Natural Computing Series

Series Editors: G. Rozenberg
Th. Bäck A.E. Eiben J.N. Kok H.P. Spaink

Leiden Center for Natural Computing

Junghuei Chen · Nataša Jonoska
Grzegorz Rozenberg (Eds.)

Nanotechnology: Science and Computation

With 126 Figures and 10 Tables

Springer

Editors
Junghuei Chen
Department of Chemistry and Biochemistry
University of Delaware
Newark, DE 19716, USA

Nataša Jonoska
Department of Mathematics
University of South Florida
4202 E. Fowler Av., PHY114
Tampa, FL 33620-5700, USA

Grzegorz Rozenberg
Leiden University
Leiden Institute for Advanced Computer Science
Niels Bohrweg 1
2333 CA Leiden, The Netherlands

Series Editors
G. Rozenberg (Managing Editor)
rozenber@liacs.nl

Th. Bäck, J.N. Kok, H.P. Spaink
Leiden Institute of Advanced
Computer Science
Leiden University
Niels Bohrweg 1
2333 CA Leiden, The Netherlands

A.E. Eiben
Vrije Universiteit Amsterdam
The Netherlands

Library of Congress Control Number: 2005936799
ACM Computing Classification (1998): F.1, G.2.3, I.1, I.2, I.6, J.3
ISBN-10 3-540-30295-6 Springer Berlin Heidelberg New York
ISBN-13 978-3-540-30295-7 Springer Berlin Heidelberg New York

Springer is a part of Springer Science+Business Media

springer.com

Cover Design: KünkelLopka, Werbeagentur, Heidelberg
Typesetting: by the Editors
Production: LE-TeX Jelonek, Schmidt & Vöckler GbR, Leipzig
Printed on acid-free paper 45/3142/YL – 5 4 3 2 1 0

This book is dedicated to Nadrian C. Seeman
on the occasion of his 60th birthday

This image was created by DADARA

Preface

Nanotechnology is slowly and steadily entering more and more aspects of our life. It is becoming a base for developing new materials as well as a base for developing novel methods of computing. As natural computing is concerned with information processing taking place in or inspired by nature, the ideas coming from basic interactions between atoms and molecules naturally become part of these novel ways of computing.

While nanotechnology and nanoengineering have flourished in recent years, the roots of DNA nanotechnology go back to the pioneering work of Nadrian (Ned) C. Seeman in the 1980s. Many of the original designs and constructions of nanoscale structures from DNA developed in Ned's lab provided a completely new way of looking at this molecule of life. Starting with the synthesis of the first immobile Holliday junction, now referred to as J1, through the double and triple cross-over molecules, Ned has shown that DNA is a powerful and versatile molecule which is ideal for building complex structures at the nanometer scale.

Through the years, Ned has used some of the basic DNA motif structures as 'tinkertoy' or 'lego' units to build a cube, two-dimensional arrays, and various three-dimensional structures, such as Borromean rings, nanomechanical devices, nano-walkers (robots), etc. All of them were designed and demonstrated originally in Ned's lab, but then all these ideas and designs were followed up by many other researchers around the world.

Adleman's seminal paper from 1994 provided a proof of principle that computing at a molecular level, with DNA, is possible. This led to a real explosion of research on molecular computing, and very quickly Ned's ideas concerning the design and construction of nanoscale structures from DNA had a profound influence on the development of both the theoretical and the experimental foundations of this research area.

Ned is a scientist and a chemist in the first place. Although Ned can be considered the founder of the DNA nanoengineering field, he has always considered himself as a chemist who is interested in basic science. Therefore, he is still very interested in the basic physical properties of DNA and enzymes that interact with nucleic acids. Ned has been continuously funded by NIH for almost 30 years and is still providing valuable insights into the DNA and RNA biophysical and topological properties as well as the mechanism of homologous recombination between two chromosomal DNAs.

Ned's enormous influence extends also to service to the scientific community. Here one has to mention that Ned is the founding president of the International Society for Nanoscale Science, Computation and Engineering (ISNSCE). The respect that Ned enjoys is also manifested through various honors and awards that he has received — among others the Feynman Prize in Nanotechnology and the Tulip Award in DNA Computing.

Besides science, Ned is very much interested in the world around him, e.g., in art. Amazingly, some of this interest has also influenced his scientific work: by studying the work of Escher he got some specific ideas for constructions of DNA-based nanostructures! Ned is an excellent lecturer and has given talks around the world, thereby instigating significant interest and research in DNA nanotechnology and computing.

With this volume, which presents many aspects of research in basic science, application, theory and computing with DNA molecules, we celebrate a scientist who has been a source of inspiration to many researchers all over the world, and to us a mentor, a scientific collaborator, and a dear friend.

December 2005

Junghuei Chen
Nataša Jonoska
Grzegorz Rozenberg

Contents

---

[1] Adapted with permission (Table 1, Figs 1–3, and associated text) from *J. Am. Chem. Soc.* **2004**, *126*, 10834–10835. Copyright 2004 American Chemical Society.

Part VI Biomolecular Computational Models

Part VII Computations Inspired by Cells

Part VIII Appendix

Scaffolded DNA Origami: from Generalized Multicrossovers to Polygonal Networks

Paul W.K. Rothemund

California Institute of Technology, Pasadena, CA 91125, USA
pwkr@dna.caltech.edu

My acquaintance with Ned Seeman began in the Caltech library sometime during 1992. At the time, I was trying to design a DNA computer and was collecting papers in an attempt to learn all the biochemical tricks ever performed with DNA. Among the papers was Ned and Junghuei Chen's beautiful construction of a DNA cube [2]. I had no idea how to harness such a marvel for computation – the diagrams explaining the cube were in a visual language that I could not parse and its static structure, once formed, did not seem to allow further information processing. However, I was in awe of the cube and wondered what kind of mad and twisted genius had conjured it.

Ned's DNA sculptures did turn out to have a relationship to computation. In 1994, Len Adleman's creation of a DNA computer [1] showed that linear DNA self-assembly, together with operations such as PCR, could tackle NP-complete computational problems. Excited by this result, Erik Winfree quickly forged an amazing link that showed how the self-assembly of geometrical DNA objects, alone, can perform universal computation [21]. The demonstration and exploration of this link have kept a small gaggle of computer scientists and mathematicians tangled up with Ned and his academic children for the last decade. At an intellectual level, the technical achievements of the resulting collaborations and interactions have been significant, among them the first two-dimensional DNA crystals [22] and algorithmic self-assembly of both linear [7] and two-dimensional [10] arrays. By various other paths, a number of physicists have joined the party, mixing their own ideas with Ned's paradigm of "DNA as Tinkertoys" to create nanomechanical systems such as DNA tweezers [26] and walkers [25, 17, 20]. DNA nanotechnology has taken on a life of its own since Ned's original vision of DNA fish flying in an extended Escherian lattice [14], and we look forward to a new "DNA world" in which an all-DNA "bacterium" wriggles, reproduces, and computes.

On a personal level, I and many others have gotten to find out *exactly* what kind of twisted genius Ned is. Ned is a singular character. He is at once gruff and caring, vulgar and articulate, stubborn and visionary. Ned is generous both with his knowledge of DNA and his knowledge of life. His life's philosophy includes a strong tension between the abysmally negative (the general state of the world) and the just tolerably positive (that which one can, with great effort, hope to achieve). To paraphrase and to whitewash, "In a world full of execrable excrescences, there is always a fetid coprostasis of an idea to make your own." Once one is correctly calibrated to Ned, this superficially gloomy counsel becomes positively bright and Ned's success with DNA nanotechnology serves as an example for the young scientist. In fact, Ned's education of young scientists reveals a latent optimism. As an advisor Ned plots a strategic course, giving graduate students projects with risks and payoffs calculated to help them succeed at every stage — from confidence builders in their first years to high-risk/high-gain projects in later years.

Ned's own relationship with science is equally telling of his character. He is healthily (and vocally) paranoid about Nature's determination to screw up his experiments. To combat this, he practices a capricious paganism, frequently switching between gods in the hope that one will answer his prayers for a highly-ordered three-dimensional DNA crystal. (A habit which he attempted unsuccessfully to break when he abandoned crystallography.) Such superstition is tongue-in-cheek, however, and Ned is one of the most careful scientists that I know. He is ever-mindful that, as Peter Medawar wrote, "research is surely the art of the soluble" and, while his highly imaginative research is constructive and nonreductionist in its goals, Ned makes sure that it rests on falsifiable Popperian bedrock.

In celebration of Ned the character, as well as the box of Tinkertoys and Legos that he has created, I cover two topics. First, I review the recent generalization of Ned's geometry of parallel crossovers to the creation of arbitrary shapes and patterns via a method called scaffolded DNA origami. I give an example pattern with roughly 200 pixels spaced 6 nm apart. Second, I propose a new method for using scaffolded DNA origami to make arbitrary polygonal networks, both two-dimensional planar stick figures and three-dimensional polyhedra.

1 Scaffolded DNA Origami for Parallel Multicrossovers

Fig. 1a,b show one of the most successful of Ned's noncanonical DNA motifs, a "double-crossover" molecule [4] fashioned from two parallel double helical domains that comprise four distinct strands of DNA. Each DNA strand winds along one helix for a number of bases before switching to the other helix by passing through a structure called a "crossover" (small black triangles). Because strands reverse direction at the crossovers, the crossovers are termed "antiparallel". It is the juxtaposition of two crossovers that holds the helices in their parallel arrangement (isolated crossovers assume an equilibrium angle of roughly 60°), and it is their juxtaposition that also holds the helices rigidly together (isolated crossovers are floppy). These properties allow double crossovers to assemble into large extended lattices [22], and nanotubes [12].

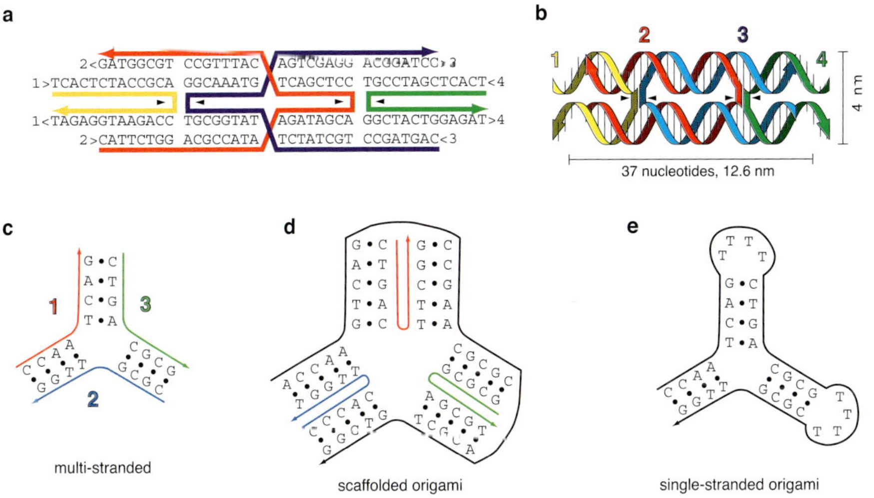

Fig. 1. Double-crossover molecules, and flavors of DNA design.

The idea of holding helical domains in a parallel arrangement via the juxtaposition of antiparallel crossovers has become a general principle in DNA nanotechnology, used in at least a dozen constructions. For example, it has been extended to molecules with three parallel helices [6], and it has been used to attach triangles rigidly to a nanomechanical device [23].

A key question is how to create generalized multicrossover molecules with parallel helices. To answer this question, it is necessary to understand the advantages and disadvantages of different approaches. Within the DNA nanotechnology paradigm, designs may be classified by how they are built up from component strands, being (1) composed entirely of short oligonucleotide strands as in Fig. 1c, (2) composed of one long "scaffold strand" (black) and numerous short "helper strands" (colored) as in Fig. 1d, or (3) composed of one long strand and few or no helpers as in Fig. 1e. Here these design approaches are termed "multistranded", "scaffolded", and "single-stranded", respectively. The last two are termed "DNA origami" because a single long strand is folded, whether by many helpers or by self-interactions.

Multistranded designs (such as Ned's original cube) suffer from the difficulty of getting the ratios of the component short strands *exactly* equal. If there are not equal proportions of the various component strands, then incomplete structures form and purification may be required. Because, for large and complex designs, a structure missing one strand is not very different from a complete structure, purification can be difficult and may have to be performed in multiple steps. Single-stranded origami such as William Shih's octahedron [19] cannot, by definition, suffer from this problem. Scaffolded origami sidesteps the problem of equalizing strand ratios by allowing an excess of helpers to be used. As long as each scaffold strand gets one of each

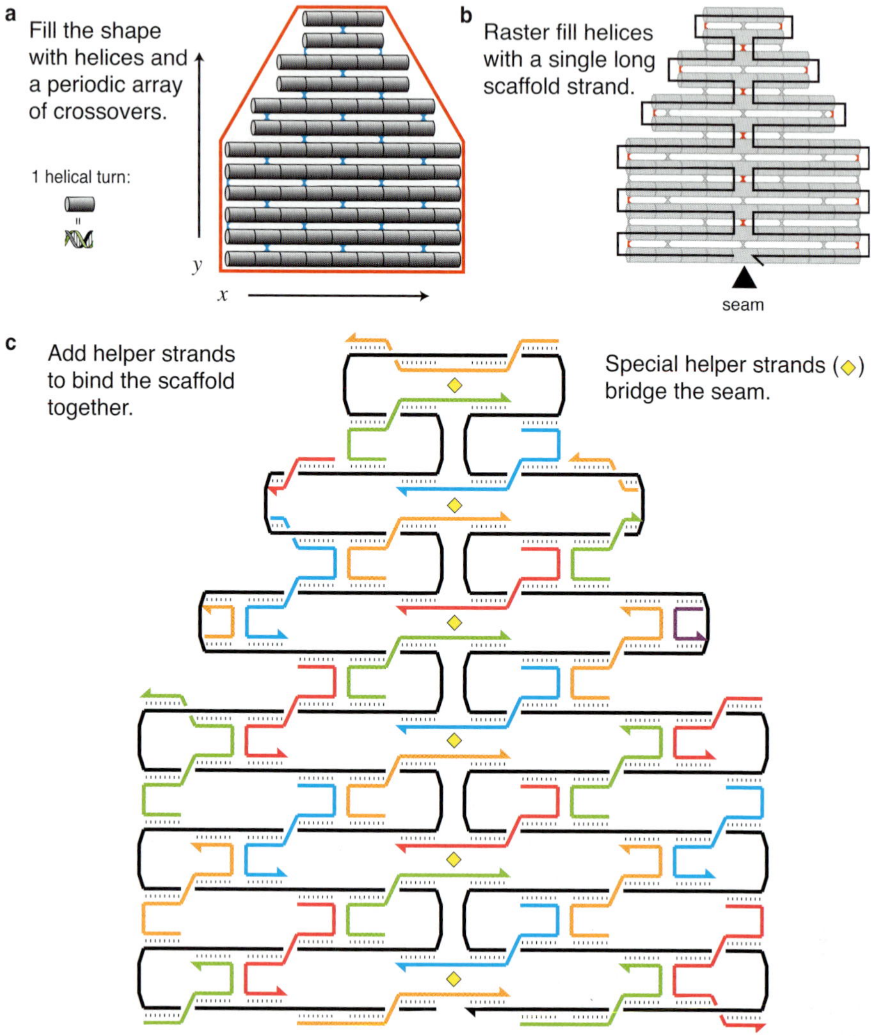

Fig. 2. Design of DNA origami.

helper, all scaffolds may fold correctly (some might get trapped in misfoldings). Because origami are easily differentiable from the helpers, separating them is not difficult (e.g. large origami stick much more strongly to mica surfaces than do tiny helpers and so excess helpers can be washed away).

Single-stranded origami and scaffolded origami thus seem the best candidates for the creation of large, complex structures. As Shih has observed (personal communication), the geometry used for the octahedron should generalize and allow the creation of arbitrary polygonal networks. However, the use of single-stranded origami to create parallel multi-crossover designs seems difficult (but perhaps only to me).

Generalization of the parallel helical geometry introduced by double-crossover molecules is simple using scaffolded DNA origami; I have recently demonstrated a technique for the creation of six arbitrary shapes and six arbitrary patterns (including the one shown here); the design method and experiments showing its generality are described in [11]. To get a feeling for the method, look at Fig. 2. Shapes are approximated by laying down a series of parallel helical domains inside the shape (Fig. 2a). Helices are cut to fit the shape, in a series of sequential pairs from top to bottom, so that the resulting geometry approximates the shape within one DNA turn (~3.6 nm) in the x-direction and two helical widths (~6 nm, including an inter-helix gap) in the y-direction. To make a molecular design, a scaffold is run exactly once through each helix; performed in a raster-fill manner, this creates a "folding path" (Fig. 2b). To hold the scaffold in this shape, helper strands are added to create a regular pattern of antiparallel crossovers (Fig. 2c).

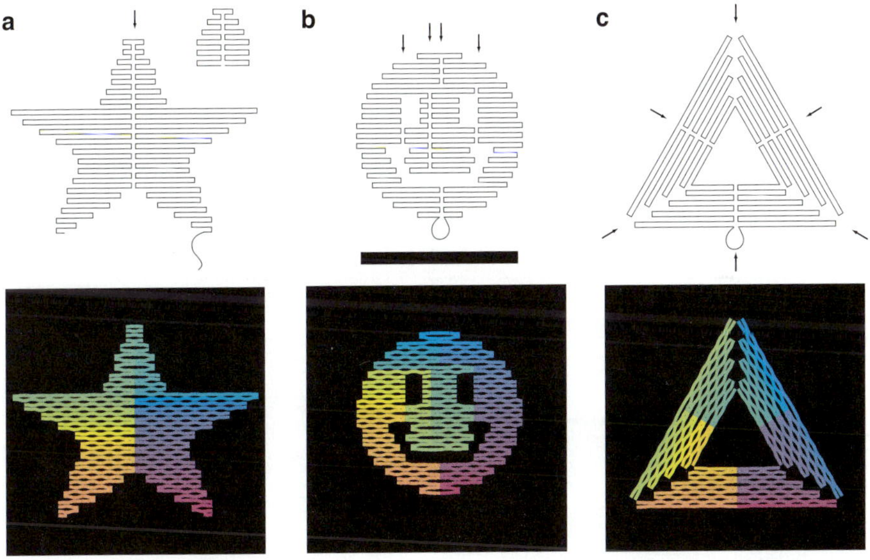

Fig. 3. Several folding paths (top) drawn without helper strands, and predicted structures (bottom) that use an ~7000-base-long scaffold. Colors indicate the base position on the scaffold, from 1 (red–orange) to 7000 (purple). Arrows indicate seams, which are bridged by helper strands for mechanical stability. Scale bar, 100 nm.

As reported in [11], the method is general and scales quite well to large origami (Fig. 3). The two shapes diagrammed in Fig. 3b,c each form in excess of a 70% yield, and each uses a 7000-base-long scaffold requiring more than

200 DNA strands for a final molecular weight of 15,000 nucleotides. Thus these DNA origami have a molecular weight 100 times that of the original double-crossover and almost 6 times larger than Ned's largest geometric construction, a truncated octahedron [27]. Further, such scaffolded origami are created in a single laboratory step: strands are mixed together in a Mg^{2+}-containing buffer and annealed from 90°C to 20°C over the course of 2 hours.

Given a shape, such as the rectangle in Fig. 4a,b, it is simple to decorate it with an arbitrary pattern of binary pixels. The position of each helper strand (of which there are roughly 200) is considered to be a pixel. The original set of helper strands is taken to represent binary '0's. To represent binary '1's a new set of labeled helper strands is constructed; so far, they have been labeled with extra DNA hairpins. To create a desired pattern (say Fig. 4c), the appropriate complementary sets of strands are drawn from the original helper strands and the labeled helper strands. Everywhere the pattern has a '0', an original helper strand is used; everywhere the pattern has a '1', a new helper strand is used. Creating the mixture of strands for a desired pattern requires about 1.5 hours of pipetting.

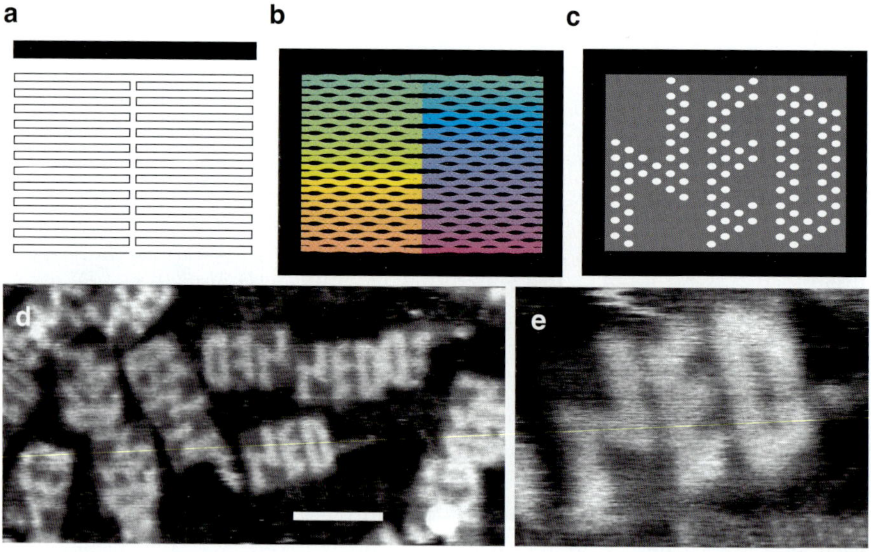

Fig. 4. An arbitrary pattern. The white features are DNA hairpins. The black scale bar in (**a**) applies to (**b,c**) and (**e**) as well. Scale bars, both black and white, 100 nm.

The pattern in Fig. 4c was made in this manner, just for this paper. Fig. 4d,e show atomic force micrographs of the result; hairpin labels appear as light dots, unlabeled positions appear gray, and the mica surface on which the sample is deposited appears black. Each letter is approximately 60 nm tall (letters half this height are shown in [11]). Roughly 50 billion copies of the pattern were made; copies stick to each other along their vertical edges via blunt-end stacking. Note that the pattern clearly shows the influence of Ned on DNA nanotechnology.

Because scaffolded DNA origami makes the creation of arbitrary shapes and patterns so simple, and because it provides the ability to pattern at the 6 nm length scale, scaffolded origami has the potential to play an important role in future lithographic techniques for nanocircuits and other nanodevices.

2 DNA Origami for Polygonal Networks

Given the ease with which scaffolded origami generalizes parallel crossovers, the question becomes, "what other general methods of creating shapes might there be?" The first thing that would probably spring to a geometer's mind is the use of polygons. Indeed an attempt to create polygonal networks – DNA stick figures – was where Ned began his quest for 3D structure [14, 15]. His original vision was to "trash the symmetry" of DNA branch junctions to create immobile motifs, which could then be assembled into polygonal networks via sticky ends (Fig. 5a,b). Unfortunately, it wasn't that easy; single-branched junctions resisted crystallization into 2D lattices for many years. In general, branched junctions formed from single helices are floppy and tend to cyclize into families of trimers, tetramers, and higher macrocycles. In particular, four-armed branch junctions vacillate between one of two different "stacked-X" conformations [9, 3] and, demonstrating a mind of their own, assume a 60° angle rather than the 90° angle one might like them to. Again, by trashing symmetries, one can use specific sticky ends that force a particular connectivity, such as the DNA cube [2], but, because of uncertainty in the junction geometry, it is still unknown whether the DNA cube was a cube or some other parallelopiped.

It was out of such frustrations that the parallel helical geometry used by Ned to create the double crossovers was born [4], giving us DNA "Lego" bricks rather than the "Tinkertoy" spools and sticks originally envisioned. DNA lattices were eventually formed from unconstrained four-arm junctions either by letting the junctions have their way, to create rhomboidal lattices with 60° angles [8], or by incorporating symmetries that apparently force the junctions to crystallize into lattices of parallel helices [13]. None of these experiments, however, gets us any closer to Tinkertoys.

Recently, in an attempt to create DNA motifs with a square 1:1 aspect ratio, Hao Yan and Thom LaBean came up with what they call a "4×4" motif (Fig. 5c). By using two DNA helices rather than one for each arm of their four-arm motif, and connecting these arms with apparently floppy junctions, Yan and LaBean have created a motif that crystallizes into rectilinear domains several microns in size [24]. Chengde Mao has modified the 4×4 to create three-arm motifs (Fig. 5d), which he calls "3-point stars", that crystallize beautifully

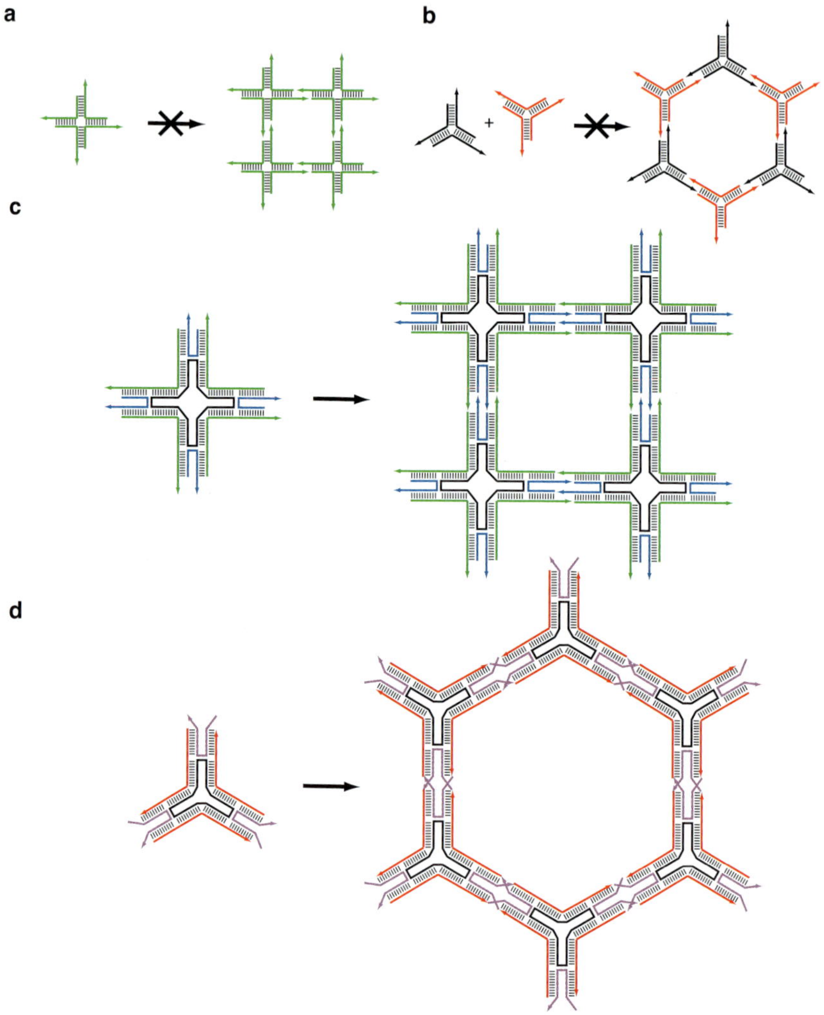

Fig. 5. Ned's original vision for branch junction lattices, and the motifs that have succeeded them. The sticky-end placement and arm lengths in (**c**) and (**d**) are not accurate; refer to [24, 5] for the actual structures.

into 30-micron hexagonal lattices [5]. It is amazing that the combination of single covalent bonds and poly-T linkers at the centers of these motifs yields structures rigid enough to form large lattices. These successes hint that the principle may be generalized to other numbers of arms — and may provide us with the sticks and spools for DNA Tinkertoys.

Here I propose a new multiarm motif, similar to the 4×4 motifs and three-point stars in that it uses two helical domains per arm, that may be used in the context of scaffolded DNA origami to create arbitrary polygonal networks. I begin by describing its use to create arbitrary pseudohexagonal networks.

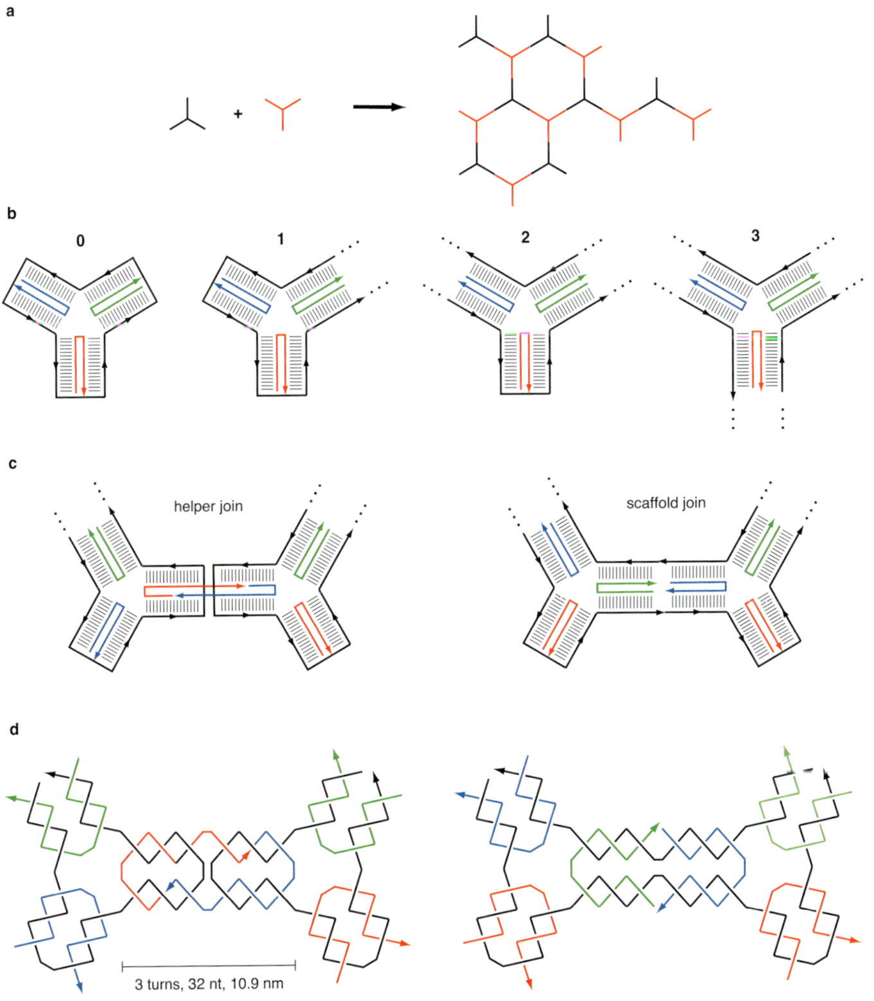

Fig. 6. A pseudohexagonal network composed of geometrical 3-stars, and the DNA 3-stars used to build a molecular approximation.

Fig. 6a shows what is meant by pseudohexagonal networks: planar figures composed from the two three-armed components at the left (which I call 3-stars) without rotation or bending. I propose that such structures can be created from scaffolded DNA origami by replacing each geometrical 3-star with one of the DNA 3-stars diagrammed in Fig. 6b.[1] In each DNA 3-star, the black strand is intended to be the scaffold strand of a DNA origami, and the colored strands are helper strands, each 32 nucleotides long. DNA 3-stars are classified by the number of "open ends" that they have, i.e. the number of breaks in the scaffold strand as it travels around the circumference of the DNA 3-star. Thus DNA 3-stars can be "type-0", "type-1", "type-2", or "type-3". The type-0 DNA 3-star is the simplest pseudohexagonal network; each arm is closed at the end by the scaffold as it crosses from one helix of the arm to the other. Note that these DNA 3-stars differ from Mao's 3-point stars (as well as the 4×4 motifs) in that they have crossovers at the junctions between arms, rather than in the middle of each arm – and thus it is uncertain how DNA 3-stars will behave in the laboratory. Let us assume for now that they will form well.

When two DNA 3-stars abut in a pseudohexagonal network, they can be joined in one of two ways: either two closed ends meet (Fig. 6c, left) or two open ends meet (Fig. 6c, right). If two closed ends meet then they are mechanically joined by modified helper strands that cross the ends closed by the scaffold strand; call this structure a "helper join".[2] On the other hand, if two open ends meet then they are joined by the scaffold strand – the scaffold strand passes along the top helix from right to left, and returns along the bottom helix from left to right. Call this structure a "scaffold join". Fig. 6d shows the helical representation of both helper and scaffold joins.

Given an arbitrary pseudohexagonal network of N 3-stars, a simple algorithm allows a molecular design M to be built up from N DNA 3-stars. Fig. 7a shows an example network; Fig. 7b shows simplified diagrams of DNA 3-stars that show only the scaffold strand and are colored according to their type. The algorithm begins by placing a type-0 DNA 3-star over a randomly chosen 3-star in the network; Fig. 7c,d show one particular choice, and Fig. 7e shows another. The algorithm proceeds by adding type-1 DNA 3-stars one at a time, until the entire network is covered (Fig. 7c–e, step 2 through step 7). Each time a type-1 DNA 3-star is added, it is positioned next to an already-placed DNA 3-star (which such a position may be chosen randomly) and it is fastened to the already-placed DNA 3-star by a scaffold join. Thus the type of the already-placed 3-star is incremented by 1 (visualized in Fig. 7 as a color change). If the type-1 DNA 3-star is placed next to two or more already-placed DNA 3-stars (Fig. 7d,e, step 7), then it is fastened to one of the DNA 3-stars (chosen randomly) by a scaffold join and to the remaining DNA 3-stars by helper joins (arrows, Fig. 7c–e). Before each addition of a type-1 DNA star, the scaffold is a single closed loop. At the end of each addition, the scaffold

[1] Technically, this motif should be called a 1.5-turn DNA 3-star; any odd number of half-turns may be used in the arm.

[2] Here each helper strand is drawn as binding to 24 bases in one DNA 3-star, and to eight bases in the other. This is by analogy with similar joints in previously created scaffolded origami; what lengths may work the best are unknown.

is still a single closed loop. Thus the algorithm always generates a design M that has a single continuous scaffold strand.

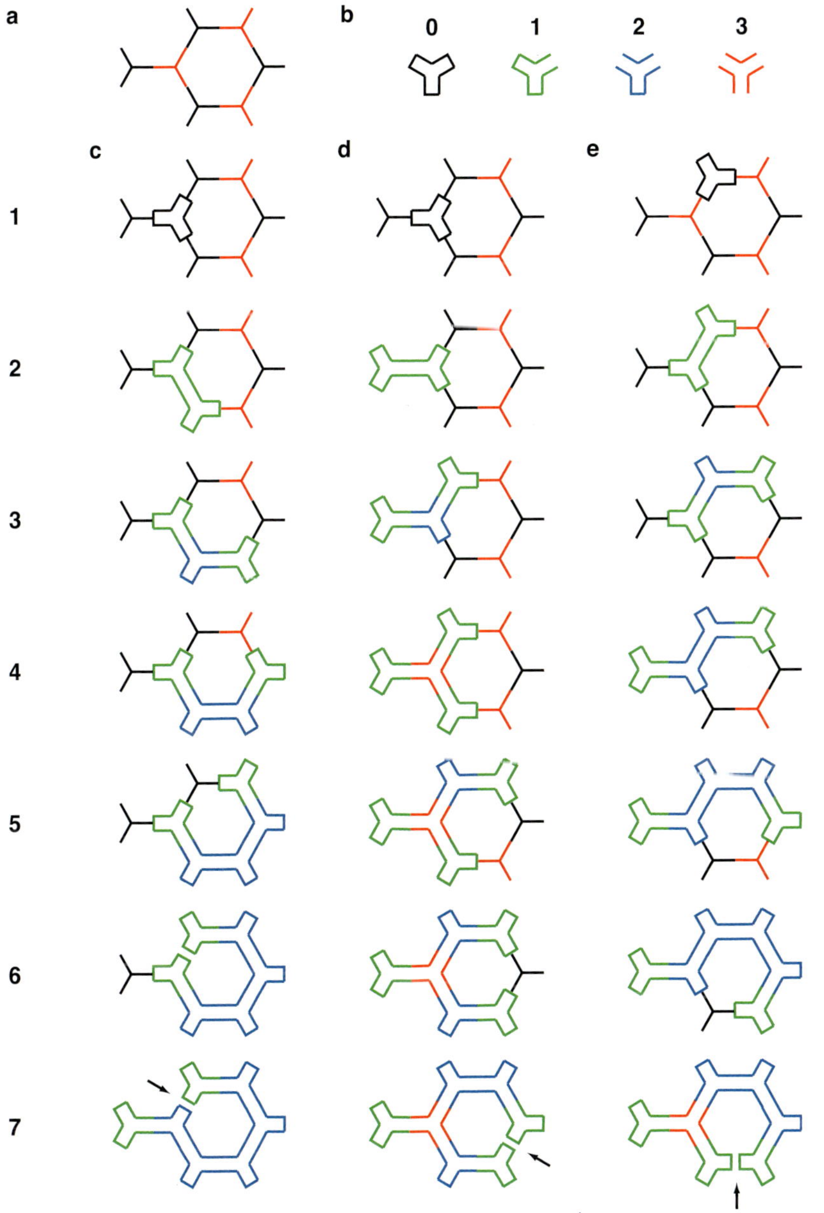

Fig. 7. A pseudohexagonal network, converted to a molecular design in three different ways. Arrows point to helper joins.

As described, the algorithm is nondeterministic and can generate different folding paths; the positions of helper and scaffold joins in M depend on the order in which 3-stars are replaced by DNA 3-stars.[3] In small designs, such as in Fig. 7, the pattern of scaffold and helper joins seems irrelevant. In large designs, however, such as those in Fig. 8, it is easy to imagine that the pattern of joins may have a bearing on whether the structures fold correctly or on their mechanical stability. For example, perhaps local folds may form faster than long-distance ones, causing short, wiggly paths to fold more reliably than long, straight ones; if this is true then the tree-like folding path of the design in Fig. 8c might fold more robustly into a triangular figure (Fig. 8a) than would the comb-like folding path of the design in Fig. 8b. Or we might expect that the folding path of Fig. 8e (for which every radius of the hexagon intersects at least two covalent scaffold bonds) would yield a more mechanically stable version of Fig. 8d than would the folding path of Fig. 8f (for which one radius of the hexagon – the dotted line – intersects only helper joins). If it is learned that the pattern of scaffold and helper joins matters, such information can be incorporated into the design algorithm.

Technically, large designs such as those in Fig. 8 seem within easy reach (at least to try). The triangular network (Fig. 8a) would require a 5856-base-long scaffold, and the hexagonal ring (Fig. 8b) a scaffold 6912 bases long (rendered using 1.5-turn DNA 3-stars).

While polygonal networks are planar graphs, the objects created with them need not be planar. Fig. 9 (top left) reproduces Ned's proposal for a single-stranded dodecahedron, drawn twisting around the Schlegel diagram[4] for a dodecahedron. In this scheme, the single blue strand that winds around the dodecahedron must leave the dodecahedron once per face, and jump to an adjacent face (Fig. 9, bottom right, makes this path clear). Ned's plan was to cut off these exocyclic arms with restriction endonucleases after the do-

[3] Note that the *number* of scaffold and helper joins in M remains the same, independent of the order in which M is built. By construction, the number of scaffold joins, S, equals $N-1$, where N is the number of 3-stars. The number of helper joins, H, is obviously $J-S$, where J is the total number of joins (determined by the network geometry). More fun (and perhaps more useful) than counting J or H is to observe that H is the number of "holes" in the network. If the network is embedded in a plane, the number of holes is the number of unconnected regions that the network divides the plane into, disregarding the region outside of the network. For example, the network in Fig. 8a has 21 holes (small hexagons), and the molecular designs in Fig. 8b,c both have 21 helper joins. The network in Fig. 8d has 19 holes (18 small hexagons and 1 large interior hexagonal void) and the designs in Fig. 8e,f both have 19 helper joins. The relationship $J = S + H = N - 1 + H$ is just a restatement of Euler's theorem for planar graphs $V - E + F = 2$, where the number of vertices V is equal to N, the number of edges E is equal to J, and the number of faces F is equal to $H + 1$ (the number of faces of a graph includes all the holes, plus the region of the plane outside the graph.)

[4] A Schlegel diagram for a polyhedron is just the planar graph associated with that polyhedron.

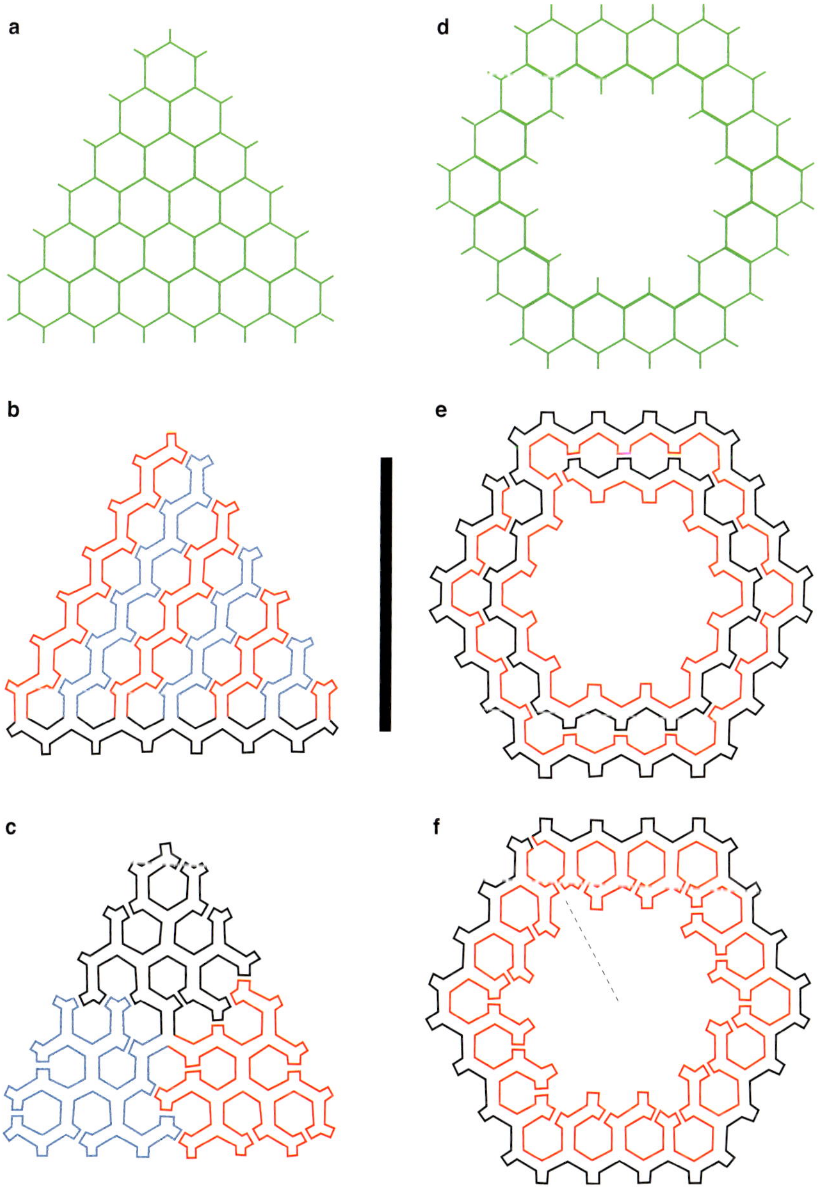

Fig. 8. Given a particular network, folding paths in molecular designs are not unique. Vertically oriented scale bar, 100 nm.

decahedron had folded. More inconvenient than the surplus arms is that this structure is a formal knot – in order for it to fold, the single strand would have

own design. As we create architectures and tools that put DNA nanotechnology into the hands of the research community at large, it will be exciting to see the legacy of Ned's flying DNA fish continue to grow.

Acknowledgments

I thank Erik Winfree and William Shih for comments and the Caltech Center for the Physics of Information for a postdoctoral fellowship.

References

1. L.M. Adleman. Molecular computation of solutions to combinatorial problems. *Science*, 266:1021–1024, 1994.
2. J. Chen and N.C. Seeman. The synthesis from DNA of a molecule with the connectivity of a cube. *Nature*, 350:631–633, 1991.
3. D.R. Duckett, A.I.H. Murchie, S. Diekmann, E. von Kitzing, B. Kemper, and D.M.J. Lilley. The structure of the Holliday junction, and its resolution. *Cell*, 55:79–89, 1988.
4. T.-J. Fu and N.C. Seeman. DNA double-crossover molecules. *Biochemistry*, 32:3211–3220, 1993.
5. Y. He, Y. Chen, H. Liu, A.E. Ribbe, and C. Mao. Self-assembly of hexagonal DNA two-dimensional (2D) arrays. *Journal of the American Chemical Society*, 10:1021, 2005.
6. T.H. LaBean, H. Yan, J. Kopatsch, F. Liu, E. Winfree, J.H. Reif, and N.C. Seeman. Construction, analysis, ligation, and self-assembly of DNA triple crossover complexes. *Journal of the American Chemical Society*, 122:1848–1860, 2000.
7. C. Mao, T.H. LaBean, J.H. Reif, and N.C. Seeman. Logical computation using algorithmic self-assembly of DNA triple-crossover molecules. *Nature*, 407(6803):493–496, 2000.
8. C.D. Mao, W.Q. Sun, and N.C. Seeman. Designed two-dimensional DNA Holliday junction arrays visualized by atomic force microscopy. *Journal of the American Chemical Society*, 121:5437–5443, 1999.
9. A.I.H. Murchie, R.M. Clegg, E. von Kitzing, D.R. Duckett, S. Diekmann, and D.M.J. Lilley. Fluorescence energy transfer shows that the four-way DNA junction is a right-handed cross of antiparallel molecules. *Nature*, 341:763–766, 1989.
10. P.W.K. Rothemund, N. Papadakis, and E. Winfree. Algorithmic self-assembly of DNA Sierpinski triangles. *PLoS Biology*, 2(12):e424, 2004.
11. P.W.K. Rothemund. Generation of arbitrary nanoscale shapes and patterns by scaffolded DNA origami. (submitted), 2005.
12. P. W. K. Rothemund, A. Ekani-Nkodo, N. Papadakis, A. Kumar, D.K. Fygenson, E. Winfree. Design and characterization of programmable DNA nanotubes. *Journal of the American Chemical Society*, 26(50):16344–16353, 2004.
13. P.W.K. Rothemund. DNA self-assembly with floppy motifs – single crossover lattices. *Foundations of Nanoscience, Self-Assembled Architectures and Devices*, Proceedings of FNANO'05 (J.H. Reif eds.) 185–186, 2005.
14. N.C. Seeman. Nucleic-acid junctions and lattices. *Journal of Theoretical Biology*, 99:237–247, 1982.
15. N.C. Seeman. Construction of three-dimensional stick figures from branched DNA. *DNA and Cell Biology*, 7(10):475–486, 1991.

16. Z.Y. Shen, H. Yan, T. Wang, and N.C. Seeman. Paranemic crossover DNA: A generalized Holliday structure with applications in nanotechnology. *Journal of the American Chemical Society*, 126:1666–1674, 2004.
17. W.B. Sherman and N.C. Seeman. A precisely controlled DNA biped walking device. *Nanoletters*, 4(7):1203–1207, 2004.
18. W.B. Sherman and N.C. Seeman. The design of nucleic acid nanotubes. *Journal of Biomolecular Structure and Dynamics*, 20(6):930–931, 2003.
19. W.M. Shih, J.D. Quispe, and G.F. Joyce. A 1.7-kilobase single-stranded DNA that folds into a nanoscale octahedron. *Nature*, 427(6453):618–621, 2004.
20. J.S. Shin and N.A. Pierce. A synthetic DNA walker for molecular transport. *Journal of the American Chemical Society*, 126(35):10834–10835, 2004.
21. E. Winfree. On the computational power of DNA annealing and ligation. In R.J. Lipton and E.B. Baum, editors, *DNA Based Computers*, DIMACS, AMS Press, Providence, RI, 27:199–221, 1996.
22. E. Winfree, F. Liu, L.A. Wenzler, and N.C. Seeman. Design and self-assembly of two-dimensional DNA crystals. *Nature*, 394:539–544, 1998.
23. H. Yan, X. Zhang, Z. Shen, and N.C. Seeman. A robust DNA mechanical device controlled by hybridization topology. *Nature*, 415:62–65, 2002.
24. H. Yan, S.H. Park, G. Finkelstein, J.H. Reif, and T.H. LaBean. DNA-templated self-assembly of protein arrays and highly conductive nanowires. *Science*, 301:1882–1884, 2003.
25. P. Yin, H. Yan, X.G. Daniell, A.J. Turberfield, and J.H. Reif. A unidirectional DNA walker that moves autonomously along a track. *Angewandte Chemie International Edition*, 43(37):4906–4911, 2004.
26. B. Yurke, A.J. Turberfield, A.P. Mills, Jr., F.C. Simmel, and J.L. Neumann. A DNA-fuelled molecular machine made of DNA. *Nature*, 406:605–608, 2000.
27. Y. Zhang and N.C. Seeman. The construction of a DNA truncated octahedron. *Journal of the American Chemical Society*, 116:1661–1669, 1994.

A Fresh Look at DNA Nanotechnology

Zhaoxiang Deng, Yi Chen, Ye Tian, and Chengde Mao

Department of Chemistry, Purdue University, West Lafayette, Indiana 47907, USA
mao@purdue.edu

Synthetic DNA structures for nanotechnological applications have experienced substantial success during the past decades benefiting from Seeman and his coworkers' pioneering work. In the last few years, some new branches have been emerging in this field. This review will summarize some recent progress in the authors' group.

1 Two-Dimensional DNA Triangle Arrays Designed with a Tensegrity Strategy

Forming crystalline DNA lattices in one, two and even three dimensions has long been a hot topic of DNA nanotechnology. These artificially designed lattices are the basis for a variety of applications. The first success with a 2D DNA lattice was achieved in [22] with building blocks of double-crossover (DX) DNA molecules. Following that, rhombus motifs, triple-crossover molecules, and a cross motif were also constructed from branched four-arm Holliday junctions [19]. Here we present a tensegrity strategy for the construction of well-structured DNA triangle molecules [14]. A DNA triangle consists of three vertices (DNA four-arm junctions) and three sides (DNA duplexes). Although individual four-arm junctions are flexible, the rigidity of the three duplex edges restricts the freedom of the component four-arm junctions and only triangles can form. The shape of such a triangle is fully defined by the lengths of the three edges. By rational use of sticky-end cohesion, we have successfully assembled triangle arrays in one and two dimensions. Fig. 1 shows the design of some such triangle arrays and some atomic force microscope (AFM) images of such triangle arrays.

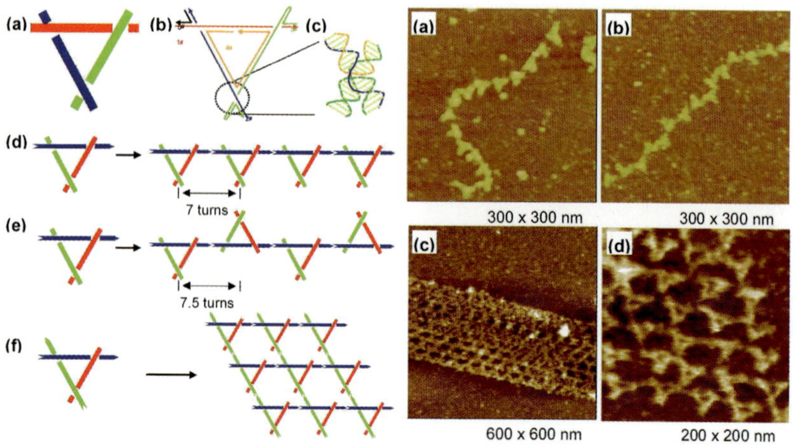

Fig. 1. DNA triangle arrays. *Left panel*: Schematic representation of one- and two-dimensional arrays. (**a**) A DNA triangle contains three DNA duplexes, shown as rods of different colors. (**b**) Strand structure of a DNA triangle. Each thin line represents a single DNA strand. An arrow indicates the 3′ end of a DNA strand. (**c**) The detailed structure of a triangle vortex. (**d**, **e**) 1D and (**f**) 2D self-assembly of DNA triangles. *Right panel*: AFM images of one-dimensional (**a**, **b**) and two-dimensional (**c**, **d**) DNA triangle arrays. (Reproduced from [14] with permission).

2 DNA Molecular Motors

Molecular motors are a very attractive topic in many scientific fields, because they are expected to be mechanical parts of future nanorobots. Complementarily to other molecular systems, DNA motors afford rational design, easy construction, and, most importantly, good control of the motions that they generate. Earlier models of DNA motors include a nanomechanical device based on B–Z transition upon a change of the ionic strength of a solution [17], and molecular tweezers with their opening and closing controlled by sequential addition of DNA strands [23]. Inspired by those early successes, the authors' group has worked intensively in this field. Comparison between cellular protein motors and macroscale man-made machines leads us to ask four questions, as listed below. Answering these questions is fundamental for the further development of molecular motors.

2.1 Can DNA Motors Perform Complicated Motions? — Modeling Gear Motion at Molecular Scale

Gears have many useful functions such as changing the direction and speed of movement, and are important parts in real machines. It is reasonable to expect that gears might play similar roles in small motor systems such nano motors and molecular motors. This notion has motivated us to model gears with circular DNA molecules, which can roll controllably against each other by use of a strand displacement strategy (Fig. 2) [21]. A DNA gear has a central, circular single strand of DNA, which base-pairs with three linear DNA strands, leaving three unpaired tails as cogs for the gears. When a linker strand L1 is added to the sample, the two gears are bridged together and become ready for rolling. Upon addition of another linker strand L2, two linkages are built between the two gears and the gears roll mutually through 60°. By use of a strand displacement method, a removal strand R1 is then added, which forms a duplex with strand L1 and strips off L1 from the gear pair. This step creates a further 60° rolling between these two gears. If the above process is repeated, a continuous rotation of the two gears is realized.

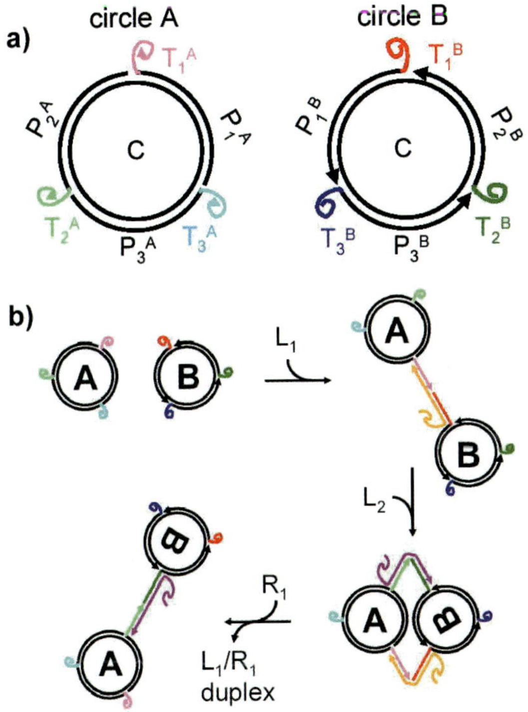

Fig. 2. Design and rolling mechanism of a pair of molecular gears. (**a**) Structures of the individual gears. C and P indicate DNA strands, and T indicates teeth. (**b**) Operation of the gears. L and R represent linker and removal strands, respectively. L_1 and R_1 are complementary to each other. Both circles remain intact during the rolling process. The only changed strands are the linker (L) and removal (R) strands. Note that no twisting motion is generated in the central strands during the rolling process. (Reproduced from [21] with permission).

2.2 Can DNA Motors Work Autonomously?

A DNA Machine Contains a DNAzyme Domain

Working autonomously is an essential feature for cellular protein motors and man-made macroscale machines. It is also desirable for nanomachines to be autonomous. This section describes our initial efforts to address this challenge. The key of our strategy is the introduction of a DNA enzyme domain, which extracts chemical energy and powers a DNA machine. We have tested this notion first with a construct that performs a simple opening and closing motion. This DNA machine undergoes continuous, autonomous motion in the presence of a fuel strand. The motion is controlled by the addition of a brake strand [7, 5]. Fig. 3 illustrates this process. This motor has a triangular shape. It contains a V-shaped dual arm spaced by a single DNA strand at the top. The single strand has a special sequence, corresponding to the core part of a DNA enzyme (E) and its flanking recognition arms on both sides. When this strand is base-paired with its substrate (S), which is a DNA strand with two RNA bases in the middle, the V-shaped arms will be opened owing to the increase in the rigidity of the single-stranded linkage after forming a duplex with its substrate. The DNAzyme then cleaves its substrate, and the cleaved products are short and dissociate from the DNAzyme, which virtually closes the two arms of the motor. If there is substrate in the solution, the above opening and closing process will continue until all substrate molecules (fuel) have been consumed. This process can be regulated by addition of a brake strand (B). The brake molecule is a DNA analog of the substrate, but has base pairs extending into the catalytic core of the enzyme. The brake strand can form a slightly longer duplex with the DNAzyme than the substrate does. The DNA machine will preferentially bind brake strands and further incorporation of fuel strands is then disabled. Therefore, the motor will be frozen in its open state. Note that the brake strand has an unpaired tail, which is designed for removal of the brake. Upon addition of a removal strand (R), the removal strand completely base-pairs with the entire brake strand. As a result, the brake will be removed from the motor system, and the motion of the machine will be resumed.

A DNAzyme-Containing DNA Walker

Recent work has shown that relatively complex motions can be realized with DNA nanocontructions, including walkers and gears. In the following, we show further that we can introduce a DNAzyme into a DNA walker. Such a DNAzyme-containing walker can move autonomously in either direction along a linear track in a controllable fashion [20]. This design takes advantage of the RNA-cleaving function of a DNAzyme. Details of this walker and its mechanism of movement are presented in Fig. 4. The enzyme strand (the red parts are base-recognizing arms, and the orange part is the catalytic core)

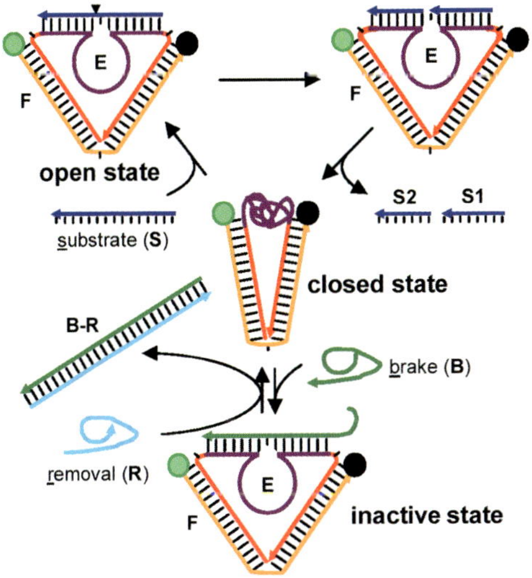

Fig. 3. Schematic of an autonomous DNA nanomotor based on a DNA enzyme. The DNA motor consists of two single strands, E and F. The strand E contains a 10–23 DNA enzyme domain, which is colored purple. The strand F has a fluorophore at the 5′ end (labeled as a solid green circle) and a quencher at the 3′ end (labeled by a solid black circle). (Reproduced from [5] with permission).

base-pairs with one of its substrates on the linear track. The blue dots on the green substrates depict the cleavage points, where the RNA bases are located. After the enzyme cleaves the substrate, the shorter product will be released into the solution owing to its relatively weaker bonding. This gives an opportunity for the exposed part of the enzyme strand to seek another substrate within its vicinity. Gradually, through branch migration, the whole enzyme strand will shift to the next neighboring substrate on the track, and the above process will be repeated until the DNA enzyme moves to the other end of the track. This process can be purposely chosen to start from either end of the track and will continue to the opposite end, but the walker cannot move backwards once it has been determined which end of the track it will start from, because the enzyme will destroy all the substrates that it has passed.

2.3 Can DNA Motors Work with Inexpensive Fuel Molecules? — pH-Switched DNA Motor Based on a Duplex–Triplex Transition

In the DNA machines described so far, DNA and RNA are used as fuel, but they quite expensive. It would be desirable to use inexpensive common chemicals as fuel. This motivation has guided us to design a pH-triggered DNA machine. Under certain conditions, a DNA triplex rather than a duplex

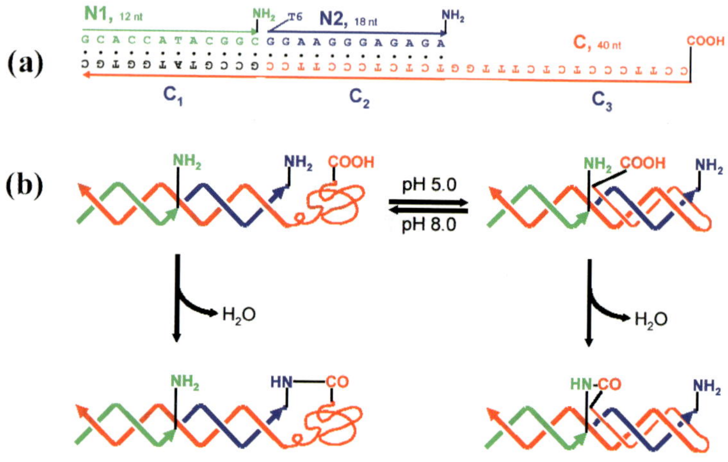

Fig. 6. Schematic illustration of the switching of a chemical reaction based on a DNA duplex–triplex transition. (**a**) DNA sequences, and the positions of the amino groups and carboxylate group of interest. Note that there is a string of unpaired T6 at the 5′ end of strand N2. Addition of the extra six bases to strand N2 causes strands N1 and N2 to have different molecular weights and electrophoretic mobilities, which allows identification of strands N1 and N2 by polyacrylamide gel electrophoresis (PAGE). (**b**) Switching of a chemical reaction by switching the location of the carboxylate group. This behavior is triggered by a change of the pH value of the solution. Note the formation and dissociation of a DNA triplex. (Reproduced from [6] with permission).

3 DNA Encoded One-Dimensional Array of Nanogold

A very challenging aspect of nanotechnology is the development of an efficient and potentially universal way to organize nanosized building blocks into designed architectures. Among the various possible materials, DNA is a superior molecule for this purpose for the following reasons: (1) DNA can be made to form well-defined nanostructures by rational design; (2) DNA can be chemically modified and operated on by enzymes; (3) DNA itself is an environmentally benign biochemical reagent. By choosing gold nanoparticles as model materials for the assembly, we have demonstrated the successful preparation of 1D gold nanoparticle arrays with lengths up to 4 _m [10]. It has been previously shown that gold nanoparticles can be assembled into small, discrete structures [1, 16, 12, 24] through hybridizing mono-DNA-modified gold particles with a DNA template. However, creating a gold nanoparticle array containing hundreds of nanoparticles does not seem to be an easy matter because of difficulties with the availability of long, single-stranded DNA templates. Fortunately, a rolling-circle DNA polymerization technique [11, 15] developed ten years ago can help. With the help of this rolling-circle polymerization, we can obtain a single-stranded DNA template with a tandemly repetitive sequence defined by the circular DNA template (Fig. 7). Also, gold nanoparticles can be modified with a thiolated single-stranded oligonucleotide, and mono-DNA-modified particles can be isolated simply by agarose gel electrohporesis using a protocol developed by Alivisatos et al. [1, 16, 12, 24]. After combining the mono-DNA-modified gold particles with the rolling-circle-synthesized long DNA template, 1D gold nanoparticle linear arrays with lengths up to several micrometers can be obtained.

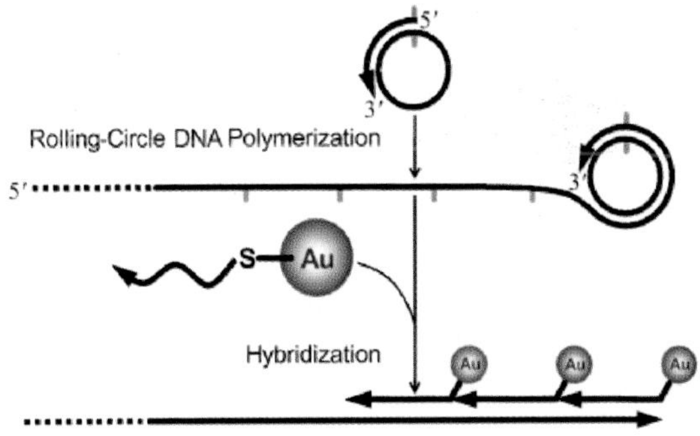

Fig. 7. Synthesis of an extended gold nanoparticle array by combining DNA-encoded self-assembly and rolling-circle polymerization of DNA. (Reproduced from [10] with permission).

4 DNA as Templates for Nanofabrication

4.1 Oriented Metallic Nanowire Networks Templated by DNA

Besides the use of self-assembly to form various structures, DNA molecules can also be used as scaffolds for nanofabrication. The first example that we have demonstrated is related to DNA metallization (Fig. 8). It is known that lambda-phage DNA, a linear DNA with a natural length of 16μm, can be aligned on a surface [2]. It is also known that DNA strands can be metallized through electroless metal reduction in solution or on a surface [3, 18]. This provides a fundamental possibility of fabricating 1D or 2D oriented metal wire networks. The method described here integrates a molecular combing technique and DNA metallization [8]. In the first step, DNA is aligned on a mica surface by a fluid flow in the presence of magnesium ions, which enhance the binding between the DNA and the mica surface and thus minimize DNA detachment during the metallization process. Note that both the DNA and the mica surface are negatively charged. After alignment, the DNA sample is then used for metal deposition. Palladium was chosen as the metal for this purpose. To avoid the formation of nanowires with abundant branches, removal of the palladium solution from the surface before adding the reduction bath solution is helpful. The incubation time for the reduction process must be controlled within a range of several minutes to as short as tens of seconds, otherwise DNA will begin to detach from the surface, and the originally created DNA network structures will be partially destroyed.

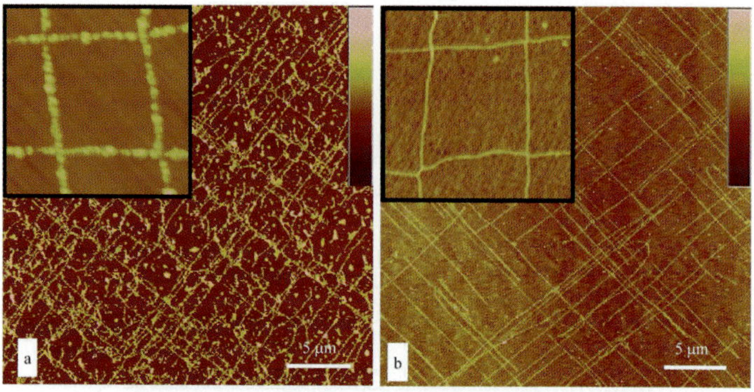

Fig. 8. AFM images of 2D aligned Pd nanowires (**a**) and the corresponding precursor DNA molecules (**b**). The insets in (**a**) and (**b**) give closer views of a 2D square of metal nanowires and of DNA molecules, respectively. Height scale: (**a**) 30 nm and (**b**) 3.0 nm. (Reproduced from [8] with permission).

4.2 Molecular Lithography with DNA Nanostructures

Another example of DNA-templated nanofabrication is the molding of DNA patterns with a metal film, resulting in a negative replica of the DNA structure (Fig. 9) [9]. The first step in realizing the replication of DNA nanostructures is to deposit DNA samples onto a mica surface. Since mica has an atomically flat surface, DNA structures on the mica surface show significant topographic patterns even though they are only about 1 nm high. Immediately after the sample has been deposited and the surface has been dried, a layer of gold metal is thermally evaporated onto the mica surface until a continuous film with a thickness of about 20 nm is formed. The weak bonding between the gold film and the mica surface offers the possibility to easily peel off the gold film to release the replica. By this strategy, DNA structures, both one-dimensional and two-dimensional can be successfully transferred to a metal substrate. Since good control over the DNA structure could be achieved by rational design, it is not a dream that in the future we might use the replicated patterns for various purposes, from ultrafine display panels and integrated sensors to nanoelectronic applications.

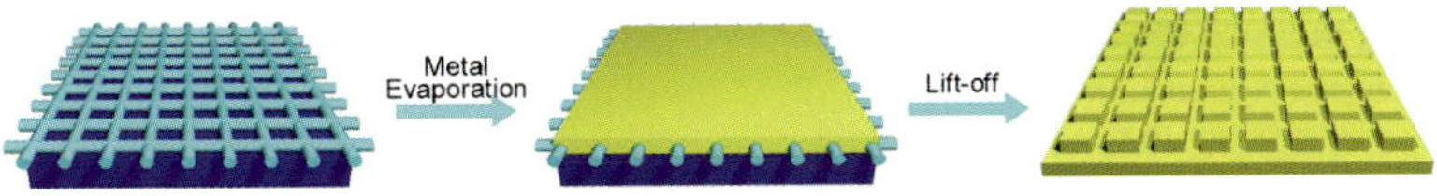

Fig. 9. A molecular lithography technique that transfers DNA nanostructures into metallic structures. (Reproduced from [9] with permission).

5 Final Remarks

This review has summarized our recent research results regarding DNA nanotechnology. DNA nanotechnology has clearly been developed into various research fields. Despite some difficulties that lie ahead, DNA, as a unique material, is paving the way towards molecular motors, directed material self-assembly, and nanofabrication, which may, potentially, solve the challenges in nanotechnology. We are expecting to see more and more breakthroughs in the coming years that will bring new content to this field, towards more practical applications of DNA nanostructures.

References

1. A.P. Alivisatos, K.P. Johnsson, X.G. Peng, T.E. Wilson, C.J. Loweth, M.P. Bruchez, and P.G. Schultz. Organization of 'nanocrystal molecules' using DNA. *Nature*, 382:609–611, 1996.
2. A. Bensimon, A. Simon, A. Chiffaudel, V. Croquette, F. Heslot, and D. Bensimon. Alignment and sensitive detection of DNA by a moving interface. *Science*, 265:2096, 1994.
3. E. Braun, Y. Eichen, U. Sivan, and G. Ben-Yoseph. DNA-templated assembly and electrode attachment of a conducting silver wire. *Nature*, 391:775, 1998.
4. Y. Chen, S.H. Lee, and C.D. Mao. A DNA nanomachine based on a duplex–triplex transition. *Angew. Chem.-Int. Ed.*, 43:5335, 2004.
5. Y. Chen and C.D. Mao. Putting a brake on an autonomous DNA nanomotor. *J. Am. Chem. Soc.*, 126:8626, 2004.
6. Y. Chen and C.D. Mao. Reprogramming DNA-directed reactions on the basis of a DNA conformational change. *J. Am. Chem. Soc.*, 126:13240, 2004.
7. Y. Chen, M.S. Wang, and C.D. Mao. An autonomous DNA nanomotor powered by a DNA enzyme. *Angew. Chem.-Int. Ed.*, 43:3554, 2004.
8. Z.X. Deng and C.D. Mao. DNA-templated fabrication of 1D parallel and 2D crossed metallic nanowire arrays. *Nano Lett.*, 3:1545, 2003.

9. Z.X. Deng and C.D. Mao. Molecular lithography with DNA nanostructures. *Angew. Chem.-Int. Ed.*, 43:4068, 2004.
10. Z.X. Deng, Y. Tian, S.H. Lee, A.E. Ribbe, and C.D. Mao. DNA-encoded self-assembly of gold nanoparticles into one-dimensional arrays. *Angew. Chem. Int. Ed.*, 44:3582–3585, 2005.
11. A. Fire and S.Q. Xu. Rolling replication of short DNA circles. *Proc. Natl. Acad. Sci. USA*, 92:4641–4645, 1995.
12. H. Fu, C.M. Micheel, J. Cha, H. Chang, H. Yang, and A.P. Alivisatos. Discrete nanostructures of quantum dots/Au with DNA. *J. Am. Chem. Soc.*, 126:10832–10833, 2004.
13. X. Li and D.R. Liu. DNA-templated organic synthesis: Nature's strategy for controlling chemical reactivity applied to synthetic molecules. *Angew. Chem.-Int. Ed.*, 43:4848–4870, 2004.
14. D.G. Liu, M.S. Wang, Z.X. Deng, R. Walulu, and C.D. Mao. Tensegrity: construction of rigid DNA triangles with flexible four-arm DNA junctions. *J. Am. Chem. Soc.*, 126:2324, 2004.
15. D.Y. Liu, S.L. Daubendiek, M.A. Zillman, K. Ryan, and E.T. Kool. Rolling circle DNA synthesis: Small circular oligonucleotides as efficient templates for DNA polymerases. *J. Am. Chem. Soc.*, 118:1587–1594, 1996.
16. C.J. Loweth, W.B. Caldwell, X.G. Peng, A.P. Alivisatos, and P.G. Schultz. DNA-based assembly of gold nanocrystals. *Angew. Chem.-Int. Ed.*, 38:1808–1812, 1999.
17. C.D. Mao, W.Q. Sun, Z.Y. Shen, and N.C. Seeman. A DNA nanomechanical device based on the B–Z transition. *Nature*, 397:144, 1999.
18. J. Richter, R. Seidel, R. Kirsch, M. Mertig, W. Pompe, J. Plaschke, and H.K. Schackert. Nanoscale palladium metallization of DNA. *Adv. Mater.*, 12:507, 2000.
19. N.C. Seeman. DNA in a material world. *Nature*, 421:427, 2003.
20. Y. Tian, Y. He, Y. Chen, P. Yin, and C.D. Mao. A DNAzyme that walks processively and autonomously along a one-dimensional track. *Angew. Chem.-Int. Ed.*, 2005. In press.
21. Y. Tian and C.D. Mao. Molecular gears: a pair of DNA circles continuously rolls against each other. *J. Am. Chem. Soc.*, 126:11410, 2004.
22. E. Winfree, F. R. Liu, L. A. Wenzler, and N.C. Seeman. Design and self-assembly of two-dimensional DNA crystals. *Nature*, 394:539, 1998.
23. B. Yurke, A.J. Turberfield, A.P. Mills, F.C. Simmel, and J.L. Neumann. A DNA-fuelled molecular machine made of DNA. *Nature*, 406:605, 2000.
24. D. Zanchet, C.M. Micheel, W.J. Parak, D. Gerion, and A.P. Alivisatos. Electrophoretic isolation of discrete Au nanocrystal/DNA conjugates. *Nano Lett.*, 1:32–35, 2001.

DNA Nanotechnology: an Evolving Field

Hao Yan and Yan Liu

Department of Chemistry and Biochemistry and The Biodesign Institute,
Arizona State University, Tempe, AZ 85287, USA
hao.yan@asu.edu

1 Introduction

In the fifty years of history since the structure of DNA was first revealed, what was once a Nobel Prize-winning research discovery has become an omnipresent cultural icon. The familiar DNA double helix is now serving as a microscopic trellis for further advances in nanotechnology. The beginnings of DNA nanotechnology can be traced back to Ned Seeman's ingenious vision of using flying-fish-like, six-arm, branched DNA to self-align into an ordered three-dimensional lattice [46] and his earlier construction of a DNA cube [5]. I can still remember what Ned has told us many times in his group meetings: "If it doesn't work, use a hammer; if it still doesn't work, use a bigger hammer." After about 20 years of hammering, DNA nanotechnology has evolved into a unique interdisciplinary field, crossing the paths of chemistry, physics, computer science, biology and materials science. This burgeoning field is producing new information as prolifically as bamboo generates new shoots from the ground. We would like to devote this chapter to Ned to celebrate his 60th birthday. Most of the work reviewed here is based on that of the present authors and their collaborators. While other prominent work in this field will also be briefly mentioned, more thorough surveys of DNA nanotechnology can be found elsewhere [48, 47].

1.1 Why Is DNA Good for Nanotechnology?

DNA, the well-known double-helix structure, offers the following advantageous features as building blocks for nanotechnology. (1) B-DNA has a diameter of about 2 nm and a helical repeat of about 3.4 nm. Its conformation is predictable with atomic precision. (2) DNA has the most predictable and programmable intramolecular and intermolecular interactions, from the perspectives of both Watson–Crick base pairing and its resulting conformation. A DNA sequence with N bases gives 4^N unique sequence compositions.

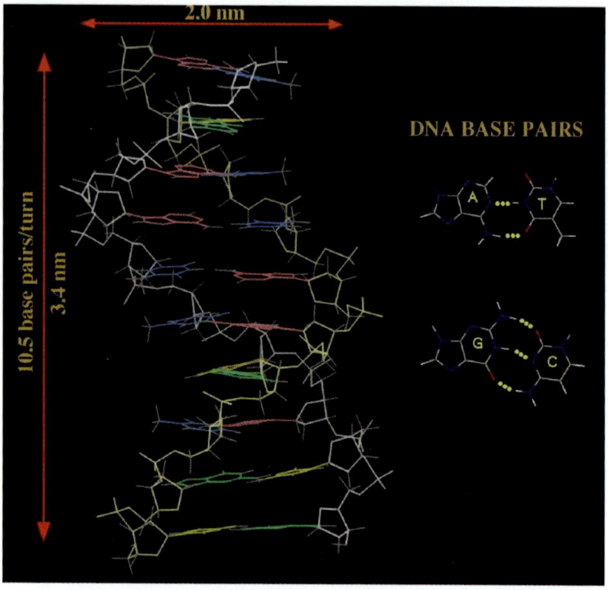

Fig. 1. B-DNA structure and base pairs.

Hybridization energies between complementary DNA sequences can be calculated. Sticky-end cohesion can serve as a "smart glue" to bring together two double-helical DNA fragments with complementary overhanging sequences. (3) Double-helix DNA has a persistence length of 50 nm under conventional conditions, and thus it is considered a stiff molecule in the context of use as a nanoscale building block. (4) DNA can be conveniently synthesized by automated phosphoramidite chemistry and functionalized with various chemical groups such as –NH2, –COOH and –SH and various fluorescent dyes. Other DNA derivatives such as peptide nucleic acid [36] and locked nucleic acid [20] can also be produced, adding different flavors to the library of nucleic acid-based nanostructures. (5) DNA structures are complemented by a sophisticated array of tools developed for DNA biotechnology: for example, DNA can be manipulated using commercially available enzymes for site-selective DNA cleavage (restriction), ligation, labeling, transcription, replication, kination and methylation [52]. DNA nanotechnology is further empowered by well-established methods for purification and structural characterization, so that essentially any designer DNA strand can be constructed.

The conventional DNA structure is a linear molecule. It has a valence of 2, so that end-to-end joining between multiple DNA double helices can result only in longer lines or topologically circularized objects. To construct interesting two-dimensional (2D) and three-dimensional (3D) nanostructures from DNA, it is necessary to design DNA structures containing branches. In 1982, Seeman first presented the idea of combining sticky-end cohesion and

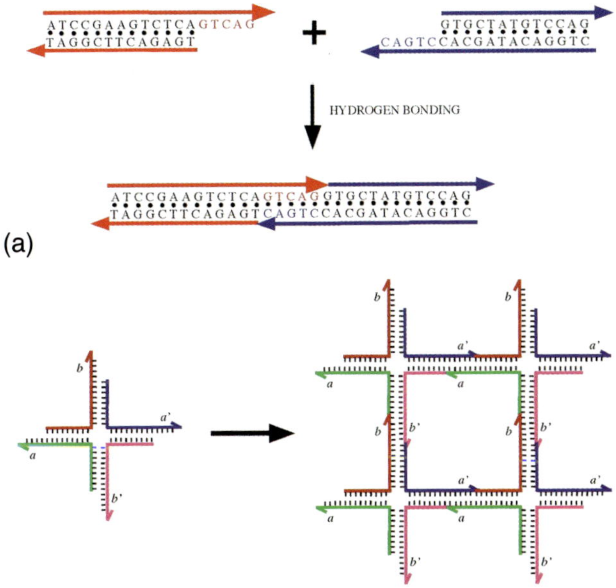

Fig. 2. (**a**) Sticky-end cohesion. (**b**) Sticky-ended assembly of branched DNA junctions into a 2D square lattice.

branched DNA to make objects and lattices [46]. This idea is illustrated in Fig. 2, which shows that a four-arm branched DNA junction with sticky ends could self-assemble into square-like lattice in two dimensions. While not appreciated by many people in the early days, this idea turned out to be a paradigm-shifting proposal that truly opened up a new field. Ned spent the next 15 years designing, constructing and characterizing a variety of branched DNA molecules, such as three- [31], four- [19], five- and six-armed DNA junctions [54], DNA triangles [43, 60], and double-crossover (DX) molecules [17]. Finally, in 1998, Seeman's group succeeded in using the DNA DX molecule to make 2D lattices by self-assembly [56]. This first example of a self-assembling 2D DNA lattice generated a lot of excitement and laid a solid foundation for subsequent advancements in this field. Substantial progress has been made in recent years with various DNA nanostructures in the construction of patterned arrays. These include triple-crossover (TX) molecules [21], parallelogram DNA lattices [32], 4×4 DNA nanogrids [58] and DNA lattices made from triangular motifs [27, 4, 13]. All these examples demonstrate the versatility and programmability of DNA-based self-assembly.

1.2 What Is DNA Nanotechnology Good for?

One of the key motivations for using branched DNA to achieve patterned lattices is the desire to use them as scaffolds to organize macromolecules into

3D crystals, thus helping to overcome the bottleneck of protein crystallization experiments. Beyond organizing macromolecules, one can also imagine organizing nanoelectronic components into functional device architectures. Indeed, the programmed and efficient self-assembly of rationally defined nanoarchitectures from nanoscale building blocks, such as nanoparticles and nanowires, is presently one of the outstanding challenges in nanotechnology. Another important application of DNA nanotechnology is the construction of nanomechanical devices based on conformational changes of DNA. Can such devices be used to control protein–protein interactions or metabolic pathways? Can one use 2D DNA nanoscaffolds to template addressable protein arrays for single-molecule proteomics, artificial organelles, enzymatic networks or even a molecular assembler? These are all open questions and will represent the next wave of breakthroughs in DNA nanotechnology.

2 Programmable Self-assembly of 2D DNA Lattices

The unique features of DNA and branched DNA nanostructures make them excellent materials of choice for creating sophisticated nanometer-scale patterns with symmetric, asymmetric or even aperiodic structures. Here we review some examples that demonstrate the programmability of DNA self-assembly for constructing patterned 2D lattices.

2.1 DNA Nanotubes and Nanogrids

We have recently constructed a DNA nanostructure [58] (referred to as 4×4 tile), which can be controlled to self-assemble into two distinct lattice morphologies: DNA nanotubes or 2D DNA nanogrids. Fig. 2a shows a schematic drawing of the strand structure of the 4×4 tile. This DNA tile consists of four four-armed junctions pointing in four directions, which flank a central cavity. The special features of this tile structure include a square aspect ratio, which helps to regularize lattice growth by balancing helix stacking, and sticky-end connections in all four directions within the lattice plane. Note also that a central strand weaves through all four four-armed junctions. Though each arm consists of a flexible four-armed junction structure, when combined with junctions on neighboring tiles they are able to form reasonably rigid nanostructures. Owing to an intrinsic curvature of the tile plane, DNA nanotubes are formed when similar faces of all constituent tiles point in the same direction in the lattice; planar nanogrids, in contrast, are formed when similar faces of adjacent tiles point up and down alternately, which cancels the curvature. The controlled self-assembly of different lattice morphologies with only a slight modification of a single DNA tile unit attests to the statement that DNA self-assembly is programmable. It is worth noting that nanotubes made from other types of DNA tile structures have also been reported and characterized [35, 45, 26, 15, 42, 34]. These groups of DNA tubes may find applications as encapsulants for the controlled release of nanomaterials.

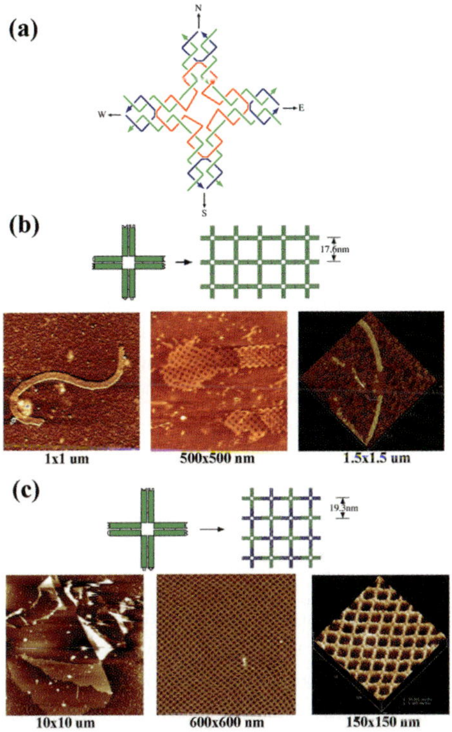

Fig. 3. Self-assembly of DNA nanoribbons and nanogrids using 4×4 DNA tiles. (**a**) Strand structure of 4×4 tile. The tile contains nine oligonucleotides, shown as simplified backbone traces. The four-arm junctions are oriented in each of the directions (N, S, E, W); the red strand participates in all four junctions and contains T_4 loops connecting adjacent junctions. (**b**) Self-assembly of nanoribbons with an original design. *Upper left*: double-helical domains are illustrated as rectangles, and paired rectangles represent four-arm junctions. Complementary sticky-ends are shown as matching geometric shapes. *Upper right*: designed structure of self-assembled lattice. *Bottom*: AFM images of the nanoribbons. Scales are shown below each image. (**c**) Self-assembly of 2D nanogrids with a corrugated design. *Upper left*: the component tile is drawn similarly to that in Fig. 2b; note that the positions of the sticky ends have changed. The tiles have two surfaces, one facing out of the plane and the other facing into the plane. Here the surface facing out of the plane is indicated in green; the other side (when visible) is colored blue. *Upper right*: corrugated self-assembly. *Bottom*: AFM images of the 2D lattices (nanogrids).

2.2 Self-assembly of Patterns with Reduced Symmetry

Future practical applications of nucleic acid nanotechnology will rely on our ability to efficiently self-assemble complex patterns with reduced symmetry. A recent publication by Jaeger and coworkers [9] has described exciting progress toward this goal. By designing 3D artificial RNA motifs called tecto-RNA, these authors programmed the building blocks to self-assemble into a variety of nanoscopic fabrics with increasing complexity and addressability. 2D lattices with predefined finite, addressable patterns can be easily achieved using this strategy.

How complex can the patterns achieved through DNA self-assembly be? Winfree has proposed that if self-assembly proceeds by cooperative binding at multiple weak binding domains, then it should be possible to encode any desired computational rules in a set of "molecular tiles" that will self-assemble into a (possibly quite complex) pattern defined by those algorithmic rules. Recent work [44] by his group on the construction of a Sierpinski triangle pattern using algorithmic DNA self-assembly has demonstrated that engineered DNA self-assembly can potentially be used to construct complex networks.

Yan et al. [57] have recently reported the construction of an aperiodic patterned DNA lattice (a barcode lattice) formed by a self-assembly process via the directed nucleation of DNA DX tiles around a longer DNA scaffolding strand. The long DNA strand not only encoded the barcode information 01101, but also served as the nucleation point for the assembly of DX tiles, with each bit represented by one DX tile. To aid in the visual readout of the encoded information, each bit-1 tile was modified with two stem loops perpendicular to the tile plane, one protruding upward and the other downward; each bit-0 tile, in contrast, was represented by the absence of such stem loops.

Multiple layers of such structure will associate with each other via sticky-end pairing and form a 2D lattice displaying the barcode information, which can be easily detected by AFM (Fig. 4). The nucleated self-assembly can be reprogrammed to achieve another patterning; an inverted barcode pattern 10010 was achieved by modifying the scaffold strands and one of the strands making up each DX tile. The directed-nucleation method described here provides another unique way to implement complex patterns.

3 DNA Lattices as Nanoscaffolds for Templated Self-assembly

A diversity of functional groups and components can be chemically at-tached to DNA. This considerably enhances the attractiveness of DNA self-assembled nanostructures as excellent templates for spatially positioning other functional molecules with subnanometer precision and programmability.

3.1 DNA Scaffolds for Protein Arrays

The 2D DNA nanogrids self-assembled from the 4×4 DNA tiles have a large cavity size, which makes them ideal to serve as scaffolds to organize other molecular components. We have used such lattices to template the self-assembly of streptavidin proteins into periodic 2D arrays [58]. In this experiment, one of the T_4 loops at the center of the 4×4 tile was modified

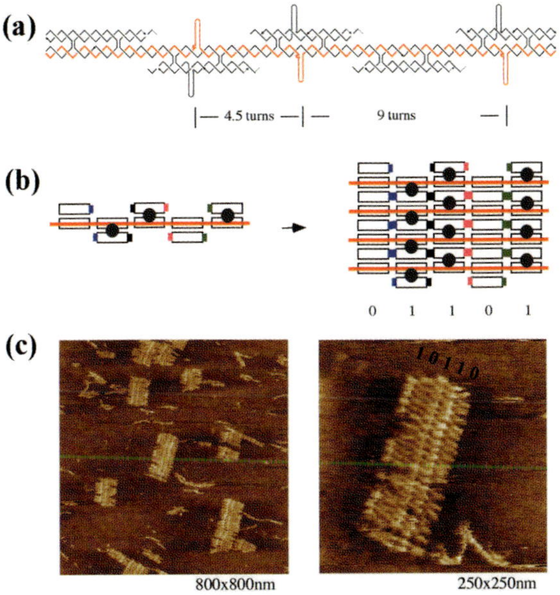

Fig. 4. Self-assembly of 01101 barcode lattice around DNA scaffold strand. (**a**) Strand structure of one barcode layer. This layer represents the barcode information 01101. The red strand is the scaffold strand required for tile assembly. The distances between adjacent hairpin loops are indicated by the number of helical turns. (**b**) Schematic illustration of self-assembly of barcode lattice layers based on DX tiles around a scaffold strand. On the *left*, a five-tile crenellated horizontal layer is shown with an input scaffold strand running through the layer (red). The scaffold strand is required for the tiles to assemble. On the *right*, a lattice of four layers is illustrated (note that sticky ends are still available on the upper and lower layers for the addition of more layers). The sticky ends are represented by different, matching colored areas. The barcode information (01101) is represented by either the presence (1) or the absence (0) of a stem loop (shown as a black circle) protruding out of the tile plane. (**c**) Atomic-force-microscopy visualization of DNA barcode lattice (01101). The scale of each image is indicated at its lower right corner. Up to 24 layers of DNA have been self-assembled; the desired stripe pattern is clearly visible. Each layer contains five DX tiles and is about 75 nm wide. The distance between the two closer adjacent stripes is ~ 16 nm. The distance between the two more distant adjacent stripes is ~ 31 nm.

with a biotin group. When streptavidin was added to the solution containing the self-assembled 4×4 DNA nanogrids, the interaction of streptavidin with biotin led to periodic streptavidin arrays (Figs. 4a,b). We have also demonstrated that these 2D nanogrids can be used to program the directed assembly of streptavidin protein arrays with a controlled spatial separation and density of the proteins [41]. These programmable protein assemblies utilized a two-tile system, where "A tiles" and "B tiles" associated with each other alternately through rationally designed sticky ends, and self-assembled into 2D nanogrids. The A tiles and B tiles can be selectively modified such that either one tile type or both carry biotin groups. Consequently, the combination of selectively modified A and B tiles in the self-assembly and the subsequent binding of streptavidin to biotin leads to a varied periodic spacing of the protein molecules on the DNA lattices. This result clearly shows that DNA can be used as a programmable nanoscaffold.

So far, the noncovalent streptavidin–biotin interaction has been the most explored method to create networks [58, 41, 40, 38, 37] of streptavidin molecules by modifying DNA oligos with biotin groups. However, this method is limited in that only one type of protein–ligand interaction exists. To realize the full potential of DNA scaffolding for protein assembly, the challenge is to develop programmable methods of using DNA nanostructures to direct the assembly of any proteins of interest. To meet this challenge, we have recently demonstrated the use of selective DNA aptamer binding as a robust platform for linking proteins to periodic sites of a self-assembled DNA array [29]. This aptamer-directed self-assembly system employs three components: (1) a rationally designed DNA nanostructure that self-assembles into highly ordered spatial lattices by virtue of specific annealing of complementary sticky ends; (2) a DNA docking site containing an aptamer sequence which will tether the protein of interest to the DNA lattice; (3) the protein to be displayed on the self-assembled DNA lattice. Aptamers are DNA or RNA molecules that can be selected from random pools on the basis of their ability to bind other molecules. Aptamers that exhibit subnanomolar affinities for a wide range of protein targets have been identified. It is possible to generate a virtually unlimited number of specific ligand–aptamer pairs, such that each class of spatially displayed aptamer will interact with high affinity with its specific ligand. The aptamer-directed self-assembly method possesses the following advantageous features. (1) The DNA-tiling self-assembly has been demonstrated to be programmable. (2) Both DNA tiles and aptamers are composed of oligonucleotides, and thus they are compatible with each other. (3) New protein-binding aptamers can be generated through automated processes. (4) DNA-nanostructure-displayed aptamers are not just limited to proteins; they can be extended to bind other ligands. As a proof-of-concept experiment, we have demonstrated the incorporation of the thrombin-binding aptamer sequence into a triple-crossover DNA tile and used the aptamer-bearing TX tiles for the directed assembly of thrombin protein linear arrays (Fig. 6).

3.2 DNA Scaffolds for Nanoelectronics

Metallic, semiconductor and magnetic nanostructures are being actively developed as electronic-device building blocks and chemical sensors. When such nanomaterials are organized into well-defined ensembles, their collective properties depend critically on the interparticle spacing and hierarchical organization. However, methods to control these parameters are still scarce. DNA

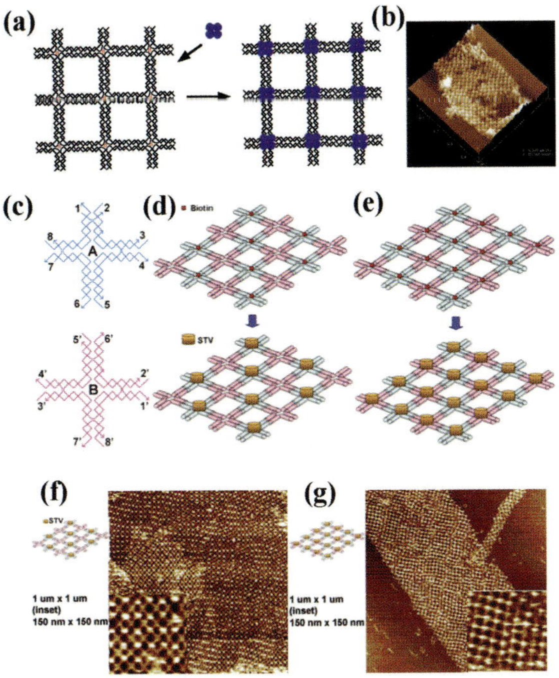

Fig. 5. Programmable self-assembly of streptavidin protein arrays templated by 4×4 DNA nanogrids. (**a**) Schematic drawing of the DNA-nanogrid-scaffolded assembly of streptavidin. *Left*: the DNA nanogrids; a biotin group, labeled by a red letter B, is incorporated into one of the T_4 loops at the center of each tile. *Right*: binding of streptavidin (represented by a blue tetramer) to biotin will lead to a protein nanoarray on a DNA lattice. Note that this is a one-tile system, so that every tile in the lattice contains a biotin group. (**b**) AFM image of a self-assembled protein array; a scale bar is shown at the bottom of the image. (**c**) Strand structure of 4×4 tiles A (blue) and B (pink) for the construction of the 2D nanogrids of a two-tile system. Complementary sticky-end pairs are labeled as n and n'. (**d**, **e**) Schematic drawings of the AB tile self-assembly and subsequent binding of streptavidin onto an A*B lattice and an A*B* lattice, respectively. Biotin groups and streptavidin molecules are represented by smaller red dots and larger yellow cylinders, respectively. (**f**, **g**) AFM images obtained after binding of streptavidin to the bare DNA nanogrids A*B and A*B*, respectively. The scan sizes of both AFM images are 1 μm × 1 μm, with 150 nm × 150 nm zoom-in insets.

self-assembly may provide a viable solution. The periodicities and interparticle spacings defined by DNA nanostructures are readily adjustable, with subnanometer spatial resolution. This level of precision provides exquisite control in the construction of rationally defined 2D and 3D nanoscale assemblies.

The development of self-assembled DNA nanoscaffolds to organize functional nanomaterials is still in a very early stage. Recently, Kiehl and coworkers [22] reported the use of 2D DX lattice as a template to organize 5 nm

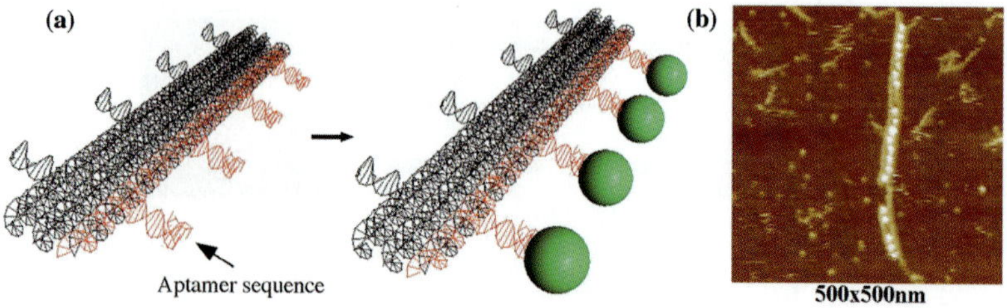

Fig. 6. Aptamer-directed self-assembly of thrombin protein on a triple-crossover DNA tile. (**a**) The red strand contains a thrombin-binding aptamer sequence. The TX tiles self-assemble into a linear DNA nanoarray. The periodic spacing between adjacent pairs of aptamer loops corresponds to 5 turns of DNA, $\sim$ 17 nm. Binding of thrombin protein (green balls) to aptamers on the linear array leads to a linear protein array. (**b**) An AFM image of the aptamer-directed self-assembly of thrombin protein linear arrays. Brighter spots show the thrombin proteins.

Au nanoparticles into arrays. In this strategy, the self-assembly of periodic Au nanoparticle arrays was achieved in three steps. First, a 2D DX array containing single-strand A_{15} at periodic sites was formed in solution. Second, this DNA scaffolding suspension was deposited on a mica surface and allowed to adsorb on the surface. Finally, a solution of DNA–Au conjugate, prepared by attaching a dense layer of thiolated T_{15} oligo to the Au NPs, was deposited on the mica substrate and allowed to hybridize with the A_{15} sequence in the 2D DX array. An alternative strategy for directing the assembly of Au nanoparticles on a DNA scaffold can be achieved through the hybridization of DNA–Au with only one DNA oligo conjugated to the Au nanoparticle [63]. Alivisatos and coworkers have used this method to construct dimeric and trimeric nanoparticle ensembles using linear and branched DNA scaffolds [30, 10]. Most recently, Mao's group has extended this method and used longer linear DNA templates containing repeating sequences to template a linear array of Au nanoparticles [11]. Another interesting study, concerning the reversible switching of DNA–gold nanoparticle aggregations, has been reported by Niemeyer's group [18]. In this design, the distance between neighboring nanoparticles can be controlled by shortening or stretching the DNA fragments connecting them.

Another general strategy for using DNA to scaffold Au nanoparticles relies on biotin–streptavidin binding [39]. For example, we have utilized a linear DNA array composed of TX molecules that are functionalized with biotinylated DNA strands and incubated with streptavidin-coated Au nanoparticles to assemble nanoparticle arrays with controlled spacings [23]. The biotin–streptavidin binding directs the placement of the Au nanoparticles on the TX array. The spacing between neighboring particles is dictated by the periodicity of the nanostructured lattice (Fig. 7). On the basis of the same strategy, Simmel and coworkers [3] have obtained linear chains of nanoparticles on longer scaffolds generated by rolling-circle amplification.

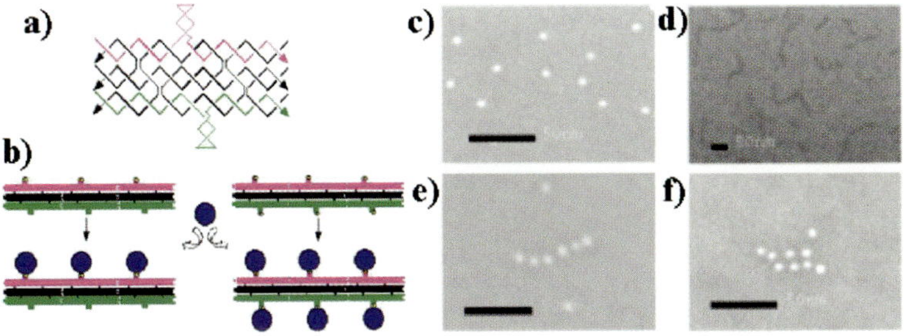

Fig. 7. DNA templated self-assembly of nanoparticle linear arrays. (**a**) The unit triple-crossover DNA tile carrying protruding loops, which can have biotin on one or both sides of the tile. (**b**) Binding of 5 nm gold nanoparticles (blue balls) coated with streptavidin (STV) to biotin-modified TX can lead to two different forms of gold arrays: single- or double-layer gold linear arrays. (**c**)–(**f**) SEM images of: (**c**) STV–gold alone; (**d**) bare TX DNA arrays; (**e**) single-layer STV–gold arrays; (**f**) double-layer STV–gold arrays (scale bars: 50 nm).

4 DNA Nanomechanical Devices

Controlled mechanical movement in molecular-scale devices is one of the key goals of nanotechnology. DNA is an excellent candidate for the construction of such devices owing to the specificity of base pairing and its robust physicochemical properties. A variety of DNA-based nanomechanical devices have recently been demonstrated. In general, the current DNA nanomechanical devices can be classified into three different categories. (1) Devices based on structural transitions triggered by the addition of small molecules to the solution. Examples include the B–Z transition device [33] and pH-driven devices [6, 28]. (2) Devices that are sequence-dependent, where their conformational states are switched by addition and removal of DNA strands that set a particular device state. Examples in this category include DNA tweezers [62], the PX-JX2 device [59], devices based on duplex/G-quadruplex structural transitions [24, 1], and two-legged DNA walkers [49, 51]. (3) Autonomous devices that do not require outside intervention during the conformational transition between states. These include autonomous DNA tweezers [8] based on DNAzyme cleavage, an autonomous unidirectional walker [61] based on ligation and asymmetric endonuclease cleavage, the DNA crawler [53], and another DNA walking device based on reactions of a DNA nicking enzyme [2].

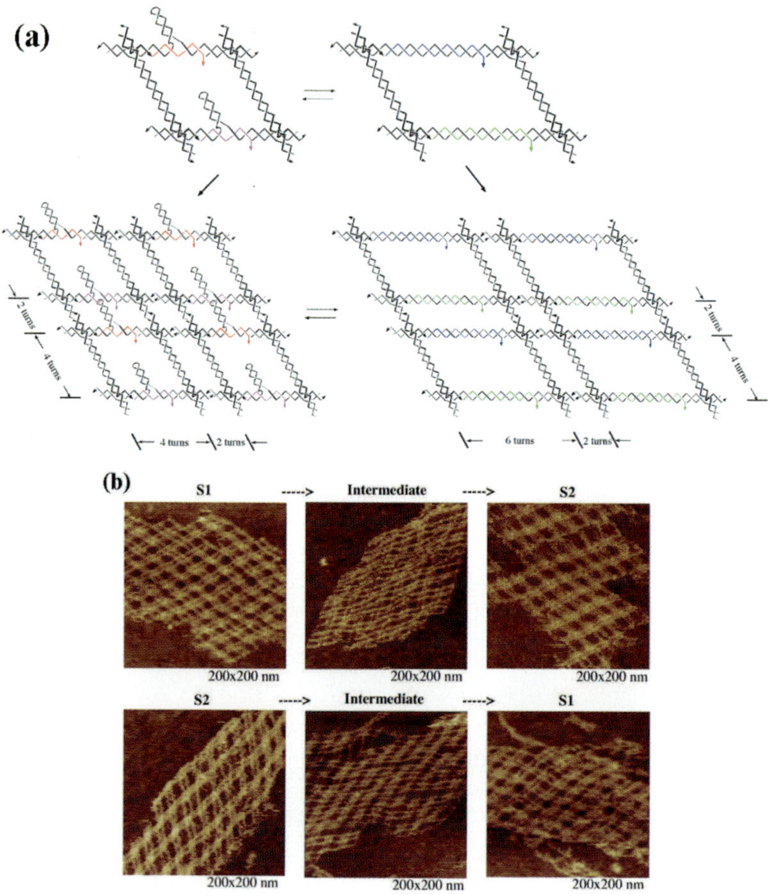

Fig. 8. A 2D DNA lattice capable of changing cavity dimensions. (**a**) Incorporation of DNA nanoactuator devices into a 2D lattice. Two nanoactuator devices, with different base sequences, are incorporated into opposite arms of a rhombus-like motif. The two states of these nanoactuator devices result in two different lattices. (**b**) AFM images illustrating reversible transitions between the two different lattices. Two well-defined end states, S1 (smaller cavity) and S2 (larger cavity), are clearly evidenced. An intermediate state shown in the middle shows a less regular lattice owing to the removal of the set strands.

4.1 Sequence-Dependent DNA Nanomechanical Devices

The DNA-based devices in the first category have the drawback that they are triggered by a single species and are not individually addressable. This means that it is not possible to construct a nanorobotic device containing multiple subunits that can be operated independently. This problem was overcome in a recent breakthrough utilizing a strand displacement technique. Yurke and Turberfield [62] have developed sequence-dependent DNA tweezers. The tweezers assume a contracted state when a particular "set" strand is added to the solution and then assume an open state when the "set" strand is removed by a "fuel" strand that is fully complementary to the "set" strand. Because the "set" strand can be designed to contain different sequences, N different types of tweezers can be constructed using N different "set" strand sequences. Seeman's group has applied the strand displacement technique in the construction of a sequence-dependent, robust rotary device [59]. In this device, reversible rotations between two well-defined states were clearly visible by AFM as the device rotated a larger DNA marker attached to it.

4.2 Incorporating DNA Devices into DNA Lattices

We have recently reported the construction of DNA nanoactuator and incorporated it into a 2D parallelogram DNA lattice [16]. This device can exist in two states. State 1 is the contracted state, with a bulging three-arm DNA branch junction; state 2 is the expanded state, with two perfectly complementary strands of DNA. Operation of the device in the 2D lattice was used to control the lattice morphology. Large alterations in lattice dimensions due to additive changes from each unit cell were observed. Fig. 8a shows the design of this nanoactuator and the lattice in which this device is embedded. The AFM images shown in Fig. 8b demonstrate the interconversion of the two states of the rhombic lattice actuated by the two-state devices. The sizes of the cavities in the rhombic lattice were switched from $\sim$ 14 nm $\times$ 14 nm (the left image) to $\sim$ 14 nm $\times$ 20 nm (the right image). The reverse process from an extended lattice to a contracted lattice was also observed.

4.3 An Autonomous Unidirectional DNA Walker

Most molecular machines that execute cellular functions in the human body are autonomous and in many cases unidirectional, which makes the construction of such autonomous unidirectional devices in artificial systems promising and attractive. We have recently reported the design and experimental construction of an autonomous unidirectional DNA walker on a DNA track [61] (Fig. 9).[1] This walker device has the following features: the device is free from any external environmental mediation, and hence is autonomous; it is powered

[1] Fig. 9: (**a**) The device contains two parts: the track and the walker. The track consists of three evenly spaced duplex DNA anchorages, A, B, and C, each linked to the backbone via a hinge, a four-nucleotide flexible, single-stranded DNA fragment. The walker is a six-nucleotide DNA fragment (colored red and indicated by *) initially positioned at anchorage A. The numbers give the lengths of the DNA fragments, in bases. (**b**) Recognition sites and restriction patterns of PflM I and BstAP I. Bases that are important for PflM I and BstAP I recognition are shown in bold green and pink type, respectively. N indicates the position of a base that does not affect recognition. (**c**). Operation of the device. The left portion shows the sequence of structural changes that occur during the device's operation; the

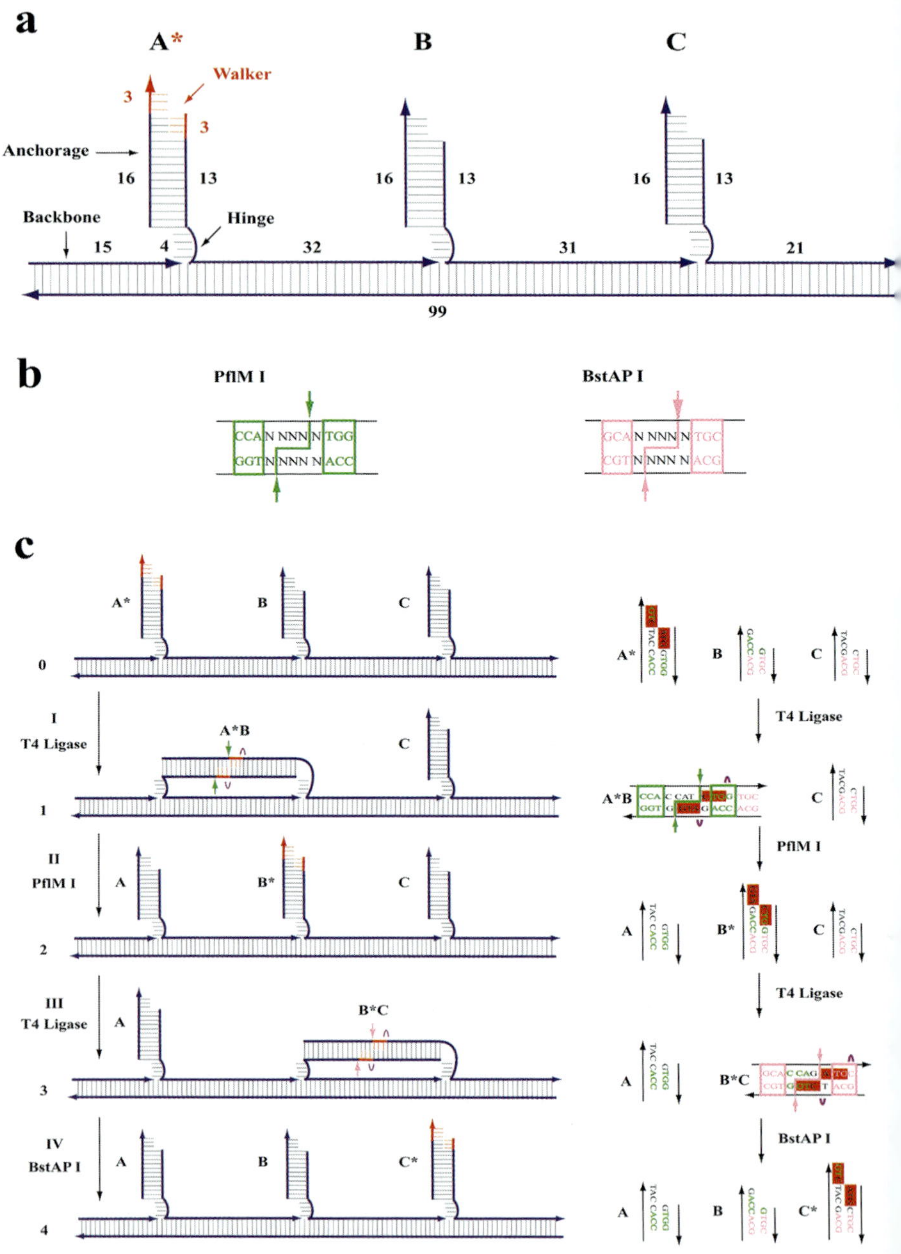

Fig. 9. The structural design and operation of an autonomous unidirectional device.

right portion describes the accompanying enzyme actions and shows how they affect the ends of the anchorages. Panel 0 depicts the device in its initial state.

by the hydrolysis of ATP consumed by T_4 ligase; the motion of the device is unidirectional; and the device executes motion along a DNA track and results in a DNA fragment moving unidirectionally from one end of the track to the other. The operation of the walker device was verified via careful tracking of the radioactively labeled walker using gel electrophoresis.

4.4 Can DNA-Based Nanomechanical Devices Perform the Desired Functions?

The most recent developments by Seeman's and Mao's groups give positive answers to this question. The ribosome-like DNA constructed by Seeman and coworkers [25] provides an exciting example of how a DNA-based nanomechanical device can perform useful functions. This device mimics the translational capabilities of a ribosome by directing positional synthesis of polymers. Different pairs of DNA "set strands" can be added or removed to bring the device into any one of four conformational states. Each state allows the positional alignment of a specific pair of DNA motifs, analogous to an aminoacyl-tRNA, selected from a pool. The pairs bear polymer components that can then be fused in a specific order. As a proof of principle, Seeman's group chose DNA as the polymer being aligned and used enzymatic ligation to fuse the polymers. Consequently, positional synthesis using the prototype device resulted in four different DNA polymers, each containing a defined sequence. Mao's group has utilized a pH-driven duplex–triplex transition device to control the distance-dependent coupling reactions between –NH2- and –COOH-modified DNA oligos [7]. Although the above experiments are only the first steps, more sophisticated nanorobotic applications can be expected in the near future.

5 Conclusions and Outlook

In this review, the use of DNA as designer building blocks for constructing self-assembled DNA lattices and nanomechanical devices has been discussed. The experimental demonstrations described here attest to DNA's role as one of the leading materials for nanoconstruction and nanoscaffolding. The fascinating potential applications of self-assembled DNA structures in nanoelectronics, nanorobotics, and nanocomputation are waiting for us to explore. As nucleic acid-based nanotechnology reaches a new crossroads and more

---

Process I is the ligation of anchorages A* and B, which have complementary sticky ends. This creates a PflM I recognition site. In process II, the device is cleaved by PflM I, transferring the walker to B (panel 2). The new sticky end of B* is complementary to that of C. In process III, B* and C hybridize with each other, and are ligated by T_4 ligase to create a recognition site for endonuclease BstAP I. In process IV, B*C is cleaved into B and C*, transferring the walker to C. This completes the motion of the walker, and the final product is shown in panel 4.

progress is made in constructing molecular devices and patterned superstructures, we need to consider carefully how to reduce errors in self-assembly, how to template functional nanoelectronic device architectures on more complex DNA nanoscaffolds, how to extend 2D self-assembly to 3D and how to use DNA self-assembly to bridge the gap between "top-down" and "bottom-up" construction. Recent progress in designing error-correcting mechanisms [55], DNA-based molecular lithography [12], chemical copying of branched DNA nanostructures [14] and a replicable 3D DNA nanoobject [50] promises a great future for DNA nanotechnology. I would like to end this chapter with a quote from Confucius: "A journey of a thousand miles begins with a single step". The journey to a greater DNA nanotechnology has just begun.

Acknowledgments

The authors thank the National Science Foundation for kind financial support. Hao Yan would like to thank Professors John Reif, Thom LaBean and Andrew Turberfield for their collaboration in some of the research reviewed here.

References

1. P. Alberti and J.L. Mergny. DNA duplex–quadruplex exchange as the basis for a nano-molecular machine. *Proceedings of the National Academy of Sciences of the United States of America*, 100:1569–1573, 2003.
2. J. Bath, S. Green, and A.J. Turberfield. A free-running DNA motor powered by a nicking enzyme. *Angewandte Chemie International Edition*, 2005. In press. Preprint DOI: 10.1002/anie.200501262.
3. S. Beyer, P. Nickels, and F.C. Simmel. Periodic DNA nanotemplates synthesized by rolling circle amplification. *Nano Letters*, 5:719–722, 2005.
4. N. Chelyapov, et al. DNA triangles and self-assembled hexagonal tilings. *Journal of the American Chemical Society*, 126:13924–13925, 2004.
5. J.H. Chen and N.C. Seeman. Synthesis from DNA of a molecule with the connectivity of a cube. *Nature*, 350:631–633, 1991.
6. Y. Chen, S.H. Lee, and C. Mao. A DNA nanomachine based on a duplex–triplex transition. *Angewandte Chemie International Edition*, 43:5335–5338, 2004.
7. Y. Chen and C. Mao. Reprogramming DNA-directed reactions on the basis of a DNA conformational change. *Journal of the American Chemical Society*, 126:13240–13241, 2004.
8. Y. Chen, M.S. Wang, and C.D. Mao. An autonomous DNA nanomotor powered by a DNA enzyme. *Angewandte Chemie International Edition*, 43:3554–3557, 2004.
9. A. Chworos, et al. Building programmable jigsaw puzzles with RNA. *Science*, 306:2068–2072, 2004.
10. S.A. Claridge, et al. Directed assembly of discrete gold nanoparticle groupings using branched DNA scaffolds. *Chemistry of Materials*, 17:1628–1635, 2005.
11. Z. Deng, Y. Tian, S.-H. Lee, A.E. Ribbe, and C. Mao. DNA-encoded self-assembly of gold nanoparticles into one-dimensional arrays. *Angewandte Chemie International Edition*, 44:3582–3585, 2005.

12. Z.X. Deng and C.D. Mao. Molecular lithography with DNA nanostructures. *Angewandte Chemie International Edition*, 43:4068–4070, 2004.
13. B. Ding, R. Sha, and N.C. Seeman. Pseudohexagonal 2D DNA crystals from double crossover cohesion. *Journal of the American Chemical Society*, 126:10230–10231, 2004.
14. L.H. Eckardt, et al. DNA nanotechnology: Chemical copying of connectivity. *Nature*, 420:286–286, 2002.
15. A. Ekani-Nkodo, A. Kumar, and D.K. Fygenson. Joining and scission in the self-assembly of nanotubes from DNA tiles. *Physical Review Letters*, 93, 2004.
16. L.P. Feng, S.H. Park, J.H. Reif, and H. Yan. A two-state DNA lattice switched by DNA nanoactuator. *Angewandte Chemie International Edition*, 42:4342–4346, 2003.
17. T.J. Fu and N.C. Seeman. DNA double-crossover molecules. *Biochemistry*, 32:3211–3220, 1993.
18. P. Hazarika, B. Ceyhan, and C.M. Niemeyer. Reversible switching of DNA–gold nanoparticle aggregation. *Angewandte Chemie International Edition in English*, 43:6469–6471, 2004.
19. N.R. Kallenbach, R.I. Ma, and N.C. Seeman. An immobile nucleic-acid junction constructed from oligonucleotides. *Nature*, 305:829–831, 1983.
20. A.A. Koshkin, et al. LNA (locked nucleic acid): An RNA mimic forming exceedingly stable LNA: LNA duplexes. *Journal of the American Chemical Society*, 120:13252–13253, 1998.
21. T.H. LaBean, et al. Construction, analysis, ligation, and self-assembly of DNA triple crossover complexes. *Journal of the American Chemical Society*, 122:1848–1860, 2000.
22. J.D. Le, et al. DNA-templated self-assembly of metallic nanocomponent arrays on a surface. *Nano Letters*, 4:2343–2347, 2004.
23. H.Y. Li, S.H. Park, J.H. Reif, T.H. LaBean, and H.Yan. DNA-templated self-assembly of protein and nanoparticle linear arrays. *Journal of the American Chemical Society*, 126:418–419, 2004.
24. J.W.J. Li and W.H. Tan. A single DNA molecule nanomotor. *Nano Letters*, 2:315–318, 2002.
25. S.P. Liao and N.C. Seeman. Translation of DNA signals into polymer assembly instructions. *Science*, 306:2072–2074, 2004.
26. D. Liu, S.H. Park, J.H. Reif, and T.H. LaBean. DNA nanotubes self-assembled from triple-crossover tiles as templates for conductive nanowires. *Proceedings of the National Academy of Sciences of the United States of America*, 101:717–722, 2004.
27. D. Liu, M.S. Wang, Z.X. Deng, R. Walulu, and C.D. Mao. Tensegrity: Construction of rigid DNA triangles with flexible four-arm DNA junctions. *Journal of the American Chemical Society*, 126:2324–2325, 2004.
28. D.S. Liu and S. Balasubramanian. A proton-fuelled DNA nanomachine. *Angewandte Chemie International Edition*, 42:5734–5736, 2003.
29. Y. Liu, C. Lin, H. Li, and H. Yan. Aptamer-directed self-assembly of protein arrays on a DNA nanostructure. *Angewandte Chemie International Edition*, 2005. In press. Preprint DOI: 10.1002/anie.200501089.
30. C.J. Loweth, W.B. Caldwell, X.G. Peng, A.P. Alivisatos, and P.G. Schultz. DNA-based assembly of gold nanocrystals. *Angewandte Chemie International Edition*, 38:1808–1812, 1999.

31. R.I. Ma, N.R. Kallenbach, R.D. Sheardy, M.L. Petrillo, and N.C. Seeman. Three-arm nucleic acid junctions are flexible. *Nucleic Acids Research*, 14:9745–9753, 1986.
32. C.D. Mao, W.Q. Sun, and N.C. Seeman. Designed two-dimensional DNA holliday junction arrays visualized by atomic force microscopy. *Journal of the American Chemical Society*, 121:5437–5443, 1999.
33. C.D. Mao, W.Q. Sun, Z.Y. Shen, and N.C. Seeman. A nanomechanical device based on the B–Z transition of DNA. *Nature*, 397:144–146, 1999.
34. F. Mathieu, et al. Six-helix bundles designed from DNA. *Nano Letters*, 5:661–665, 2005.
35. J.C. Mitchell, J.R. Harris, J.Malo, J. Bath, and A.J. Turberfield. Self-assembly of chiral DNA nanotubes. *Journal of the American Chemical Society*, 126:16342–16343, 2004.
36. P.E. Nielsen, M. Egholm, and O. Buchardt. Peptide nucleic-acid (PNA) – a DNA mimic with a peptide backbone. *Bioconjugate Chemistry*, 5:3–7, 1994.
37. C.M. Niemeyer. DNA–protein nanostructures, In C.M. Niemeyer and C.A. Mirkin, editors, *Nanobiotechnology: Concepts, Applications and Perspectives*, pages 227–243. Wiley-VCH, Weinheim, 2004.
38. C.M. Niemeyer, M. Adler, S. Gao, and L. Chi. Supramolecular DNA–streptavidin nanocircles with a covalently attached oligonucleotide moiety. *Journal of Biomolecular Structure Dynamics*, 20:223–230, 2002.
39. C.M. Niemeyer and B. Ceyhan. DNA-directed functionalization of colloidal gold with proteins. *Angewandte Chemie International Edition in English*, 40:3685–3688, 2001.
40. C.M. Niemeyer, T. Sano, C.L. Smith, and C.R. Cantor. Oligonucleotide-directed self-assembly of proteins: semisynthetic DNA–streptavidin hybrid molecules as connectors for the generation of macroscopic arrays and the construction of supramolecular bioconjugates. *Nucleic Acids Research*, 22:5530–5539, 1994.
41. S.H. Park, et al. Programmable DNA self-assemblies for nanoscale organization of ligands and proteins. *Nano Letters*, 5:729–733, 2005.
42. S.H. Park, et al. Three-helix bundle DNA tiles self-assemble into 2D lattice or 1D templates for silver nanowires. *Nano Letters*, 5:693–696, 2005.
43. J. Qi, X.J. Li, X.P. Yang, and N.C. Seeman. Ligation of triangles built from bulged 3-arm DNA branched junctions. *Journal of the American Chemical Society*, 118:6121–6130, 1996.
44. P.W.K. Rothemund, N. Papadakis, and E. Winfree. Algorithmic self-assembly of DNA Sierpinski triangles. *PLoS Biology*, 2:2041–2053, 2004.
45. P.W.K. Rothemund, et al. Design and characterization of programmable DNA nano-tubes. *Journal of the American Chemical Society*, 126:16344–16352, 2004.
46. N.C. Seeman. Nucleic acid junctions and lattices. *Journal of Theoretical Biology*, 99:237–247, 1982.
47. N.C. Seeman. From genes to machines: DNA nanomechanical devices. *Trends in Biochemical Sciences*, 30:119–125, 2005.
48. N.C. Seeman and P.S. Lukeman. Nucleic acid nanostructures: bottom-up control of geometry on the nanoscale. *Reports on Progress in Physics*, 68:237–270, 2005.
49. W.B. Sherman and N.C. Seeman. A precisely controlled DNA biped walking device. *Nano Letters*, 4:1203–1207, 2004.
50. W.M. Shih, J.D. Quispe, and G.F. Joyce. A 1.7-kilobase single-stranded DNA that folds into a nanoscale octahedron. *Nature*, 427:618–621, 2004.

51. J.S. Shin and N.A. Pierce. A synthetic DNA walker for molecular transport. *Journal of the American Chemical Society*, 126:10834–10835, 2004.
52. L.Stryer. *Biochemistry*. W. H. Freeman, New York, 1995.
53. Y. Tian, Y. He, Y. Chen, P. Yin, and C. Mao. A DNAzyme that walks processively and autonomously along a one-dimensional track. *Angewandte Chemie International Edition*, 2005. In press. Preprint DOI: 10:1002/anie.200463101.
54. Y.L. Wang, J.E. Mueller, B. Kemper, and N.C. Seeman. Assembly and characterization of five-arm and six-arm DNA branched junctions. *Biochemistry*, 30:5667–5674, 1991.
55. E. Winfree and R. Bekbolatov. Proofreading tile sets: Error correction for algorithmic self-assembly. *DNA Computing*, 2943:126–144, 2004.
56. E. Winfree, F.R. Liu, L.A. Wenzler, and N.C. Seeman. Design and self-assembly of two-dimensional DNA crystals. *Nature*, 394:539–544, 1998.
57. H. Yan, T.H. LaBean, L.P. Feng, and J.H. Reif. Directed nucleation assembly of DNA tile complexes for barcode-patterned lattices. *Proceedings of the National Academy of Sciences of the United States of America*, 100:8103–8108, 2003.
58. H. Yan, S.H. Park, G. Finkelstein, J.H. Reif, and T.H. LaBean. DNA-templated self-assembly of protein arrays and highly conductive nanowires. *Science*, 301:1882–1884, 2003.
59. H. Yan, X. Zhang, Z. Shen, and N.C. Seeman. A robust DNA mechanical device controlled by hybridization topology. *Nature*, 415, 2002.
60. X.P. Yang, L.A. Wenzler, J. Qi, X.J. Li, and N.C. Seeman. Ligation of DNA triangles containing double crossover molecules. *Journal of the American Chemical Society*, 120:9779–9786, 1998.
61. P. Yin, H. Yan, X.G. Daniell, A.J. Turberfield, and J.H. Reif. A unidirectional DNA walker that moves autonomously along a track. *Angewandte Chemie International Edition*, 43:4906–4911, 2004.
62. B. Yurke, A.J. Turberfield, A.P. Mills, Jr., F.C. Simmel, and J.L.A. Newmann. DNA-fuelled molecular machine made of DNA. *Nature*, 406, 2000.
63. D. Zanchet, C.M. Micheel, W.J. Parak, D. Gerion, and A.P. Alivisatos. Electrophoretic isolation of discrete Au nanocrystal/DNA conjugates. *Nano Letters*, 1:32–35, 2001.

Self-healing Tile Sets

Erik Winfree

California Institute of Technology, Pasadena, CA 91125, USA
winfree@caltech.edu

1 Algorithmic Crystal Growth

Biology provides the synthetic chemist with a tantalizing and frustrating challenge: to create complex objects, defined from the molecular scale up to meters, that construct themselves from elementary components, and perhaps even reproduce themselves. This is the challenge of bottom-up fabrication. The most compelling answer to this challenge was formulated in the early 1980s by Ned Seeman, who realized that the information carried by DNA strands provides a means to program molecular self-assembly, with potential applications including DNA scaffolds for crystallography [19] or for molecular electronic circuits [15]. This insight opened the doors to engineering with the rich set of phenomena available in nucleic acid chemistry [20].

This chapter focuses on what might be considered the most elementary phenomenon, the self-assembly of macromolecular crystals. As commonly practiced today, DNA self-assembly is a two-stage process. In the first stage, which typically occurs at elevated temperatures, DNA oligonucleotides self-assemble into well-defined molecular complexes often known as DNA tiles (e.g. [9]). In the second stage, which typically occurs at substantially lower temperatures, the DNA tiles stick to each other and form crystalline arrays [25]. This self-assembly is mediated by single-stranded "sticky ends" with complementary sequences that allow tiles to stick to each other by forming a double-helical domain. The situation becomes particularly interesting when there are multiple types of DNA tiles containing multiple types of sticky ends. Under such circumstances, the ground state of the crystal might not be a periodic arrangement of the molecular units [13], which motivates generalized concepts of crystalline order [11]. Since in general a physical system may take exponentially long to reach its ground state (meaning that large perfect structures will effectively never form), the use of DNA self-assembly for bottom-up fabrication requires an understanding of the kinetics of crystal growth processes – and the means to control them.

Inspired by Seeman's work, Len Adleman's work on DNA computing [1], and Hao Wang's work on the mathematical theory of tilings [24], algorithmic self-assembly [26] provides a mechanism whereby crystal growth can do information processing. A crystal of DNA tiles can store information in the spatial arrangement of the different sticky-end types exposed on its surface or along its perimeter. When a DNA tile binds to a particular sticky-end combination, it covers them up and simultaneously exposes new sticky-ends – thus effectively modifying the information presented by the crystal. A set of DNA tiles with particular input and output sticky ends therefore corresponds to a program that leaves the trace of its operations embedded in the crystal. As one application, the program can direct the construction of a shape [16]; in fact, in an "error-free" model, self-assembly is universal for the construction of arbitrary shapes [21]. From this perspective, algorithmic self-assembly presents us with an extremely simplified model of morphogenesis based on elementary crystalline growth mechanisms. The most interesting case occurs when the information present in a small "seed assembly" directs the growth of a specific shape or pattern much larger than the seed. Thus algorithmic self-assembly may be compared to biological development, a process that operates robustly over 24 orders of magnitude in volume from the information encoded in DNA to the mature organism.

For algorithmic self-assembly to direct growth at such a large scale, error-free assembly cannot be assumed and fault-tolerance becomes a central issue. Previous work suggested that physically reversible self-assembly can perform "proofreading" on redundantly encoded information [27], that by preventing undesired nucleation on growth facets exponentially low error rates can be achieved [5], and that spontaneous nucleation of undesired assemblies unrelated to the seed can be made arbitrarily rare [17] – all with only a modest increase in the complexity of the tile set. Considered together, this work appears to solve (at least theoretically) the basic issues for fault-tolerant self-assembly according to reversible, error-prone growth processes.

Here, we consider a new model: repair of a self-assembled structure after gross damage – be it destruction by cosmic rays, fragmentation by ripping, or attack by an adversary. We call a tile set **self-healing** if, at any point during error-free growth, when any n tiles (not including the seed) are removed, subsequent error-free growth will perfectly repair the damage in average time $O(n)$. Although hints of self-healing were seen in prior work [27], large damaged areas almost always healed imperfectly. In fact, no previously considered tile sets for 2D algorithmic patterns are self-healing according to this formal definition. Nonetheless, we present a construction that transforms a tile set of interest into a self-healing tile set (containing 25 times more tile types) that performs the same assembly task but at a five-fold larger scale. This transformation works for a wide class of original tile sets, including all widely studied examples.

Since the self-assembled pattern was originally produced by algorithmic growth in the forward direction, the information required for repairing the hole is already present along the perimeter of the hole, and forward growth will rebuild the correct structure – unless backwards growth (which is generally not guaranteed to be correct) gets there first. The key to our construction is to prevent holes (caused by damage) from filling in backwards. This is done by replacing each tile in the original tile set with a 5×5 block of tiles; however, each block is designed such that it can grow in only one direction, "forward." To achieve this, we rely on the technique developed in [5], wherein a pattern of strong bonds, weak bonds, and null bonds within each block controls the order in which tiles can be attached. A simplified 3×3 transformation (not general, but sufficient for transforming a tile set that constructs an infinite Sierpinski triangle pattern) is also shown, as well as a 7×7 transformation that has additional robustness to a type of spurious nucleation error.

1.1 Models for Algorithmic Self-assembly

The abstract Tile Assembly Model (aTAM) provides a rigorous framework for analyzing algorithmic self-assembly. In the formulation used here, *tiles* are considered to be unit squares with each side labeled by a *bond type*. Tiles cannot be rotated. Each bond type has an associated *strength*, which may be 0, 1, or 2 (respectively called null, weak, and strong bonds). A *tile set* is a finite set of tile types, which may be used with replacement during the assembly process. Assembly begins with a specified *seed tile*. A tile may be legally added to an assembly whenever it may be placed so as to match one or more sides with a total bond strength greater than or equal to 2 (i.e., if it either forms at least two weak bonds or one strong bond). Mismatches neither help nor hinder. Assemblies that can be created from the seed tile via a sequence of legal tile additions are called the *produced* assemblies. (For a more formal description, see [21].)

Fig. 1 gives three examples of algorithmic self-assembly from prior work [28, 29, 27]. The Sierpinski tile set, for example, consists of seven tile types: four *rule tiles* (each with four weak bonds), two *boundary tiles* (each with two strong bonds, a weak bond, and a null bond), and one *seed tile* (with two strong bonds and two null bonds). Each tile type is given a distinct color. Growth is unbounded. In the limit, the pattern formed by the pink and yellow tiles give the positions of 0's in a discrete Sierpinski fractal; the other tiles represent 1's. The red arrows, which are not part of the tiles per se, describe the forward growth process by which the assemblies were formed: diagonal arrows point away from two weak input sides (i.e., sides by which the tile attaches to the crystal), while horizontal and vertical arrows point away from a strong input side. At each forward growth site, there is a unique tile type that matches sufficiently many sides to be legally added according to the aTAM; in fact, the rule tiles implement the logic of XOR. The binary counter tile set is similar but uses different logic. Here, tile type colors are chosen to create a derivative pattern. The positions of black tiles in the nth row above the blue tiles correspond to 1's in the binary expansion of the integer n. In general,

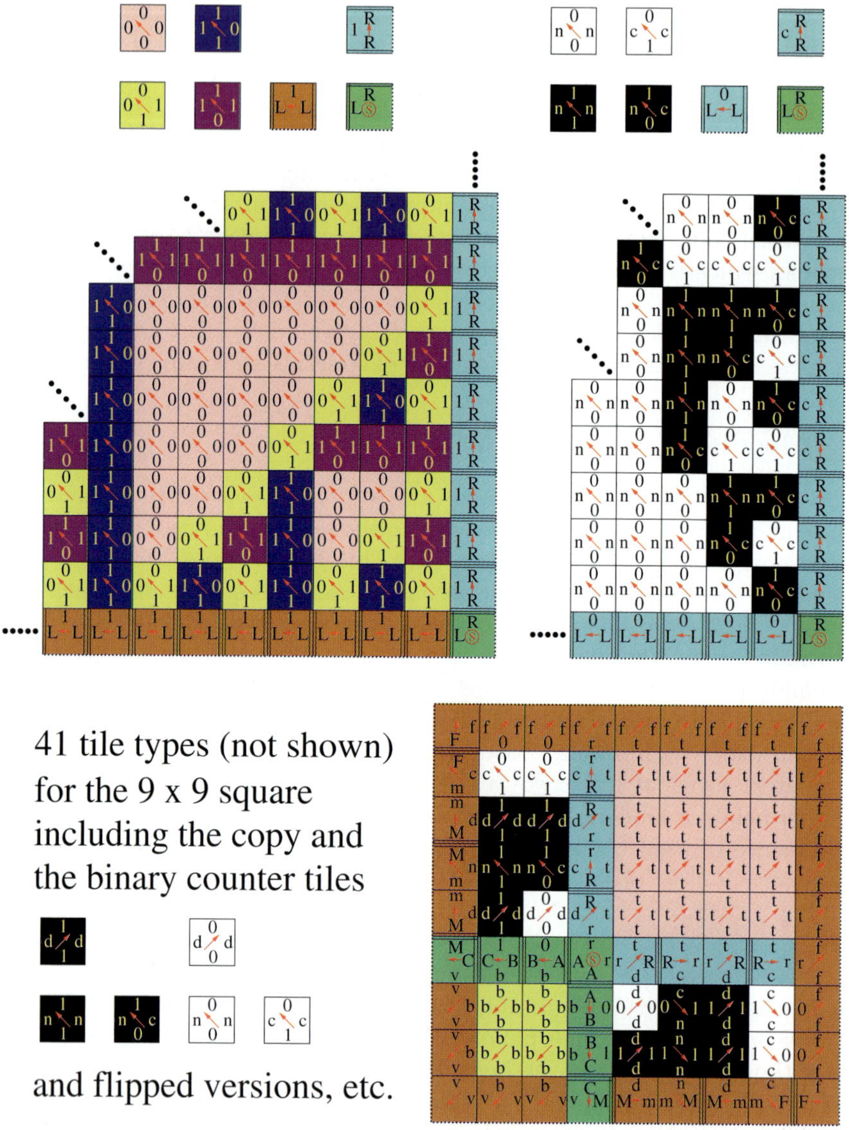

Fig. 1. Three tile sets demonstrating algorithmic growth. Bond types are indicated by letters or digits. Double lines indicate a strong bond, dotted lines indicates a null bond, all other bonds (single lines) are weak. **s** indicates the seed tile. Top left: the Sierpinski tile set. Top right: the binary counter tile set. Bottom: a tile set for constructing a 9×9 square.

by endowing each tile type with a color, the assembled tiling may produce a pattern with less complexity than the tiling itself, since the information processing required to construct the pattern is hidden. In some cases, in fact, a single color is used and one is interested only in the shape of the assembled structure. This is the case for the third tile set, which forms a finite square. (Tile types are colored just to aid understanding of the growth logic.) Here, two orthogonally oriented binary counters count down from an initial number encoded in the green tiles; when they reach zero, the growth is terminated. The 41 tile types (not all shown) may be inferred as the set of distinct tile types that appear in the assembly. This technique can be used to construct an $N \times N$ square by replacing just the green tiles with $O(\log N)$ new tiles encoding the size of the square to be constructed.

Tile additions in the aTAM are non-deterministic, in the sense that at any given moment there are typically several locations where a tile may be legally added; for some tile sets there may also be locations at which more than one tile type may be legally added. Therefore, many tile sets will produce different assemblies dependent upon the order in which tiles are added. The three examples of Fig. 1, however, uniquely produce the assemblies shown. How do we know this? Thankfully, there is a simple yet powerful technique for establishing that this is so for a tile set of interest. Consider an *assembly sequence* of legal tile additions in a particular order. For each tile, we define the *input sides* to be the sides that created weak or strong bonds when the tile was added; the *propagation sides* to be those that serve as the input sides for subsequent tile additions; and the remaining sides (if any) are called *terminal sides*. An assembly sequence is *locally deterministic* if (1) every tile addition makes *exactly* either two weak bonds or one strong bond (i.e., a strength-2 addition), and (2) if the tile at location (i, j) and all tiles abutting its propagation sides are removed from the final assembly, then there is exactly one tile type that can be legally added at (i, j). This is easy to check, tile by tile. Furthermore, if a tile set has a locally deterministic assembly sequence, then the same final assembly is produced regardless of the order in which tiles are added legally. (For a more formal description and a proof, see Theorem 2.3 of [21]. Here, we also allow infinite assembly sequences, which poses no problems for the proof.) Locally deterministic tile sets include the majority of examples considered in the literature.[1] Furthermore, the definitions of self-healing tile sets and transformable tile sets introduced later in this chapter use ideas similar to local determinism.

The aTAM is considered an "error-free" model because perfect assembly can be guaranteed, despite the asynchronous and non-deterministic order of tile additions. This is the appropriate level of abstraction for reasoning about how to program algorithmic self-assembly. However, considering how algorithmic growth can occur in a physical system, such as DNA tiles in solution, requires more realistic models that admit a variety of error modes expected to be present in any real chemistry. For example, the kinetic Tile Assembly Model (kTAM) describes physically reversible assembly as a continuous-time Markov

[1] Not all tile sets that uniquely produce an assembly are locally deterministic; it is a sufficient but not necessary condition.

process in which tiles may be added at a location at a rate proportional to their concentration ($k_f = k[\text{tile type}] = ke^{-G_{mc}}$) regardless of how well they match their neighbors, but tiles also fall off at a rate determined by the total strength, b, of bonds holding them to their neighbors ($k_{r,b} = ke^{-bG_{se}}$). Thus, tile additions that are illegal in the aTAM will sometimes occur in the kTAM and may persist due to the addition of subsequent tiles that stabilize them – resulting in assemblies containing errors. However, if $G_{mc} \approx 2G_{se}$, then exactly the legal tile additions have $k_f \geq k_{r,b}$ (favorable growth) while exactly the illegal tile additions have $k_{r,b} \geq k_f$ (unfavorable growth). Thus, error rates can be reduced to arbitrarily low values by simultaneously decreasing tile concentration and the temperature [28]. This improvement in fidelity comes at the expense of speed: an m-fold reduction of errors requires m^2-slower growth conditions.

Errors can be reduced dramatically without significant slow-down using the technique of block transforms of tile sets that increase their robustness [27]. The transformed tile set contains more tile types but produces the same pattern as the original tile set, although at a larger scale.[2] The basic principle is to make assembly steps cooperative, so that multiple mistakes must occur before erroneous information can be used in subsequent steps; in physically reversible assembly, this gives the erroneous tiles ample opportunity to dissociate before becoming embedded in the crystal – a simple form of "proofreading". Specifically, robustness is achieved by using redundant or distributed information encoded in the bond types and by controlling the growth path by clever placement of strong bonds and null bonds [5]. These techniques can produce tile sets that are robust to several distinct types of errors that can occur in the kTAM: *growth errors* in which a weakly-binding tile attaches at a location where another tile could and should have been added; *facet nucleation errors* in which a weakly-binding tile attaches at a location where no tile should yet be added; and *spontaneous nucleation errors* in which a large assembly grows in the absence of a seed tile. By replacing each tile in the original tile set by a $k \times k$ block of new tiles, growth and facet nucleation errors [5] can be reduced exponentially (in k) with only moderate slow-down. A similar exponential reduction can be achieved for nucleation errors in a mass-action variant of the kTAM [17]. Each of these works addresses only certain error types and provides a construction that works for a limited class of tile sets. Therefore, the outstanding issue for fault-tolerant algorithmic self-assembly according to reversible, error-prone growth processes is whether these methods can be combined into a single transformation that works for a wide class of tile sets and simultaneously solves all three types of errors. Although we do not yet have a definitive answer to this question, it appears that the basic principles have been identified and the foundation has been laid for a complete solution.

[2] Transformations that don't increase the scale also exist, but they come at the cost of a dramatic increase in the number of tile types for most patterns [14, 22].

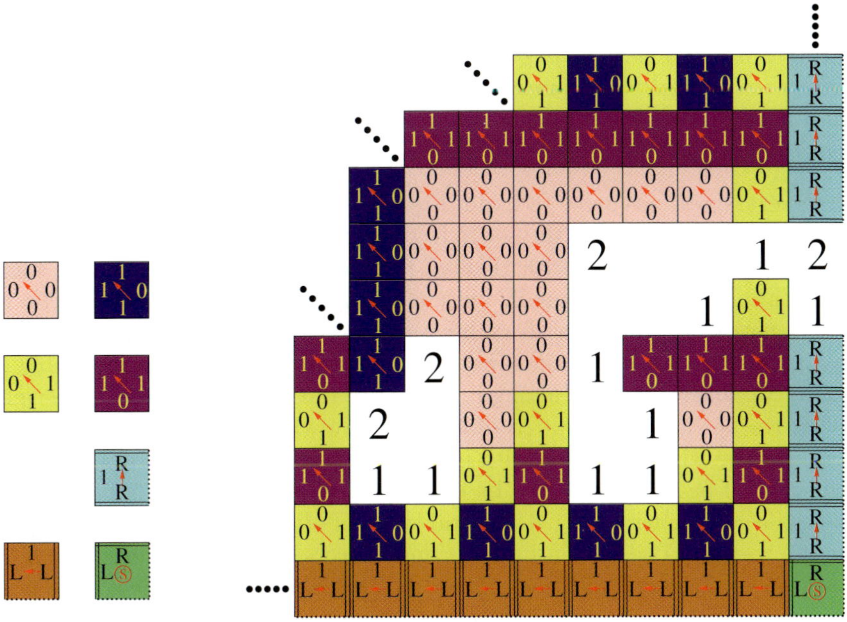

Fig. 2. Erroneous regrowth of a Sierpinski assembly after damage that removed $n = 19$ tiles. Numbers at empty sites indicate the number of distinct tile types that could be legally added at a site according to aTAM growth; where more than one tile may be added at a location, correct regrowth is no longer guaranteed. Note that the positions where incorrect regrowth can occur are dependent upon the tile set; the binary counter tiles, for example, exhibit non-deterministic regrowth in different growth directions (regrowth from north and west inputs is sometimes ambiguous, as is regrowth from east and west inputs) than the Sierpinski tiles (regrowth from north and west inputs is always ambiguous).

1.2 The Challenge of Self-healing Crystals

Here we consider a qualitatively new type of error: gross damage to an assembly, such as a puncture, that removes a region containing many tiles. Such events are so rare in the kTAM as to be effectively non-existent, yet it is easy to imagine physical circumstances that would result in gross damage, such as fragmentation and ripping induced by fluid flow or interaction with other objects in solution. The question is whether an algorithmic crystal subject to such misfortune will be capable of healing the damage correctly. This self-healing behavior was observed to occur frequently, but not always, in kTAM simulations of proofreading tile sets [27]. Can self-healing behavior be *guaranteed* for some tile sets? We formulate this question with respect to the aTAM, so as to focus on the information-propagation aspects of the problem rather than on the probabilistic aspects.

Definition 1. *We call a tile system* **self-healing** *(in the aTAM) if the following property holds for any produced assembly: If any number of tiles are removed such that all remaining tiles are still connected to the seed tile, then subsequent growth is guaranteed to eventual restore every removed tile without error.*

Several questions come to mind. First, why the restriction that remaining tile are still connected to the seed? If gross damage breaks an assembly into several pieces, we might wish that *all* fragments regrow properly. But some fragments could be very small – just a few tiles – and it is unreasonable to expect correct regrowth in all such cases. On the other hand, since we know that growth from the seed tile is capable of constructing the entire assembly, it is also capable of *re*-constructing it, at least *if* tile additions occur in the right order. Rather than attempt to discern exactly which fragments can support regrowth, we will be satisfied with just the seed fragment.[3] So, are our favorite tile sets self-healing? This can be quickly answered, in the negative, for the three tile sets shown in Fig. 1. Several types of erroneous regrowth are shown for the Sierpinski tile set in Fig. 2; similar errors occur in the other tile sets. Then do any self-healing tile sets exist? Yes; the simplest example is a periodic crystal in which every bond type is unique to the two tile types it joins, and all bonds are strong. Uniquely-addressed finite assemblies can also be self-healing. This is, however, an extremely limited class of self-assembled patterns. Can algorithmic self-assembly be self-healing? It is far from obvious.

A first hope might be that robustness-enhancing tile set transformations, such as the original [27] and snaked [5] proofreading schemes, already provide self-healing properties. While kTAM simulations do show improved ability to regrow into punctures, it is not perfect, and in the aTAM errors are even more frequent. Examination of those block transformations suggests that typically both proofreading approaches will result in new tile sets that suffer the same regrowth problems as the original tile sets.

The remainder of this chapter shows that self-healing is possible for algorithmic self-assembly. We first present a 3×3 block transformation that can be applied to tile sets, like the Sierpinski and binary counter tile sets, that grow within a quarter-plane from an L-shaped boundary. A proof technique is developed for showing that the resulting tile sets are indeed self-healing. The simplicity of these techniques makes it straightforward, then, to design and test block transformations that work for a wider class of tile sets. We present a 5×5 scheme that works for many (though not all) locally deterministic tile sets, including all three examples from Fig. 1. Finally, we ask how these results are affected if regrowth occurs not one tile at a time (as in the aTAM), but by the addition of strongly-connected chunks of tiles that may have formed on their own without the seed (which we call the polyomino aTAM). Under these more challenging conditions, self-healing is still possible, but our construction uses 7×7 blocks. We conclude with a discussion of open questions.

[3] Salamanders can regrow their tails, but their tails can't regrow the salamander.

2 Self-healing Transformations for Quarter-Plane Patterns

What makes self-healing hard? The problem is that whereas the original tile set may have been deterministic when growing in the expected directions, with the expected input sides and propagation sides, regrowth may occur in any direction and tile additions may no longer be deterministic. For example, in the case of quarter-plane growth from an L-shaped boundary, the rule tiles have four weak bonds, two of which serve as input and two as output; while there must be a unique tile for any input pair, there may be multiple tiles that have the same output pair. If such a tile is removed, the other (incorrect) tile could be added during regrowth, binding by the two weak bonds on its output sides. Incorrect regrowth could also occur if two tiles share some combination of an output side and an input side (either adjacent or opposing). Between the Sierpinski tile set and the binary counter tile set, all such situations occur.

Quarter-plane growth from an L-shaped boundary is a rich class of tile sets, capable of creating a great variety of patterns. In fact, it is sufficient for universal computation by simulation of blocked cellular automata or Turing machines [26, 28]. In general we may wish to use rule tiles simulating a blocked cellular automaton that outputs $\langle f(x,y), g(x,y)\rangle$ for input $\langle x,y\rangle$ where x and y are from some possibly large finite alphabet and $f(\cdot,\cdot)$ and $g(\cdot,\cdot)$ are arbitrary functions.[4] Input is provided by the boundary tiles that create the L; locally deterministic growth allows each arm of the L to consist of a finite initial sequence of boundary tiles followed by a finite repeating sequence. We need a block transformation that works for all such tile sets, which we call L-BCA tile sets.[5]

There are two options: either to make sure that in the transformed tile set sufficient information is present for any direction of regrowth, or else to ensure that regrowth in the wrong direction is impossible. The former seeming impossible, we take the second approach. The only way to prevent backward and sideways regrowth is for the transformed tile set to contain null bonds at key positions that control the growth path; the principle here is adapted from the mechanism that prevents facet nucleation in the snaked proofreading construction [5]. A 3×3 self-healing block transformation is shown in Fig. 3, wherein each original tile produces nine new tiles with labels and bond strengths according to a template that depends upon the original tile's bond strength pattern. (Rotated tiles use rotated templates.) For each tile type $t = \langle a, b, c, d\rangle$ in the original tile set, nine tile types are included in the new

[4] Reversible 1D cellular automata, for which the "inputs" $\langle x,y\rangle$ are also a function of the "outputs" $\langle f,g\rangle$, are a widely studied class that includes Turing-universal computation [12]. Tile sets directly simulating these cellular automata would be immune from ambiguity in the backward regrowth direction, but the possibility for problems in other directions remains.

[5] These are quarter-plane tile sets discussed in [22], but augmented by a seed tile and the boundary tiles.

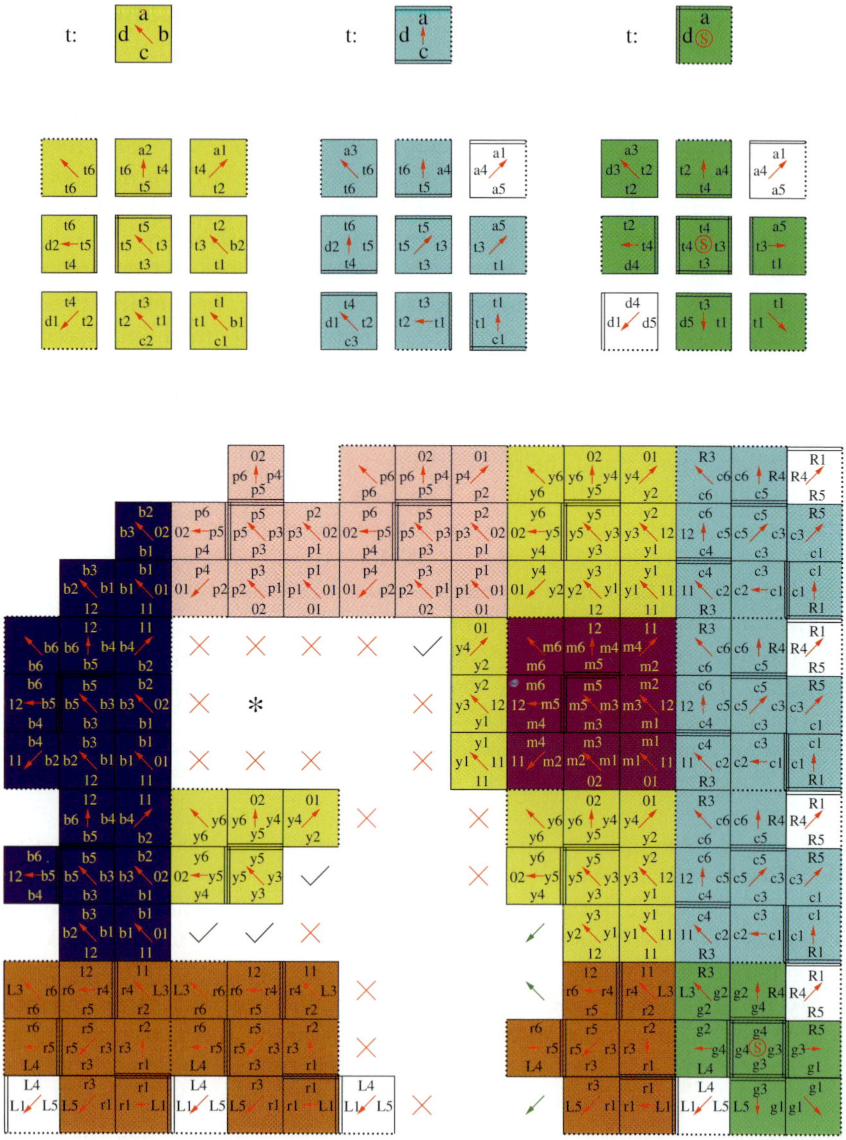

Fig. 3. Top: templates for the 3 × 3 self-healing transformation for rule tiles, boundary tiles, and the seed tile. Bottom: a damaged assembly grown using the transformed Sierpinski tile set. Sites may allow no immediate regrowth (red crosses), regrowth from incorrect input sides (black checks), or regrowth from correct input sides (green arrows). Note that a series of legal tile additions allows the partial yellow blocks to regrow independent of other activity, but the upper left damaged block (*) cannot regrow until both its input blocks have formed.

tile set. The new bond types are indexed variants of the original tile type (e.g. $t3$) or bond type (e.g. $a5$). These are respectively called *tile-type bonds* and *bond-type bonds*. New tiles that use at least one tile-type bond are given the same color as the original tile and are called *block tiles*, whereas new tiles that use exclusively bond-type bonds are left uncolored and are called *bond tiles*. The same bond tile type may result from the transformation of distinct original tiles. Examination of a damaged crystal grown from the transformed Sierpinski tile set illustrates the inability to grow backwards or sideways where there is any potential ambiguity. But how can we prove that this is always the case? We use two simple lemmas, stated informally but hopefully unambiguously.

Lemma 1. *If a tile can be added at a particular site in some assembly, then it can be added at the same site (if it is open) in any larger assembly that contains all the same tiles (and then some).*

This follows immediately from the threshold condition for tile addition in the aTAM: bond strengths are non-negative and mismatches do not interfere (i.e., they contribute strength zero). □

Lemma 2. *Consider an assembly produced from a tile set according to the aTAM. Remove any single tile (not the seed), as a test. The test succeeds if there is a unique tile that can now be added at that site according to aTAM growth. The tile set is self-healing if and only if this test succeeds for every possible tile in every possible produced (i.e., correct) assembly.*

First, the easy implication is immediate: if a test fails, then the tile set is not self-healing. For the converse, now suppose that a tile set is not self-healing. That means that there is some pattern of damage, and some sequence of regrowth that leads to a first incorrect tile t. Prior to adding t, every tile that was present was correct. Add to this assembly all the other tiles (other than at t's site) that had been removed in the damage. We now have a produced (correct) assembly with a single tile removed. By Lemma 1, t can be added in this assembly too. So can the correct tile, which is different from t. We have thus identified a test that fails. □

We can now prove that the 3×3 self-healing transformation works for all L-BCA tile sets.

Theorem 1. *The 3×3 block transformation shown in Fig. 3 produces a self-healing tile set when applied to any L-BCA tile set. Furthermore, the resulting tile set will construct the same pattern as the original tile set, but at a three-fold larger scale; specifically, the majority color of each block will be identical to the corresponding tile in the original pattern.*

To prove this, we need to show (1) that aTAM growth from the seed will produce the correct pattern (at a larger scale), and (2) that every test conceivable by Lemma 2 is bound to succeed. For claim (1), all we need to do is identify a locally deterministic assembly sequence. This is easy: since all L-BCA tile sets are locally deterministic, we can start with a locally deterministic assembly sequence for the original tile set, and show that we can elaborate it into an assembly sequence for the transformed tile set that remains locally deterministic. For each tile added in the original sequence, we add a series of nine tiles for the corresponding block of the transformed tile set. Since we know which sides are the input sides for the original L-BCA tile (south and east for rule tiles, south or east for boundary tiles, and no inputs for the seed tile), it is easy to find a canonical series of tile additions for each transformed block, assuming the blocks for the corresponding inputs are already completely present. Therefore, in the blocks each tile has a canonical growth direction (as illustrated) and it can easily be verified that each tile addition is locally deterministic (exactly strength-2, and growth from input and terminal sides is unique). When at least one input side is within the block, uniqueness is automatically guaranteed; when binding via a single strong bond as the first tile in a boundary block, uniqueness follows from the uniqueness of tile addition on the boundary in the original tile set; when binding via two weak bonds as the first tile in a rule block, uniqueness follows again from uniqueness of tiles with a given input pair in the original tile set. Thus, we have constructed a locally deterministic assembly sequence for the transformed tile set. This establishes part (1) of the result.

Part (2) can also be established by local examination of the transformed blocks, using Lemma 2. For each tile within each block, we examine all possible combinations of sides that contain total bond strength at least 2, and we ask whether there is a unique tile that matches those sides. For rule blocks, eight tiles need at least one side internal to the block, which therefore establishes uniqueness; the exception is the tile in the lower right corner, which can grow from input sides on the south and east – but for L-BCA tile sets, there is a unique tile with this pair of inputs. So no test can fail within a rule block. For boundary blocks, there are three exceptions to the rule that at least one side must be internal to the block; these are (a) the bond tile, whose sides are unique to that bond type; (b) the lower right tile, which is unique because boundary growth in the original tile set is unique; and (c) the lower left tile, which is unique for the same reason. For the seed block, the only exceptions are again the bond tiles (which are unique) and the upper left tile (which is the only tile touching both a horizontal and a vertical boundary, and thus is unique). This establishes the conditions for Lemma 2, and thus completes the proof that the transformed tile set is indeed self-healing. □

This proof also helps us understand why it was necessary to include "bond tiles", which at first seem like an out-of-place hack: if both boundary blocks and the seed block had block-specific tiles in the upper right corner, then backward growth from a boundary block (with a damaged region underneath it) would no longer be unique – sometimes the seed block corner tile would attach in this position. In fact, we will see that bond tiles play an important role in the more general self-healing transformations to come.

3 A General Self-healing Transformation

We now know that it is possible to have algorithmic growth that is self-healing. Unfortunately, L-BCA tile sets, though computationally universal, do not include most examples of algorithmically generated morphology, such as the square of Fig. 1, which exhibit much greater variety in the growth path. We would therefore like a self-healing transformation that will work for any locally deterministic tile set, thus being applicable to essentially all algorithmic self-assembly tile sets considered in the literature (e.g. [16, 2, 7, 21]). However, two situations are allowed in locally deterministic tile sets that cause technical difficulties for the block transformations that are presented below: first, the same tile type might appear with different input sides and propagation sides at different locations within an assembly; and second, the correct final assembly might contain weak bonds along the outer perimeter or internal mismatches between tiles. Rather than attempt to handle these possibilities, we choose to restrict our attention to tile sets in which neither of these situations arises. Many tile sets in the literature are already of this form, and others can be converted with a little thought. In any case, we are led to the following definition.

Definition 2. *A* **transformable** *tile set is a locally deterministic tile set with the additional properties that (1) each tile type always appears with the same sides as input, propagation, and terminal sides, and (2) all non-null bonds are either input sides or propagation sides.*

This means that all final structures are "capped" by tiles with null bonds on the outside, as is the case in the square assembly of Fig. 1. It also means that tile types can be labeled with arrows indicating their input sides, and these labels are correct for all locally deterministic assembly sequences.

Rather than the three bond-strength patterns (rule tiles, boundary tiles, seed tiles) of L-BCA tile sets, transformable tile sets may have a great variety of bond-strength patterns. We can avoid having a separate block transform for every bond-strength pattern by adopting a uniform convention for the interface between blocks that is the same regardless of whether a strong bond or a weak bond is being represented. By using the same block transform for all tile types with the same input bond-strength pattern (regardless of the strengths on the non-input sides), we require only four block schemes to be specified, as shown in Fig. 4 for a 5×5 self-healing transformation. The four cases are (a) *diagonal rule blocks*, in which two adjacent weak bonds serve as the input; (b) *convergent rule blocks*, in which two weak bonds on opposite sides of the tile serve as the input; (c) *strong blocks*, in which a single strong

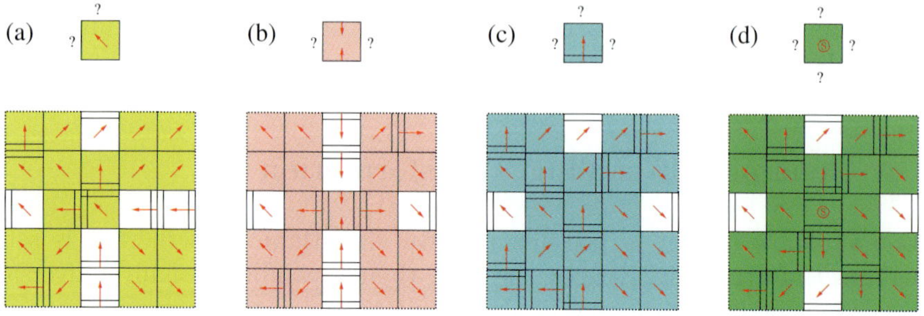

Fig. 4. A 5×5 self-healing transformation. Top: the four bond-strength patterns for tile input sides. Non-input sides (indicated by ?) may have any strength. Bottom: the corresponding block templates. Colored tiles are called *block tiles*; bonds between block tiles are tile-type bonds. Uncolored tiles are *bond tiles* and have exclusively bond-type bonds. Within each block template, each bond type appears uniquely (in any given direction). Bond-type bonds in equivalent positions of different blocks are of the same type. If the original tile had a null bond, then the corresponding bond tile in the template is replaced by a block tile with three tile-type bonds and a null bond. Original tiles with rotated input bond patterns use rotated templates.

bond serves as the input; and (d) *seed blocks*, which have no input. Note that bond tiles now play a much more prominent role in the blocks. The growth pattern within each block is designed so that output bond tiles receive their input in a clockwise growth direction; this way, rotated blocks that define the same bond tile will use it with a consistent growth direction.

Theorem 2. *The* 5×5 *block transformation shown in Fig. 4 produces a self-healing tile set when applied to any transformable tile set. Furthermore, the resulting tile set will construct the same pattern as the original tile set, but at a five-fold larger scale; specifically, the majority color of each block will be identical to the corresponding tile in the original pattern.*

The proof of this theorem follows exactly along the lines of the proof for the 3×3 transformation, but with more cases to test for Lemma 2. Tests of block tiles succeed because either at least one input is another block tile from the same block, or else all inputs (from bond tiles) specify a unique tile due to the original tile set being locally deterministic. Tests of bond tiles succeed because all their bonds are unique to the particular bond tile type. □

This transformation is now sufficient for showing that the growth of arbitrary algorithmic shapes [21] can be self-healing. Incidentally, the transformation of a transformable tile set is also a transformable tile set, an interesting closure property that could aid in combining different robustness transformations.

We can now revisit the question of why bond tiles were necessary. The essential reason is that in the original tile set, strong bonds dictate a deterministic tile choice in the forward growth direction, but may be non-deterministic in the backward growth direction. This difficulty is compounded by our choice (made for the convenience of being able to write the block transformation concisely) to treat output sides uniformly for both weak and strong outputs. Consequently, *every* output side has a strong bond, and non-deterministic backward growth could be severe. Thankfully, by padding all sides of the block with null bonds, we can prevent the backward growth from continuing for more than a single tile – the bond tile. However, all those null bonds make forward growth difficult for diagonal blocks and convergent blocks, because the two pieces of information required to know the new block's type are not co-localized. The solution in this case is to project that information into the center of the tile by a non-committal growth process (bond tiles); the actual decision is then made in the center where the information can be combined.

4 Self-healing for Polyomino Tile Sets

Tile sets produced by the 5×5 self-healing transformation have a lot of strong bonds, even when the original tile set had relatively few. This elicits some concern from those familiar with physical self-assembly, because it brings into question the assumption that growth occurs only from the seed tile, and that all subsequent steps consist of the accretion of a single isolated tile at a time, rather than by the aggregation of separately nucleated fragments. In the absence of the seed tile (for the seed block), one can consider aTAM growth from each of the other tiles in the tile set. Ideally, such growth cannot proceed far, thus supporting the accretion hypothesis in spirit if not in detail. However, we are not so lucky with this 5×5 transformation. The worst offenders here are the strong blocks: starting with first tile in the block's usual assembly sequence as a "mock seed", aTAM growth puts together the entire 25-tile block, and possibly more. This is just asking for trouble.

We therefore consider whether it is possible to create self-healing tile sets in which significant spurious nucleation does not occur, and for which aggregation of seeded assemblies with spuriously nucleated assemblies is too weak to proceed, except when it results in correct assemblies. Previous work on controlling spurious nucleation in a mass-action kTAM model made use of the principle that growth from a non-seed tile must take several unfavorable steps (which would not be allowed in the aTAM) before unbounded favorable growth (allowed in the aTAM) becomes possible [17]. Essentially, the solution presented there corresponds to a block transformation in which strong bonds are placed sufficiently far apart; in fact, instead of using tiles with strong bonds, in that work such tiles were permanently stuck together and treated as a single polyomino tile with each unit side containing a weak bond (or a null bond). The polyomino formalism provides a suitable "worst-case" framework for treating aggregation. (Our model is essentially the same as the "multiple tile" model of [3].)

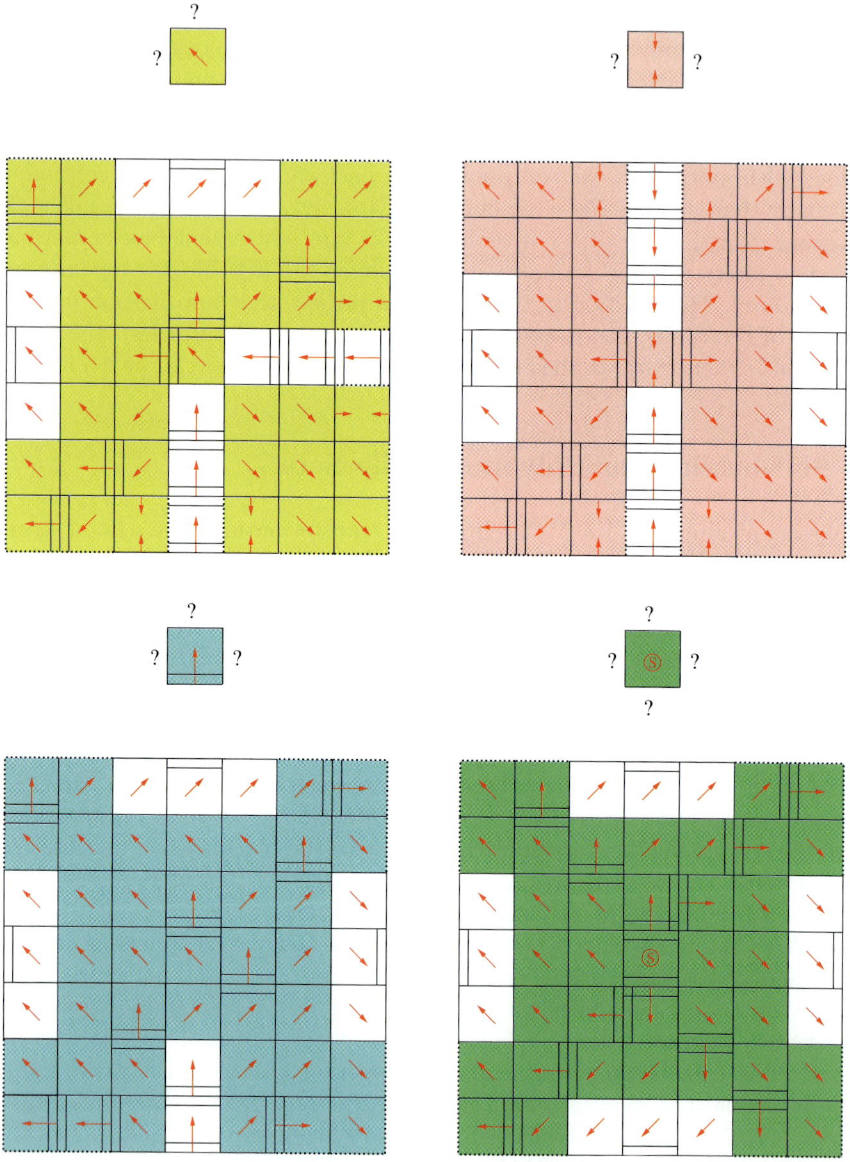

Fig. 5. A 7 × 7 self-healing transformation that yields polyomino-safe tile sets. Top: the four bond-strength patterns for tile input sides. Bottom: the corresponding block templates. Note that each side of each block now exposes one strong bond and two weak bonds.

Given a tile set that uniquely produces a target assembly under aTAM growth from the seed, we will define a corresponding set of polyominoes. Begin with the given tile set excluding the seed tile – this is the first step in the construction of all possible spuriously nucleated assemblies (here called polyominoes). Now iterate: if it is possible to place two such assemblies next to each other such that they can form bonds with a total strength at least 2, then add the resulting assembly to the set of polyominoes. If this process does not terminate or if any polyomino is not a subset of the target assembly, declare failure; the given tile set is not polyomino-safe. Otherwise, we have a finite set of polyominoes representing assemblies that have spuriously nucleated and aggregated without the seed tile.

The polyomino aTAM begins with the seed tile and allows the addition of any polyomino (in the set defined above) placed such that it can form bonds with a total strength of at least 2. Under the polyomino aTAM, any assembly that was produced by the aTAM can still be produced, since all individual tiles are also in the polyomino set (except for the seed tile itself, which is not used for growth in transformable tile sets). Possibly additional (and thus incorrect) assemblies can also be formed when polyominoes are used. For our purposes, uniqueness will follow from the polyomino-safe self-healing property – if deviations from the correct tile placement are impossible during regrowth, then it must also have been impossible during growth the first time around.

Definition 3. *We say a tile set gives rise to* **polyomino-safe self-healing** *if the following property holds for any produced assembly: If any number of tiles are removed such that all remaining tiles are still connected to the seed tile, then subsequent growth according to the polyomino aTAM with the corresponding polyomino set is guaranteed to eventually restore every removed tile without error.*

To prove that a tile set has this property, we need polyomino variants of the previous lemmas.

Lemma 3. *If a polyomino can be added at a particular site in some assembly, then it can be added at the same site (if it is open) in any larger assembly that contains all the same tiles (and then some).*

Lemma 4. *Consider an assembly produced from a tile set according to the aTAM. Choose a polyomino from the corresponding polyomino set, and choose a location where it overlaps existing tiles. (It necessarily does not overlap the seed tile.) As a test, remove all overlapped tiles. The test succeeds if either the polyomino makes no more than a single weak bond with the remaining assembly, or if all tiles in the polyomino are identical with the removed tiles. The tile set gives rise to polyomino-safe self-healing if and only if this test succeeds for every possible case.*

The proofs are straightforward adaptations of the proofs of the previous lemmas. □

It now becomes straightforward, although tedious, to verify the following.

Theorem 3. *The 7×7 block transformation shown in Fig. 5 produces a polyomino-safe self-healing tile set when applied to any transformable tile set. Furthermore, the resulting tile set will construct the same pattern as the original tile set, but at a seven-fold larger scale; specifically, the majority color of each block will be identical to the corresponding tile in the original pattern.*

The corresponding polyomino set contains only small polyominoes (no more than four tiles each) that consist of either entirely bond tiles or entirely block tiles. Bond polyominoes can only replace identical bond tiles, since their bond-type bonds are unique. Block polyominoes may have both tile-type bonds and bond-type bonds. Most block polyominoes have no more than one bond-type bond; therefore, to attach, the polyomino must make at least one tile-type bond, which uniquely positions it within the correct block. The only exceptions occur at the centers of diagonal and convergent rule blocks and at the input to strong blocks. At these sites, a block polyomino may bind by bond-type bonds with strength 2, but in these cases uniqueness is guaranteed by the original tile set being locally deterministic. □

This tile set operates on the same principles as the 5×5 tile sets, with the added precaution that in order for a strong block to grow, the central strong bond tile must be supported by tiles presenting weak bonds on either side. By distributing responsibility for propagating information through the sides of the blocks, no single tile on its own is capable of nucleating the growth of the entire block. Note that even if the original tile set was *not* polyomino-safe, the transformed tile set will be.

5 Open Questions

We now know that self-healing is possible in passive self-assembly. How good can it get?

Generality and Optimality of the Block Transformations. The first question is whether a wider class than the "transformable" tile sets can be made self-healing. Tile sets that produce a language of shapes – rather than uniquely producing a target assembly – are clearly not going to work, because self-healing can't be guaranteed at the first non-deterministic site. But might it be possible to find a transformation that works for any locally deterministic tile set?

Scale is an important issue for self-assembled objects [21, 14]. In previous work on fault-tolerant self-assembly (in the kTAM), increased robustness was achieved at the cost of increased scale [5, 17, 22]. In this work (in the aTAM), a maximal level of robustness is achieved with a constant scale-up – seven-fold, for polyomino-safe self-healing. It is intriguing to ask whether the strategies of [14, 22] can be use to produce that self-healing tile sets that incur *no* scale-up costs – although this will come at the cost of an increase in the number of

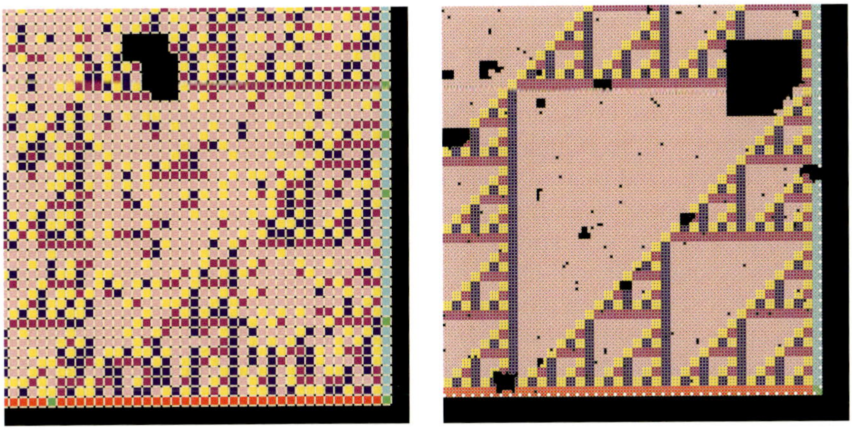

Fig. 6. Growth under a barrage of damage events. Size $k \times k$ square puncture events occur (centered at any given tile) at a rate $1000k^4$-fold less than the forward rate f for tile addition (i.e., an exactly 10×10 hole will be punctured somewhere in a 100×100 area in about the same time as it takes for 1000 tiles to visit a particular site and attempt to bind. There being 61 tile types in the assembly on the right, this corresponds to about once every 17 successful tile additions, i.e., 17 layers of tiles regrown.) Left: the original Sierpinski tile set. The target Sierpinski pattern has not yet been entirely erased and can still be discerned. Right: the 3×3 transformed tile set. The scale is reduced by a factor of 3, so that each 3×3 block is the same size as a single original tile on the left. The simulation was allowed to run four times as long (in terms of events per tile). Except for holes that are in the process of healing, the entire Sierpinski pattern is perfectly correct.

tile types. The technical challenge, in this case, concerns the bond tiles, which will not necessarily carry the color of the block they appear in.

Can self-healing be achieved without the use of extra strong bonds and null bonds, which presumably make a self-assembled molecular object more fragile? In this case, most tiles will be rule tiles (i.e., they will have four weak bonds), and therefore a puncture will be able to grow back in from any direction. The self-healing property requires, in this case, that no two rule tiles may have any pair of identically labeled sides. This seems very restrictive. How restrictive?

We chose here to define "self-healing" with respect to the fragment of a damaged assembly that contains the seed tile – we were not concerned with what happens when the other fragments regrow. In fact, there are some situations, such as when just a small region containing the seed is destroyed, for which it would be very desirable if regrowth could repair the damage. This seems in principle possible for some definition of "small", for example by having unique bonds in a region surrounding the seed. How can this robustness be quantified, and can a general construction be found that achieves arbitrary levels of robustness for a small cost?

Robustness to Continual Damage. So far, we have considered repairing an isolated damage event, and we have shown that it is possible to do so. What if there is repeated damage, with punctures of various sizes occurring at various rates? If the damage events are sufficiently far apart in space and time, then each puncture will be completely healed before any further damage occurs nearby. The expected time to repair n-tile damage is $O(n)$, since in the worst case there is a linear chain of dependencies and the n sites must be filled in that order. Thus, even if damage events have a weak power-law distribution (i.e., with a long tail), self-healing tile sets should be able to maintain the correct pattern: we have a guarantee that any tile added to the assembly will be correct, and the only question is whether tiles are being removed faster or being replaced faster. Fig. 6 shows simulations that confirm this intuition, in a variant of the aTAM in which each tile type is tested to be added at each site with forward rate f (as a continuous-time Markov process) [2].

However, there is a catch. Two catches. The first is that for many natural models of environmental damage, the distribution of event sizes has *very* long tails. This is due to the connectivity constraint: damaging or removing a small number of tiles from an assembly may result in a disconnected fragment, and thus necessitate the formal removal of a large number of additional tiles. This is particularly severe in long thin assemblies and near the corner of L-shaped assemblies. The second catch is that there is a finite rate at which either the seed tile itself will be destroyed, or barring that, a small region around the seed tile will be disconnected from the rest of the assembly. This means that every so often, the entire structure will have to regrow from the seed – a hard reboot. Is it possible that algorithmic growth can be designed to repair itself even when a region containing the seed tile is removed?

Performance in the kTAM. At the beginning of this chapter, we mentioned earlier work that addressed how to make a tile set more robust to growth errors, facet nucleation errors, and spurious nucleation errors in physically reversible models such as the kTAM. Here, we examined robustness to punctures – which seems like an error mode orthogonal to the previously examined ones – and analyzed how to achieve robustness in the aTAM, so as to focus on the new aspects of this problem. How well do our solutions work in the kTAM? Preliminary tests with the 3×3 self-healing tile set show that although it is a great improvement over the original 1×1 tile set, it does not perform dramatically better than the simpler 3×3 proofreading tile set of [27]. We can attribute this to two factors: first, the self-healing tile set uses only two sides of each block to encode information – rather than all three in the proofreading tile set – and therefore it suffers a higher rate of growth errors. Secondly, even when proofreading tiles regrow incorrectly, the growth usually does not proceed far before an inconsistency prevents further growth; this tends to stall the regrowth and allows the incorrect tiles to fall of, often, but not always. Can better performance be achieved by explicitly incorporating principles for all previously examined types of errors into the design of a block transformation that yields tile sets robust to all error types?

Experimental Practicality. The study of fault-tolerant tile sets is motivated in large part by the promise of using algorithmic self-assembly for bottom-up fabrication of complex molecular devices. Theory, however, naturally leads in directions appreciated only by theorists. How practical are the self-healing tile sets presented here? For comparison, there is already on-going experimental work investigating 2×2 proofreading systems as well as 2×6 blocks for controlling spurious nucleation. Therefore, 3×3 blocks could in principle be investigated in the near future – but I think it would be a challenging experiment! For DNA tile self-assembly, having a polyomino-safe tile set may be important to help prevent spurious nucleation, but 7×7 blocks (49-fold more tiles!) don't engender enthusiasm. Finding smaller self-healing tile sets would be a considerable advance.

A completely different approach to self-healing would be to use more sophisticated molecular components. There have already been proposals for DNA tiles that reduce self-assembly errors by means of mechanical devices (implemented by DNA hybridization and branch migration) that determine when a tile is ready to attach to other tiles or when it can be replaced by other tiles [23, 6, 10]. Although intimidating to experimentally develop such a complex tile, these approaches may ultimately have great pay-off as they can in principle reduce all the types of errors discussed in this chapter, and the resulting complex tiles are likely to be much smaller than the, e.g., 7×7 blocks presented here.

Finally, there are more serious types of physical damage that could occur. For example, within the damaged area, some tiles might be broken such that they continue to stick to the crystal, but no further tiles can stick to them. It seems that removing such tiles would require active processes.

6 Discussion

As Ned tells it, DNA nanotechnology began with a vision of an Escher print and a scheme for creating DNA crystals using six-armed junctions – which we now know won't work. Nonetheless, this vision has led to an incredible richness of experimentally demonstrated DNA structures, devices, and systems, which confirms the validity of the original insight. This gives the theorist some hope that in this field persistently pursuing a compelling idea can lead to something real – even if the original formulation is tragically flawed. Most importantly, Ned's vision has inspired new fields of research that seem to have taken on a life of their own.

Consider passive molecular self-assembly of the sort discussed in this chapter. It is a small corner of DNA nanotechnology, devoid of complicated DNA structures, nanomechanical devices, catalysts and fuels, and other sophisticated inventions. Even so, passive self-assembly has revealed itself to be more interesting than I ever would have imagined! Rather than appearing more and more like crystals (the lifeless stuff of geology), passive self-assembly now seems to be a microcosmos for the fundamental principles of biology – at least, if seen through a blurry and somewhat rose-colored lens. Specifically, passive molecular self-assembly seems to encompass several of the main aspects for how molecularly encoded information can direct the organization of matter and behavior:

Programming. How can one specify a molecular algorithm? Algorithmic self-assembly – a natural generalization of crystal growth processes – is Turing-universal [26]. The choice of a tile set is a program for self-assembly. This shows that molecularly encoded information can be very simple (just the complementarity of binding domains) and yet capable of specifying arbitrarily complex information-processing tasks.

Complexity. What kinds of structures can be self-assembled, and at what costs? In fact, any shape with a concise algorithmic description can be constructed by a concise tile set – at some increase in scale [21]. There is a single tile set that acts as a universal constructor; given a seed assembly containing a program for what shape to grow (encoded as a pattern of bond types presented on its perimeter), this tile set will follow the instructions in a way vaguely reminiscent of a biological developmental program.

Fault-tolerance. Can errors in self-assembly be reduced sufficiently to approach biological complexity? Biological organisms often grow by many orders of magnitude from their seed or egg, and often the mature individual consists of over 10^{24} macromolecules. All this despite the stochastic, reversible, and messy biochemistry underlying all the molecular processes. Reducing errors in algorithmic self-assembly to this level seems quite challenging, but theoretical constructions for error-correcting tile sets [27, 5, 17] appear to do the job – at least, on paper.

Self-healing. Can severe environmental damage be repaired? The purpose of this paper has been to show that if the damage is simply the removal of tiles in the damaged region, then it is possible to design algorithmic tile sets that heal the damage perfectly.

Self-reproduction and evolution. Can algorithmic crystals have a life cycle? The copying of genetic information from layer to layer in a crystal is a simple algorithmic task. If, when haphazardly fragmented, both pieces of the original crystal contain copies of the same information, then one can say the information has been reproduced. If the information has a selective advantage, for example serving as the program for some algorithmic growth process, then Darwinian evolution can be expected to occur [18].

Remarkably, what seems to be the most elementary physical mechanism – crystallization – is already capable of exhibiting many of the phenomena commonly associated with life [4].

Acknowledgements. The author is indebted to discussions with Ashish Goel, Ho-Lin Chen, Rebecca Schulman, David Soloveichik, Matthew Cook, and Paul Rothemund. This work was partially funded by NSF award #0523761

References

1. L.M. Adleman, Molecular computation of solutions to combinatorial problems. *Science*, 266:1021–1024, November 11, 1994.
2. L.M. Adleman, Q. Cheng, A. Goel, M.-D.A. Huang, Running time and program size for self-assembled squares. In *ACM Symposium on Theory of Computing (STOC)*, pages 740–748, 2001.
3. G. Aggarwal, Q. Cheng, M.H. Goldwasser, M.-Y. Kao, P.M. de Espanes, R.T. Schweller, Complexities for generalized models of self-assembly. *SIAM Journal on Computing*, 34:1493–1515, 2005.
4. A.G. Cairns-Smith. *The Life Puzzle: on Crystals and Organisms and on the Possibility of a Crystal as an Ancestor.* Oliver and Boyd, New York, 1971.
5. H.-L. Chen, A. Goel, Error free self-assembly using error prone tiles. In Ferretti et al. eds. *DNA Computing 10*, volume LNCS 3384, Berlin Heidelberg, Springer-Verlag, pages 62–75, 2005.
6. H.-L. Chen, Q. Cheng, A. Goel, M. deh Huang, P.M. de Espanés, Invadable self-assembly: Combining robustness with efficiency. *ACM-SIAM Symposium on Discrete Algorithms (SODA)*, pages 883–892, 2004.
7. Q. Cheng, P.M. de Espanes, Resolving two open problems in the self-assembly of squares. Computer science technical report #03-793, University of Southern California, 2003.
8. C. Ferretti, G. Mauri, C. Zandron, eds., *DNA Computing 10*, volume LNCS 3384, Berlin Heidelberg, Springer-Verlag, 2005.
9. T-J. Fu, N.C. Seeman, DNA double-crossover molecules. *Biochemistry*, 32:3211–3220, 1993.
10. K. Fujibayashi, S. Murata, A method of error suppression for self-assembling DNA tiles. In Ferretti et al. eds. *DNA Computing 10*, volume LNCS 3384, Berlin Heidelberg, Springer-Verlag, pages 113–127, 2005.
11. A.L. Mackay. Generalised crystallography. *Izvj. Jugosl. Centr. Krist. (Zagreb)*, 10:15–36, 1975.
12. K. Morita, Computation-universality of one-dimensional one-way reversible cellular automata. *Information Processing Letters*, 42:325–329, 1992.
13. C. Radin, Tiling, periodicity, and crystals. *J. Math. Phys.*, 26(6):1342–1344, 1985.
14. J.H. Reif, S. Sahu, P. Yin, Compact error-resilient computational DNA tiling assemblies. In Ferretti et al. eds. *DNA Computing 10*, volume LNCS 3384, Berlin Heidelberg, Springer-Verlag, pages 293–307, 2005.
15. B.H. Robinson, N.C. Seeman, The design of a biochip: A self-assembling molecular-scale memory device. *Protein Engineering*, 1(4):295–300, 1987.
16. P.W.K. Rothemund, E. Winfree, The program-size complexity of self-assembled squares. In *Symposium on Theory of Computing (STOC)*, ACM, 2000.
17. R. Schulman, E. Winfree, Programmable control of nucleation for algorithmic self-assembly. In Ferretti et al. eds. *DNA Computing 10*, volume LNCS 3384, Berlin Heidelberg, Springer-Verlag,pages 319–328, 2005.

18. R. Schulman, E. Winfree. Self-replication and evolution of DNA crystals. To appear in the proceedings of the *VIIIth European Conference on Artificial Life (ECAL).*
19. N. C. Seeman, Nucleic-acid junctions and lattices. *Journal of Theoretical Biology*, 99(2):237–247, 1982.
20. N.C. Seeman, P.S. Lukeman, Nucleic acid nanostructures: bottom-up control of geometry on the nanoscale. *Reports on Progress in Physics*, 68:237–270, 2005.
21. D. Soloveichik, E. Winfree, Complexity of self-assembled shapes. In Ferretti et al. eds. *DNA Computing 10*, volume LNCS 3384, Berlin Heidelberg, Springer-Verlag, pages 344–354, 2005. Extended abstract; preprint of the full paper is cs.CC/0412096 on arXiv.org.
22. D. Soloveichik, E. Winfree, Complexity of compact proofreading for self-assembled patterns. Extended abstract. in *Proceedings of the 11th Meeting on DNA Based Computing*, London, Canada, Springer, LNCS 2005 (to appear).
23. A.J. Turberfield, B. Yurke, A.P. Mills, Jr., DNA hybridization catalysts and molecular tweezers. In Erik Winfree and David K. Gifford, editors, *DNA Based Computers V*, volume 54 of *DIMACS*, American Mathematical Society, Providence, RI, 2000.
24. H. Wang, An unsolvable problem on dominoes. Technical Report BL-30 (II-15), Harvard Computation Laboratory, 1962.
25. E. Winfree, F. Liu, L.A. Wenzler, N.C. Seeman, Design and self-assembly of two-dimensional DNA crystals. *Nature*, 394:539–544, 1998.
26. E. Winfree, On the computational power of DNA annealing and ligation. In Richard J. Lipton and Eric B. Baum, editors, *DNA Based Computers*, volume 27 of *DIMACS*, American Mathematical Society, Providence, RI, pages 199–221, 1996.
27. E. Winfree, R. Bekbolatov, Proofreading tile sets: Error-correction for algorithmic self-assembly. In Junghuei Chen and John Reif, editors, *DNA Computing 9*, volume LNCS 2943, Springer-Verlag, Berlin Heidelberg, pages 126–144, 2004.
28. E. Winfree, Simulations of computing by self-assembly. Technical Report CS-TR:1998.22, Caltech, 1998.
29. E. Winfree, Algorithmic self-assembly of DNA: Theoretical motivations and 2D assembly experiments. *Journal of Biomolecular Structure & Dynamics*, pages 263–270, 2000. Special issue S2.

Compact Error-Resilient Computational DNA Tilings

John H. Reif, Sudheer Sahu, and Peng Yin

Department of Computer Science, Duke University
Box 90129, Durham, NC 27708-0129, USA.
{reif,sudheer,py}@cs.duke.edu

1 Introduction

Self-assembly is a process in which simple objects associate into large (and complex) structures. The self-assembly of DNA tiles can be used both as a powerful computational mechanism [8, 13, 21, 24, 27] and as a bottom-up nanofabrication technique [18]. Periodic 2D DNA lattices have been successfully constructed with a variety of DNA tiles, for example, double-crossover (DX) DNA tiles [26], rhombus tiles [12], triple-crossover (TX) tiles [7], "4×4" tiles [30], triangle tiles [9], and hexagonal tiles [3]. Aperiodic barcode DNA lattices have also been experimentally constructed [29]. In addition to forming extended lattices, DNA tiles can also form tubes [10, 15].

Self-assembly of DNA tiles can be used to carry out computation, by encoding data and computational rules in the sticky ends of tiles [23]. Such self-assembly of DNA tiles is known as *algorithmic self-assembly* or *computational tilings*. Researchers have experimentally demonstrated a one-dimensional algorithmic self-assembly of triple-crossover DNA molecules (TX tiles), which performs a four-step cumulative XOR computation [11]. A one-dimensional "string" tiling assembly was also experimentally constructed that computes an XOR table in parallel [28]. Recently, two-dimensional algorithmically self-assembled DNA crystals were constructed that demonstrate the pattern of Sierpinski triangles [16] and the pattern of a binary counter [1]. However, these two dimensional algorithmic crystals suffer quite high error rates. Reducing such errors is thus a key challenge in algorithmic DNA tiling self-assembly.

How do we decrease such errors? There are primarily two approaches. The first one is to decrease the intrinsic error rate ϵ by optimizing the physical environment in which a fixed tile set assembles [27], by improving the design of the tile set using new molecular mechanisms [4, 6], or by using novel materials. The second approach is to design new tile sets that can reduce the total number of errors in the final structure even with the same intrinsic error rate [5, 17, 25]. Three kinds of errors have been studied in the direction of the second approach, namely, the mismatch error, the facet error, and the

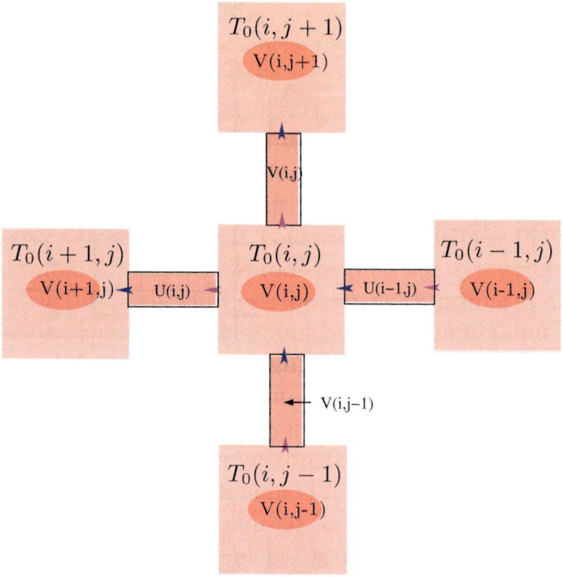

Fig. 2. Tile $T_0(i, j)$ takes input $U(i-1, j)$ and $V(i, j-1)$; determines $V(i, j) = U(i-1, j)$ $\mathtt{OP}_1$ $V(i, j-1)$ and $U(i, j) = U(i-1, j)$ $\mathtt{OP}_2$ $V(i, j-1)$; displays $V(i, j)$.

binding site for an internal 1 tile that has label 1 on its left side and label 0 on its bottom side. And this 1 tile can attach itself at this site. This in turn produces further growing sites for 0 internal tiles on top of and to the left of this just assembled 1 tile. Thus the growth can go on inductively by the accretion of appropriate individual tiles. It is straightforward to verify that the accretion of the tiles forms a binary counter with each row representing a binary number. As another concrete example, the tile set in Fig. 1b forms a Sierpinski triangle [2]. Though the above two examples appear simple, it has been proven that algorithmic assembly of tiles holds universal computing power by simulating a one-dimensional cellular automaton [22].

Note that each internal tile performs two computations: the right pad and bottom pad of each pad serve as two input bits; the left pad represents an output bit as the result of binary `AND` of the two input bits; the upper pad represents the result of the binary `XOR` operation of the two input bits Recall that `XOR` is exclusive `OR`, a binary operator that outputs bit 1 if the two input bits are different and 0 otherwise) .

By modifying the internal computational tiles and letting the left pad represent an output bit as the result of binary `XOR` of the two input bits, we obtain a set of tiles that can self-assemble into a Sierpinski triangle [2] (Fig. 1b).

The above two assemblies serve as illustrative examples for the general algorithmic assembly problem considered in this chapter, the assembly of a *Boolean array*. A Boolean array assembly is an $N \times M$ array, where the elements of each row are indexed over $\{0, \ldots, N-1\}$ from right to left and the elements of each column are indexed over $\{0, \ldots, M-1\}$ from bottom to top. The bottom row and right-most column both have some given values. Let $V(i,j)$ be the value of the i-th (from the right) bit on the j-th row (from the bottom) displayed at position (i,j) and communicated to the position $(i,j+1)$. Let $U(i,j)$ be a Boolean value communicated to the position $(i+1,j)$. For $i = 1, \ldots, N-1$ and $j = 1, \ldots, M-1$, we have $V(i,j) = U(i-1,j)\ \texttt{OP}_1\ V(i,j-1)$ and $U(i,j) = U(i-1,j)\ \texttt{OP}_2\ V(i,j-1)$, where $\texttt{OP}_1$ and $\texttt{OP}_2$ are two Boolean functions, each with two Boolean arguments and one Boolean output. See Fig. 2 for an illustration.

The binary counter shown in Fig. 1a is an $N \times 2^N$ Boolean binary array. In a binary counter, the bottom row has all 0s and the j-th row (from the bottom) is the binary representation of counter value j, for $j = 0, \ldots, 2^N - 1$. Note that the i-th bit is i-th from the right – this is in accordance with the usual left-to-right binary notation of lowest precision bits to highest precision bits. $V(i,j)$ represents the value of the i-th (from the right) counter bit on the j-th row (from the bottom), and $U(i,j)$ is the value of the carry bit from the counter bit at position (i,j). In the binary counter, we have $V(0,j) = V(0,j-1)\ \texttt{XOR}\ 1$; $V(i,j) = U(i-1,j)\ \texttt{XOR}\ V(i,j-1)$ for $i = 1, \ldots, N-1$; $U(i,j) = U(i-1,j)\ \texttt{AND}\ V(i,j-1)$. Hence $\texttt{OP}_1$ is the $\texttt{XOR}$ operation and $\texttt{OP}_2$ is the $\texttt{AND}$ operation. The Sierpinski triangle shown in Fig. 1b is an $N \times N$ Boolean binary array, where the bottom row and right-most column all have 1s; its $\texttt{OP}_1$ and $\texttt{OP}_2$ operators are both $\texttt{XOR}$.

To construct a Boolean array assembly, we make each side of each tile, denoted $T_0(i,j)$, a binary valued pad. The bottom, right, top, and left pads of tile $T_0(i,j)$ represent the values of $V(i,j-1)$ (as communicated from the tile below $T_0(i,j-1)$), $U(i-1,j)$ (as communicated from the tile on its right $T_0(i-1,j)$), $V(i,j)$ (as computed by $V(i,j-1)\ \texttt{OP}_1\ U(i-1,j)$), and $U(i,j)$ (as computed by $V(i,j-1)\ \texttt{OP}_2\ U(i-1,j)$), respectively. In the practical context of DNA tiling assemblies, a determined value $V(i,j) = 1$ can be displayed by the tile $T_0(i,j)$ using, for example, an extruding stem loop of single-strand DNA. Note that such assembly requires only four tile types in addition to three frame tiles, but results in rather small scale error-free assemblies (with the actual size contingent on the probability of single pad mismatch between adjacent tiles).

2.2 Thermodynamic Error Analysis in the Kinetic Tile Assembly Model

Experimental construction of Boolean array assemblies has shown that such algorithmic assemblies are error prone. In particular, the experimental construction of Sierpinski triangles suffers a pad mismatch rate ϵ of $1-10\%$ [16]. To analyze the error rate, Winfree further extended the above aTAM model to a kinetic Tile Assembly Model (kTAM), which includes rates both for tiles to associate to (forward rate) and to dissociate from (reverse rate) growing assemblies [25]. We reproduce Winfree's kTAM model below for completeness.

Winfree's kTAM model computes the forward and reverse rates as thermodynamic parameters. The forward rate is determined solely by the concentration of tiles, but not the type of the tiles. When the concentration of the tiles is fixed, the absolute forward rate is given by

$$r_f = k_f[\text{monomer tile}] = k_f e^{-G_{mc}},$$

where $G_{mc} = -\ln[\text{monomer}]/M$ is a unitless free energy that measures the monomer, i.e., tile, concentration in the system.

In contrast, the reverse reaction rate depends inversely exponentially on the number of base pair bonds that must be broken for the tile to dissociate from the assembly. It is given by

$$r_{r,b} = k_{r,b} = k_f e^{-bG_{se}},$$

where $G_{se} = \Delta G/RT$ is a unitless free energy corresponding to the dissociation of a single sticky end, and b is the number of such sticky ends.

It has been shown that when G_{mc} is a little smaller than $2G_{se}$, the algorithmic self-assembly under temperature 2 proceeds with optimal error rate. Intuitively, when $G_{mc} \approx 2G_{se}$, the assembly occurs near melting temperature of the system. Under such conditions, the self-assembly can achieve equilibrium, and the probability of observing a particular assembly $\mathcal{A}$ is given by

$$\Pr(\mathcal{A}) = \frac{1}{Z} e^{-G(\mathcal{A})} \quad \text{with} \quad Z = \sum_{\mathcal{A}'} e^{-G(\mathcal{A}')},$$

where $G(\mathcal{A}) = nG_{mc} - iG_{se}$ is the free energy of the assembly, n is the number of tiles in the assembly, i is the number of mismatches in the assembly, and Z is the partition function. As such, an n-assembly with Δi more mismatches will occur $e^{\Delta i G_{se}}$ less likely.

Now let $\mathcal{A}_i$ be the collection of assemblies with i mismatches in the assembly and let k_i be the number of the distinct types of $\mathcal{A}_i$ assemblies. In particular, $\mathcal{A}_0$ is the unique correct assembly and $k_0 = 1$. In addition, $\mathcal{A}_1$ represents the assemblies with exactly one mismatch. Since there are altogether $2n$ bonds in an n assembly, $k_1 = 2n$. Then we have

$$\begin{aligned}
\Pr(\mathcal{A}_0) &= \frac{[\mathcal{A}_0]}{\sum_{i=0}^{n}[\mathcal{A}_i]} && (1)\\
&= \frac{1}{\sum_{i=0}^{n}[\mathcal{A}_i]/[\mathcal{A}_0]} && (2)\\
&= \frac{1}{1 + k_1 e^{-G_{se}} + k_2 e^{-2G_{se}} + k_3 e^{-3G_{se}} + \ldots + k_n e^{-nG_{se}}} && (3)\\
&\approx \frac{1}{1 + 2ne^{-G_{se}}} && (4)\\
&\approx 1 - 2ne^{-G_{se}}. && (5)
\end{aligned}$$

On the other hand, since it takes n error-less steps to assembly A_0, we have

$$\Pr(\mathcal{A}_0) = ((1-\epsilon)^2)^n \approx 1 - 2n\epsilon, \tag{6}$$

where ϵ is the pad mismatch rate. Comparing (5) and (6), we have $\epsilon = e^{-G_{se}}$.

Under equilibrium conditions, Winfree further showed that the net growth rate of the assembly is given by

$$r_0 = r_f - r_r \approx \beta e^{-G_{mc}} \approx \beta e^{-2G_{se}} = \beta\epsilon^2,$$

where β is a constant reflecting the small difference between G_{mc} and $2G_{se}$. Now, based on the above formula, a straightforward method to reduce the error rate ϵ is to reduce the growth rate r_0. However, since r_0 depends quadratically on ϵ, a small decrease in error rate may entail dramatic decrease in the growth rate.

3 Error-Resilient Assembly Using Two-Way Overlay Redundancy

Let ϵ be the probability of a single pad mismatch between adjacent assembling DNA tiles, and assume that the likelihood of a pad mismatch error is independent for distinct pairs of pads as long as they do not involve the binding of the same two tiles. As such, a pad mismatch rate of $\epsilon = 5\%$ would imply an error-free assembly with an expected size of only 20 tiles, which is disappointingly small. Thus, a key challenge in experimentally demonstrating large-scale algorithmic assemblies is to construct error-resilient tiles. Winfree's construction is an exciting step towards this goal [25]. However, to reduce the error rate to ϵ^2 (resp. ϵ^3), his construction replaces each tile with a group of $2\times 2 = 4$ (resp. $3 \times 3 = 9$) tiles and hence increases the size of the tiling assembly by a factor of 4 (resp. 9). Our construction described below, in contrast, reduces the tiling error rate without scaling up the size of the final assembly. This would be an attractive feature in the attempt to obtain assemblies with large computational capacity. We call our constructions *compact error-resilient assemblies* and describe them below in detail.

3.1 Construction

To achieve the goals stated above, we propose the following error resilient tiling scheme. Our *Error-Resilient Assembly I* (using two-way overlay redundancy) uses only eight computational tile types plus four frame tile types. This drops the probability of assembly error to $3\epsilon^2$, which is 1.5% for $\epsilon = 5\%$, potentially allowing for error-free assemblies of expected size in hundreds of tiles.

The construction is depicted in Fig. 3. Tiles in this construction are denoted as T_1 tiles (for version 1). Each pad of each tile encodes a pair of bits.

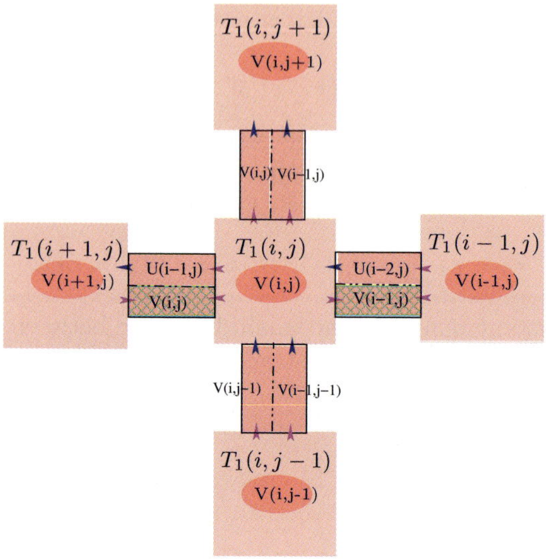

Fig. 3. Construction of compact error-resilient assembly version I. Each pad has two portions. A portion encoding an input (resp. output) value is indicated with a dark blue (resp. light pink) colored arrow head. The error checking portion is depicted as a checked rectangle. Tile $T_1(i, j)$ takes inputs $U(i-2, j)$, $V(i-1, j-1)$, and $V(i, j-1)$; determines $V(i-1, j) = U(i-2, j)$ $\texttt{OP}_1$ $V(i-1, j-1)$, $U(i-1, j) = U(i-2, j)$ $\texttt{OP}_2$ $V(i-1, j-1)$, and $V(i, j) = U(i-1, j)$ $\texttt{OP}_1$ $V(i, j-1)$; displays $V(i, j)$.

The basic idea to achieve error resiliency is to use *two-way overlay redundancy*: each tile $T_1(i, j)$ computes the outputs for its own position (i, j) and also for its right neighbor's position $(i-1, j)$; the redundant computation results obtained by $T_1(i, j)$ and its right neighbor $T_1(i-1, j)$ are compared via an additional *error checking portion* on $T_1(i, j)$'s right pad (which is the same as $T_1(i-1, j)$'s left pad). Tile $T_1(i, j)$'s right neighbor $T_1(i-1, j)$ is not likely to bind to $T_1(i, j)$ if these pad values are not consistent. Hence if only one of $T_1(i, j)$ and $T_1(i-1, j)$ is in error (incorrectly placed), the kinetics of the assembly may allow the incorrectly placed tile to be ejected from the assembly.

The four pads of $T_1(i, j)$ are constructed as follows (Fig. 3).

- The right and left portions of the bottom pad represent the value of $V(i-1, j-1)$ and $V(i, j-1)$, respectively, as communicated from the tile $T_1(i, j-1)$.
- The top portion of the right pad represents the value of $U(i-2, j)$ as communicated from the tile $T_1(i-1, j)$. The bottom portion of the right pad represents the value of $V(i-1, j)$ as determined by the tile $T_1(i, j)$. Note that the value $V(i-1, j)$ is also redundantly determined by $T_1(i-1, j)$ and hence this bottom portion performs a comparison of the two values

and is referred to as the *error checking portion*, and labeled with checked background in Fig. 3.

- The top and bottom portions of the left pad represent the values of $U(i-1,j)$ and $V(i,j)$, respectively, as determined by the tile $T_1(i,j)$. Again, the bottom portion is the error checking portion.
- The right and left portions of the top pad represent the values of $V(i-1,j)$ and $V(i,j)$, respectively, as determined by tile $T_1(i,j)$.

The above tile design allows the values $V(i-1,j-1)$ and $V(i,j-1)$ to be communicated to tile $T_1(i,j)$ from the tile $T_1(i,j-1)$ just below $T_1(i,j)$. The value $U(i-2,j)$ is communicated to tile $T_1(i,j)$ from its immediate right neighbor $T_1(i-1,j)$. These three values, $V(i-1,j-1)$, $V(i,j-1)$, and $U(i-2,j)$, can be viewed as input bits to tile $T_1(i,j)$, and the other portions of the pads as outputs. The values $V(i-1,j)$ and $U(i-1,j)$ are determined by tile $T_1(i,j)$ from $V(i-1,j-1)$ and $U(i-2,j)$: $V(i-1,j) = U(i-2,j)\ \mathtt{OP}_1\ V(i-1,j-1)$ and $U(i-1,j) = U(i-2,j)\ \mathtt{OP}_2\ V(i-1,j-1)$. The value $V(i,j)$ is determined from $V(i,j-1)$ and $U(i-1,j)$: $V(i,j) = U(i-1,j)\ \mathtt{OP}_1\ V(i,j-1)$. The determined value $V(i,j) = 1$ is displayed by the tile $T_1(i,j)$.

In this construction, each pad encodes two bits. However, since the values of the left pad, the top pad, and the bottom portion ($V(i-1,j)$) of the right pad each depend only on the values of the top portion ($U(i-2,j)$) of the right pad and the bottom pads, the tile type depends on only three input binary bits, namely, $V(i-1,j-1)$, $V(i,j-1)$, and $U(i-2,j)$. Hence only $2^3 = 8$ tile types are required. In addition, four tiles are required to assemble the frame, as described in Sect. 5.

We emphasize that though a pad has two portions, it should be treated as a whole unit. A value change in one portion of a pad changes the pad to a completely new pad. If the pad is implemented as a single strand DNA, this means that the sequence of the single-strand DNA will be a complete new sequence. One potential confusion to be avoided is mistakenly considering two pads encoding, say 00 and 01, as having the 0 portions identical or, in the context of single-strand DNA, as having half of the DNA sequences identical. To emphasize the unity of a pad, we put a box around each pad in Fig. 3.

3.2 Error Analysis

Recall that ϵ is the probability of a single pad mismatch between two adjacent DNA tiles. We further assume that the likelihood of a pad mismatch error is independent for distinct pads as long as they do not involve the binding of the same two tiles and that $\mathtt{OP}_1$ is the function $\mathtt{XOR}$.

Our intention is that the individual tiling assembly error rate (and hence the propagation of these errors to further tile assemblies) is substantially decreased, due to cooperative assembly of neighboring tiles, which redundantly compute the $V(-,-)$ and $U(-,-)$ values at their positions and at their right neighbors.

Without loss of generality, we consider only the cases where the pad binding error occurs on either the bottom pad or the right pad of a tile $T_1(i,j)$. Otherwise, if the pad binding error occurs on the left (resp. top) pad of tile $T_1(i,j)$, then use the same argument for right of tile $T_1(i+1,j)$ (resp. bottom of $T_1(i,j+1)$). We define the *neighborhood* of tile $T_1(i,j)$ to be the set of eight distinct tiles $\{ T_1(i',j') : |i'-i| < 2, |j'-j| < 2 \} \setminus \{ T_1(i,j) \}$ with coordinates that differ from (i,j) by at most 1. A neighborhood tile $T_1(i',j')$ is *dependent* on $T_1(i,j)$ if both its coordinates are equal to or greater than those of $T_1(i,j)$; otherwise $T_1(i',j')$ is *independent* of $T_1(i,j)$. Note that a neighborhood tile $T_1(i',j')$ is dependent on $T_1(i,j)$ if and only if the values $V(i',j')$ and $U(i',j')$ are determined at least partially from $V(i,j)$ or $U(i,j)$. More specifically, the neighborhood tiles dependent on $T_1(i,j)$ are $T_1(i+1,j+1)$, $T_1(i+1,j)$, and $T_1(i,j+1)$. The neighborhood tiles independent of $T_1(i,j)$ are $T_1(i+1,j-1)$, $T_1(i,j-1)$, $T_1(i-1,j+1)$, $T_1(i-1,j)$, and $T_1(i-1,j-1)$.

Lemma 1. *Suppose that the neighborhood tiles independent of tile $T_1(i,j)$ have correctly computed $V(-,-)$ and $U(-,-)$. If there is a single pad mismatch between tile $T_1(i,j)$ and another tile just below $T_1(i,j)$ or to its immediate right, then there is at least one further pad mismatch in the neighborhood of tile $T_1(i,j)$. Furthermore, given the location of the initial mismatch, the location of the further pad mismatch can be determined among at most three possible pad locations.*

Proof. Suppose that a pad binding error occurs on the bottom pad or the right pad of tile $T_1(i,j)$ but no further pad mismatch occurs between two neighborhood tiles which are independent of $T_1(i,j)$. We now consider a case-by case analysis of possible pad mismatches.

(1) First consider the case where the pad binding error occurs on the bottom pad of tile $T_1(i,j)$. Recall that the right and left portions of the bottom pad represent the values of $V(i-1,j-1)$ and $V(i,j-1)$ respectively as communicated from tile $T_1(i,j-1)$. Observe that neighborhood tiles $T_1(i,j-1)$, $T_1(i-1,j-1)$, and $T_1(i-1,j)$ are all independent of $T_1(i,j)$ and so all *correctly* compute $V(-,-)$ and $U(-,-)$ according to the assumption of the lemma.

(1.1) Consider the case where the pad binding error is due to the *incorrect* value of the right portion $V(i-1,j-1)$ of the bottom pad of tile $T_1(i,j)$ as shown in Fig. 4. Note that the left portion $V(i,j-1)$ of the bottom pad of tile $T_1(i,j)$ may also be *incorrect*. In case (i), $T_1(i,j)$ has an *incorrect* value for the $U(i-2,j)$ portion of its right pad and hence there is a further pad mismatch on the right pad of $T_1(i,j)$. In case (ii), $T_1(i,j)$ has a *correct* value for the $U(i-2,j)$ portion of its right pad. Since $T_1(i,j)$ uses the formula $V(i-1,j) = U(i-2,j) \ \mathtt{OP}_1 \ V(i-1,j-1)$ to compute $V(i-1,j)$ and $\mathtt{OP}_1$ is assumed to be the XOR function, it will determine an *incorrect* value for $V(i-1,j)$, which is distinct from the *correct* value of $V(i-1,j)$ determined by its (independent) right neighbor tile $T_1(i-1,j)$. This again implies a further pad mismatch on the right pad of tile $T_1(i,j)$.

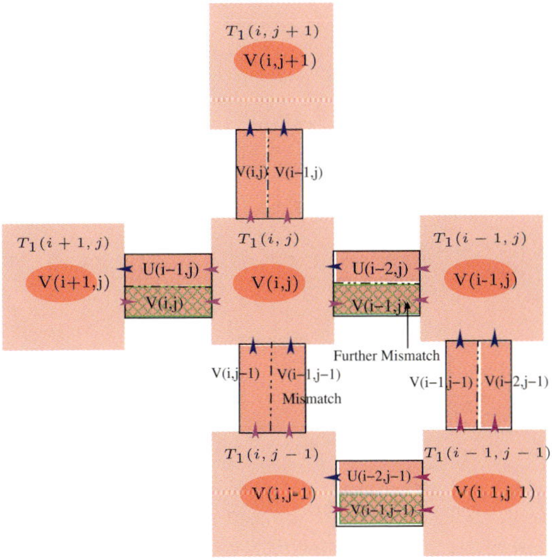

Fig. 4. Case 1.1 in the proof of Lemma 1: error in right portion $V(i-1, j-1)$ of the bottom pad of tile $T_1(i, j)$ causes a further mismatch on the right pad of tile $T_1(i, j)$.

(1.2) Next consider the case in Fig. 5 where the pad binding error is due to the wrong value of the left portion $V(i, j-1)$ of the bottom pad of tile $T_1(i, j)$. However, there is a *correct* match in the right portion $V(i-1, j-1)$ of the bottom pad of tile $T_1(i, j)$. In case (i), $T_1(i, j)$ has an *incorrect* value for the top portion $U(i-2, j)$ of its right pad, then there will be a mismatch on the right pad of $T_1(i, j)$. In case (ii), $T_1(i, j)$ has a *correct* value for the top portion $U(i-2, j)$ of its right pad, then it will further determine a *correct* value for $U(i-1, j)$, since $U(i-1, j) = U(i-2, j)\ \mathtt{OP}_2\ V(i-1, j-1)$ and both $U(i-2, j)$ and $V(i-1, j-1)$ have correct values. Since $V(i, j) = U(i-1, j)\ \mathtt{OP}_1\ V(i, j-1)$, $U(i-1, j)$ is correct and $V(i, j-1)$ is incorrect, $T_1(i, j)$ will determine an *incorrect* value for $V(i, j)$.

Note that the neighborhood tiles $T_1(i-1, j-1)$, $T_1(i, j-1)$, and $T_1(i+1, j-1)$ are independent of $T_1(i, j)$ and so both *correctly* compute $V(-,-)$ and $U(-,-)$. However, $T_1(i, j)$'s immediate left neighbor $T_1(i+1, j)$ is dependent both on the *incorrect* value communicated by the pad of $T_1(i, j)$ and the *correct* values communicated by the pad of $T_1(i+1, j-1)$. So in case (ii) there must be a further pad mismatch at tile $T_1(i+1, j)$ as argued below. In case (iia) there is pad mismatch on the right pad of $T_1(i+1, j)$ either due to a mismatch on the portion of $U(i-1, j)$ or on the portion of $V(i, j)$. Otherwise, in case (iib) there is no mismatch on either the $U(i-1, j)$ or the $V(i, j)$ portion of the pad between $T_1(i, j)$ and $T_1(i+1, j)$. This implies that $V(i, j)$ is *incorrectly* computed by $T_1(i+1, j)$ (since $T_1(i, j)$ has incorrectly computed $V(i, j)$), but $T_1(i+1, j)$ has a correct value of $U(i-1, j)$. However,

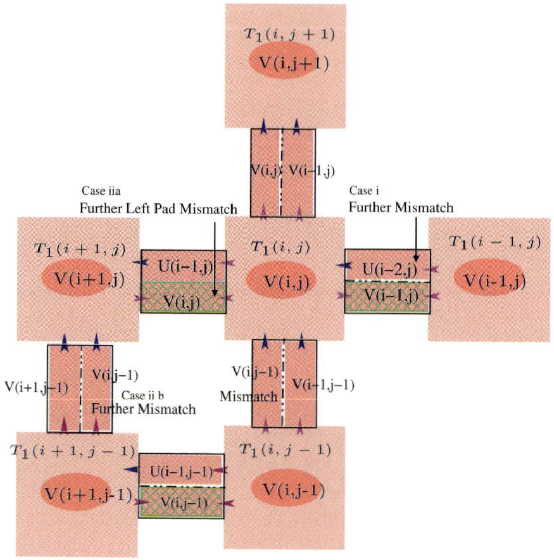

Fig. 5. Case 1.2 in the proof of Lemma 1: a further mismatch is caused by an error in the $V(i, j-1)$ portion of the bottom pad of tile $T_1(i, j)$

$V(i, j) = U(i-1, j)\ \mathtt{OP}_1\ V(i, j-1)$ and $\mathtt{OP}_1$ is XOR; this implies that the right portion $V(i, j-1)$ of the bottom pad of $T_1(i+1, j)$ has an *incorrect* value, and hence there is a mismatch between $T_1(i+1, j)$ and $T_1(i+1, j-1)$.

(2) Next consider the case where the pad binding error occurs on the right pad of tile $T_1(i, j)$, but there is no error on the bottom pad of $T_1(i, j)$. We first note that the value of the top portion $U(i-2, j)$ of the right pad of $T_1(i, j)$ must have an *incorrect* value. Assume the opposite case where $U(i-2, j)$ is correct. But the $V(i-1, j-1)$ portion of $T_1(i, j)$'s bottom pad must also have a correct value (no mismatch on the bottom pad), and this results in a further correct value for the $V(i-1, j)$ portion of $T_1(i, j)$'s right pad. Thus both $U(i-2, j)$ and $V(i-1, j)$ portions of $T_1(i, j)$'s right pad are correct and there must be no mismatch on the right pad. A contradiction. Therefore, $U(i-2, j)$ must have an *incorrect* value, and hence we only need to consider this case.

(2.1) Now consider the case where the pad binding error is due to the *incorrect* value of the top portion $U(i-2, j)$ of the right pad of tile $T_1(i, j)$ as shown in Fig. 6. We note that $T_1(i, j)$ will compute an *incorrect* value for the right portion $V(i-1, j)$ of its top pad, according to the formula $V(i-1, j) = U(i-2, j)\ \mathtt{OP}_1\ V(i-1, j-1)$. Note that $T_1(i, j+1)$ is dependent on $T_1(i, j)$. In case (i), tile $T_1(i, j+1)$ has a *correct* value of $V(i-1, j)$. There must be a pad mismatch on $V(i-1, j)$ between $T_1(i, j+1)$ and $T_1(i, j)$, since the value of $V(i-1, j)$ determined by $T_1(i, j)$ is incorrect. In case (ii), tile $T_1(i, j+1)$ has an *incorrect* value of $V(i-1, j)$, and using similar argument as

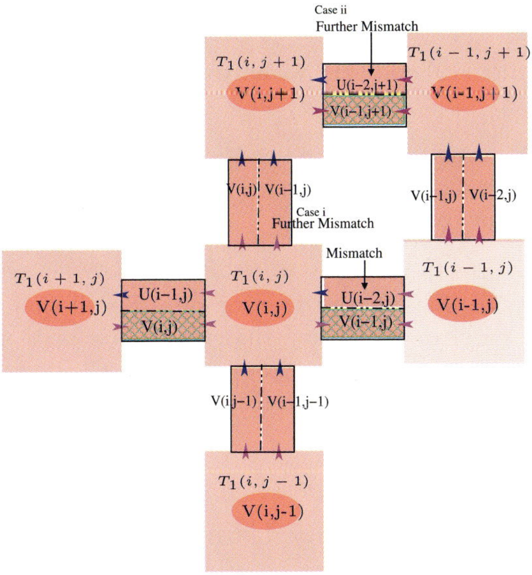

Fig. 6. Case 2.1 in the proof of Lemma 1: a further mismatch is caused by an error in the $U(i-2,j)$ portion of the right pad of tile $T_1(i,j)$.

in case 1.1 we can show that there must be a pad mismatch on the $U(i-2,j+1)$ portion of $T_1(i,j+1)$'s right pad.

Hence we conclude that in each case, there is a further pad mismatch between a pair of adjacent tiles in the neighborhood of tile $T_1(i,j)$. Furthermore, we have shown in each case that given the location of the initial mismatch, the location of the further pad mismatch can be determined among at most three possible pad locations.

Using the analytical methodology described in Sect. 2.2, we next calculate the error rate ϵ_1 in our two-way overlay construction. The key observation here is that the number of assemblies with one mismatch is $k_1 = 0$. In addition, since one pad mismatch is linked with one of three possible further mismatches, we have $k_2 = 2n * 3 = 6n$. This gives us

$$\Pr(\mathcal{A}_0) = \frac{1}{1 + k_1 e^{-G_{se}} + k_2 e^{-2G_{se}} + k_3 e^{-3G_{se}} + \ldots + k_n e^{-nG_{se}}} \tag{7}$$

$$\approx \frac{1}{1 + k_2 e^{-2G_{se}}} \tag{8}$$

$$= \frac{1}{1 + 6ne^{-2G_{se}}} \tag{9}$$

$$\approx 1 - 6ne^{-2G_{se}}, \tag{10}$$

where $\mathcal{A}_0$ is the unique error-less assembly.

Again, we also have

$$\Pr(\mathcal{A}_0) = ((1 - \epsilon_1)^2)^n \approx 1 - 2n\epsilon_1. \qquad (11)$$

Putting together (10) and (11), we have $\epsilon_1 = 3e^{-2G_{se}} = 3\epsilon^2$. Thus we have shown,

Theorem 1. *The error rate ϵ_1 for assemblies constructed from version 1 error resilient tiles is $3\epsilon^2$, where ϵ is the error rate for the corresponding assembly system with no error correction.*

Note that the growth rate is $r_1 \approx \beta e^{-G_{mc}} \approx \beta e^{-2G_{se}} = \frac{\beta}{3}\epsilon_1$. Hence the growth rate depends linearly on the error rate. Recall that, in contrast, in the system with no error correction, the growth rate is $r_0 \approx \beta\epsilon^2$. As such, compared with the system with no error correction, the decreasing error rate in our version 1 error-resilient system results in a much lower decrease in the speed of assembly.

4 Error-Resilient Assembly Using Three-Way Overlay Redundancy

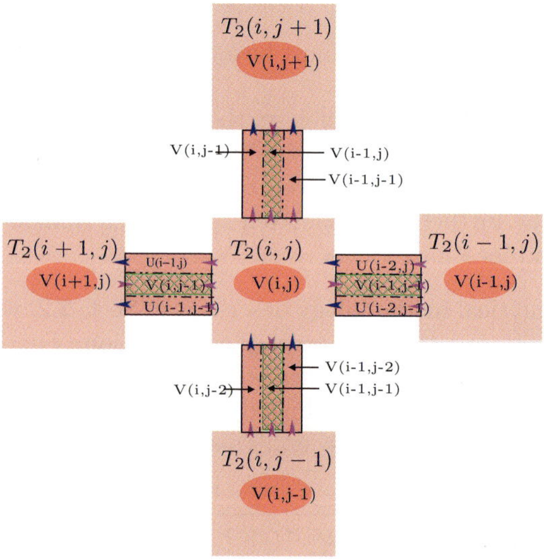

Fig. 7. Tile T_2 takes inputs $U(i-2,j)$, $U(i-2,j-1)$, $V(i-1,j-2)$ and $V(i,j-2)$; determines $V(i-1,j-1) = U(i-2,j-1)$ $\texttt{OP}_1$ $V(i-1,j-2)$, $U(i-1,j-1) = U(i-2,j-1)$ $\texttt{OP}_2$ $V(i-1,j-2)$, $V(i,j-1) = U(i-1,j-1)$ $\texttt{OP}_1$ $V(i,j-2)$, $U(i-1,j) = U(i-2,j)$ $\texttt{OP}_2$ $V(i-1,j-1)$, $V(i,j) = U(i-1,j)$ $\texttt{OP}_1$ $V(i,j-1)$ and $V(i-1,j) = U(i-2,j)$ $\texttt{OP}_1$ $V(i-1,j-1)$; displays $V(i,j)$.

4.1 Construction

We next extend the design of our scheme to a three-way overlay scheme. The *Error-Resilient Assembly version 2* (using three-way overlay redundancy) uses 16 computational tile types and five frame tile types. One mismatch on a tile forces two more mismatches in its neighborhood. This property further lowers the assembly error.

The basic construction is shown in Fig. 7. In this construction, each pad encodes a tuple of three bits and hence is an eight-valued pad. The basic idea of this error-resilient assembly is to have each tile $T_2(i,j)$ compute error checking values for positions $(i-1,j)$, $(i,j-1)$, $(i+1,j)$, and $(i,j+1)$, which are compared with corresponding error checking values computed by $T_2(i,j)$'s four neighbors. Again, the neighbors are unlikely to bind with $T_2(i,j)$ if such error checking values are inconsistent, and the kinetics of the assembly will allow these tiles to dissociate from each other, as in version 1 (two-way overlay redundancy). However, instead of introducing just one additional mismatch in $T_2(i,j)$'s neighborhood, the three-way overlay redundancy (version 2) forces two mismatches, and hence we have a further lowered error rate.

4.2 Error Analysis

For error analysis, in addition to the assumptions made in Sect. 3.2, we require that $\mathtt{OP}_2$ can detect incorrect value of input 1 regardless of the correctness of input 2. This property seems essential to guarantee two further mismatches in a tile's neighborhood when there is an initial mismatch on one of the tile's four pads. One example instance of $\mathtt{OP}_2$ is given in Table 1.

Table 1. An instance of $\mathtt{OP}_2$. This binary operation can detect the incorrect value of input 1, regardless of the correctness of input 2.

Input 1	Input 2	Output
0	0	0
1	0	1
0	1	0
1	1	1

The middle portions of all four pads (top, right, left, bottom) are computed as described in the caption of Fig. 7 and serve as the part to redundantly compute and compare the outputs of two neighboring tiles as shown in the figure.

Without loss of generality, we again consider only the cases where the pad binding error occurs on either the bottom pad or right pad of a tile $T_2(i,j)$. Otherwise, if the pad binding error occurs on the left pad of tile $T_2(i,j)$, then use the same argument for tile $T_2(i+1,j)$; likewise if the pad binding error occurs on the top pad of tile $T_2(i,j)$, use the same argument for tile $T_2(i,j+1)$.

Lemma 2. *Suppose that the neighborhood tiles independent of tile $T_2(i,j)$ have correctly computed $V(-,-)$ and $U(-,-)$. If there is a single pad mismatch between tile $T_2(i,j)$ and another tile just below or to its immediate right, then there are at least two further pad mismatches between pairs of adjacent tiles in the immediate neighborhood of tile $T_2(i,j)$. Furthermore, given the location of the initial mismatch, the location of the second mismatch can be determined among at most three locations in the neighborhood of $T_2(i,j)$; given the location of the initial and the second mismatches, the location of the third mismatch can be determined among at most five locations.*

Proof. Suppose a pad binding error occurs on a bottom pad or right pad of tile $T_2(i,j)$ but no further pad mismatch occurs between two neighborhood tiles which are independent of $T_2(i,j)$. We now consider a case-by-case analysis of possible pad mismatches.

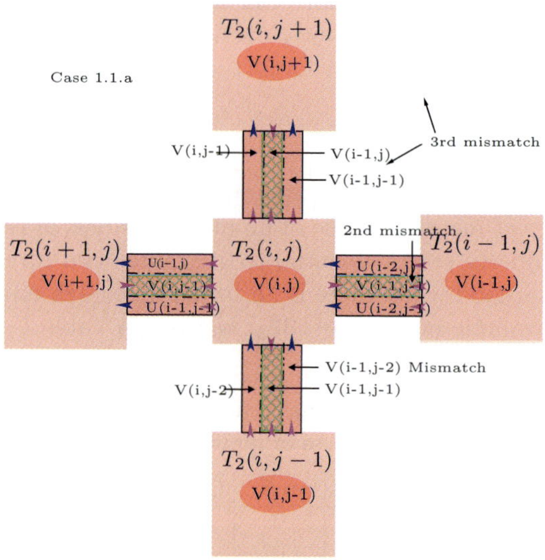

Fig. 8. Case 1.1.a in the proof of Lemma 2.

(1) First consider the case where the pad binding error occurs on the $V(i-1,j-2)$ portion of the bottom pad of tile $T_2(i,j)$.

(1.1) Consider the case where the pad binding error is due to the *incorrect* value of the right portion $V(i-1,j-2)$ of the bottom pad of tile $T_2(i,j)$ (there may also be the *incorrect* value of the other portions of the bottom pad of tile $T_2(i,j)$). Further consider case **(1.1a)** (Fig. 8) when there is no mismatch on the bottom portion $U(i-2,j-1)$ of the right pad. Immediately, we have a mismatch on the portion $V(i-1,j-1)$ of the right pad of $T_2(i,j)$, since $V(i-1,j-1) = U(i-2,j-1)$ $\texttt{OP}_1$ $V(i-1,j-2)$ and $\texttt{OP}_1$ is $\texttt{XOR}$. Furthermore, tile $T_2(i,j)$ will determine an incorrect value for the $V(i-1,j-1)$ portion of

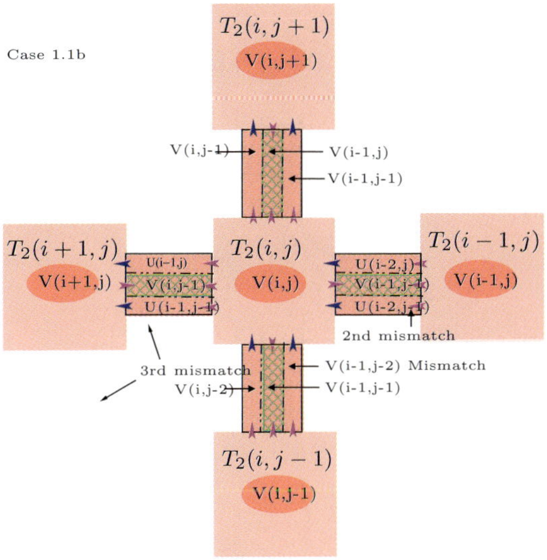

Fig. 9. Case 1.1.b in the proof of Lemma 2

its top pad, resulting in a mismatch either on the bottom or on the right pad of $T_2(i, j+1)$. Next consider case **(1.1b)** (Fig. 9) when there is a mismatch on the $U(i-2, j-1)$ portion of the right pad of $T_2(i, j)$. This will result in an incorrect value of the $U(i-1, j-1)$ portion of $T_2(i, j)$'s left pad (since $\mathtt{OP}_2$ can detect the incorrect value of $U(i-2, j-1)$), leading to a further mismatch either on the right pad or on the bottom pad of $T_2(i+1, j)$.

(1.2) (Fig. 10) Consider the case where the pad binding error is due to the *incorrect* value of the middle portion $V(i-1, j-1)$ of the bottom pad of tile $T_2(i, j)$, but there is a *correct* match in the right portion $V(i-1, j-2)$ of tile $T_2(i, j)$ (there may also be the *incorrect* value of the left portion $V(i, j-2)$ of the bottom pad of tile $T_2(i, j)$). Since the value of $V(i-1, j-2)$ is correct and $V(i-1, j-1)$ is determined by $U(i-2, j-1)$ $\mathtt{OP}_1$ $V(i-1, j-2)$ and $\mathtt{OP}_1$ is $\mathtt{XOR}$, we immediately have that there must be a mismatch on the $U(i-2, j-1)$ portion of $T_2(i, j)$'s right pad, due to the incorrect value of the $U(i-2, j-1)$ portion of this pad. However, since $V(i-1, j-1) = U(i-2, j-1)$ $\mathtt{OP}_1$ $V(i-1, j-2)$, the value of $V(i-1, j-1)$ (right portion of its top pad) computed by $T_2(i, j)$ must be incorrect, resulting in a further mismatch either on the bottom or on the right pad of $T_2(i, j+1)$.

(1.3) Consider the case where the pad binding error is due to the *incorrect* value of the left portion $V(i, j-2)$ of the bottom pad of tile $T_2(i, j)$, but there are both *correct* matches in the right portion $V(i-1, j-2)$ and middle portion $V(i-1, j-1)$ of the bottom pad of tile $T_2(i, j)$. Further consider case **(1.3a)** (Fig. 11) when there is no mismatch on the $U(i-2, j-1)$ portion of the right pad. Then $T_2(i, j)$ must compute a correct value for $U(i-1, j-1) = U(i-2, j-1)$ $\mathtt{OP}_2$ $V(i-1, j-2)$. $T_2(i, j)$ further computes both an incorrect

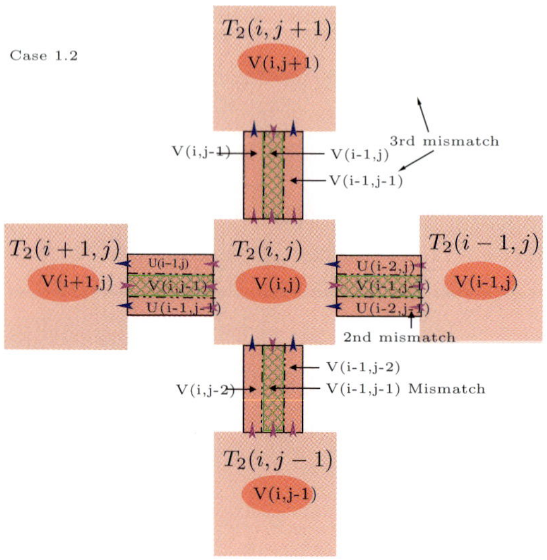

Fig. 10. Case 1.2 in the proof of Lemma 2.

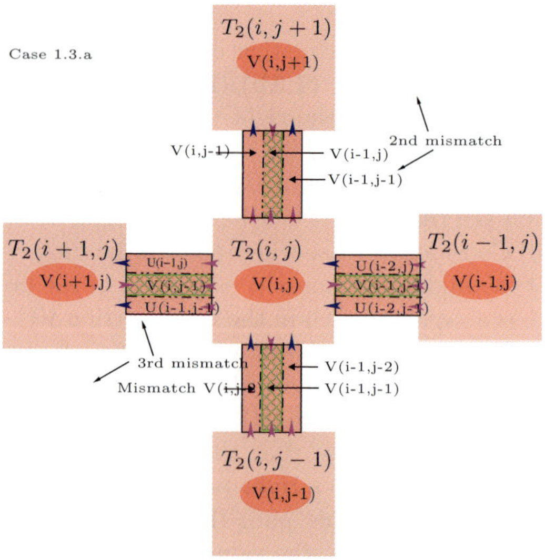

Fig. 11. Case 1.3.a in the proof of Lemma 2.

value of the $V(i,j-1)$ portion of its top pad (since $V(i,j-1) = U(i-1,j-1)\ \texttt{OP}_1\ V(i,j-2)$) and an incorrect value for the $V(i,j-1)$ portion of its left pad. The first incorrect value will result in a mismatch either on the bottom or on the right pad of $T_2(i,j+1)$. The second incorrect value will result in a mismatch either on the right or on the bottom pad of $T_2(i+1,j)$. Next consider case **(1.3b)** when there is a mismatch on the $U(i-2,j-1)$ portion of the right pad of $T_2(i,j)$. But this case cannot occur since both the $V(i-1,j-1)$ and $V(i-1,j-2)$ portions of $T_2(i,j)$'s bottom pad are correct, and $V(i-1,j-1) = U(i-2,j-1)\ \texttt{OP}_1\ V(i-1,j-2)$, where $\texttt{OP}_1 = \texttt{XOR}$.

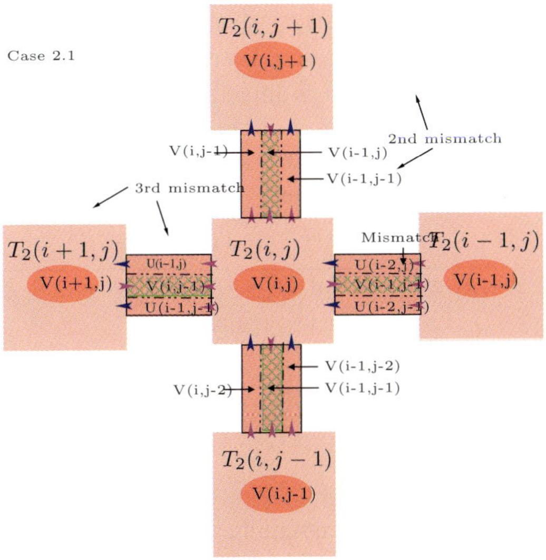

Fig. 12. Case 2.1 in the proof of Lemma 2.

(2) Now consider the case where the pad binding error occurs on the right pad of $T_2(i,j)$, but there is no binding error on the bottom pad of $T_2(i,j)$.

We note that since both the $V(i-1,j-2)$ and $V(i-1,j-1)$ portions of the bottom pad are correct, the $U(i-2,j-1)$ and $V(i-1,j-1)$ portions of the right pad must also be correct, so we only need to consider the case **(2.1)** (Fig. 12) where the binding error is due to an incorrect value of the top portion $U(i-2,j)$ of the right pad of $T_2(i,j)$, but there is no mismatch on other portions of the right pad of $T_2(i,j)$. First note an incorrect value of $U(i-2,j)$ will result in an incorrect value of the right portion $V(i-1,j-1)$ of the top pad of $T_2(i,j)$. And this will lead to a further mismatch either between $T_2(i,j)$ and $T_2(i,j+1)$ or between $T_2(i,j+1)$ and $T_2(i-1,j+1)$. Next note that $T_2(i,j)$ must compute an incorrect value for the $U(i-1,j)$ portion of its left pad, resulting in yet another mismatch either between $T_2(i,j)$ and $T_2(i+1,j)$ or between $T_2(i+1,j)$ and $T_2(i+1,j+1)$.

We have thus proven that a mismatch in the right or bottom pad of $T_2(i,j)$ results in at least two further mismatches. And given the location of the first mismatch, the location of the second mismatch can be determined among at most three locations (between $T_2(i,j)$ and $T_2(i-1,j)$, or between $T_2(i,j)$ and $T_2(i,j+1)$, or between $T_2(i,j+1)$ and $T_2(i-1,j+1)$). Furthermore, given the locations of the first two mismatches, the location of the third mismatch can be determined among at most five locations (between $T_2(i,j)$ and $T_2(i+1,j)$, between $T_2(i+1,j)$ and $T_2(i+1,j-1)$, between $T_2(i,j)$ and $T_2(i,j+1)$, between $T_2(i,j+1)$ and $T_2(i-1,j+1)$, or between $T_2(i+1,j)$ and $T_2(i+1,j+1)$).

We again calculate the error rate ϵ_2 for our version 2 construction using thermodynamic analysis. The key observation here is that the number of assemblies with exactly one mismatch or exactly two mismatches is 0. In addition, since one pad mismatch is linked with a second mismatch at one of three possible locations, and each of these three second mismatches is in turn linked with a third mismatch at one of five possible locations, we have $k_3 = 2n * 3 * 5 = 30n$. As such, we have

$$\begin{aligned}
\Pr(\mathcal{A}_0) &= \frac{1}{1 + k_1 e^{-G_{se}} + k_2 e^{-2G_{se}} + k_3 e^{-3G_{se}} + \ldots + k_n e^{-nG_{se}}} && (12)\\
&\approx \frac{1}{1 + k_3 e^{-3G_{se}}} && (13)\\
&= \frac{1}{1 + 30ne^{-3G_{se}}} && (14)\\
&\approx 1 - 30ne^{-3G_{se}}, && (15)
\end{aligned}$$

where $\mathcal{A}_0$ is the unique error-less assembly.

Again, we also have

$$\Pr(\mathcal{A}_0) = ((1-\epsilon_2)^2)^n \approx 1 - 2n\epsilon_2. \quad (16)$$

Putting together (15) and (16), we have $\epsilon_2 = 15e^{-3G_{se}} = 15\epsilon^3$. Thus we have shown,

Theorem 2. *The error rate ϵ_2 for assemblies constructed from version 2 error-resilient tiles is $15\epsilon^3$, where ϵ is the error rate for the corresponding assembly system with no error correction.*

Note that the growth rate is $r_2 \approx \beta e^{-2G_{se}} \approx (1/15)^{2/3}\beta(\epsilon_2)^{2/3}$.

Note that each pad encodes a tuple of three bits, and the values of the left pad, the top pad, the middle portion of the right pad, and the middle portion of the bottom pad each depend only on the values of the top portion and the bottom portion of the right pad and the right and left portion of the bottom pad. As such, the tile type depends on only four binary bits, and hence only $2^4 = 16$ tile types are required in addition to the initial frames at the bottom and to the right (requiring five additional tiles).

5 Computer Simulation

We first give below the construction of a Sierpinski triangle using our error resilient assembly version 1, and then perform an empirical study of the error rates using computer simulation of an assembly of the Sierpinski triangle and compare the results with those of Winfree [25].

We show below the construction of a binary counter and a Sierpinski triangle. For each of them we use a total of 12 tiles, including 8 counter tiles and 4 frame tiles as shown in Fig. 13 and Fig. 14.

We would like to emphasize again that although we gave the construction of the tiles in previous sections with each pad having two or three distinct portions, a mismatch on any portion of a pad results in a *total* mismatch of the whole pad instead of a partial mismatch of only that portion. Hence, in Fig. 13 and Fig. 14, we use a distinct label for each pad, emphasizing the wholeness of the pad.

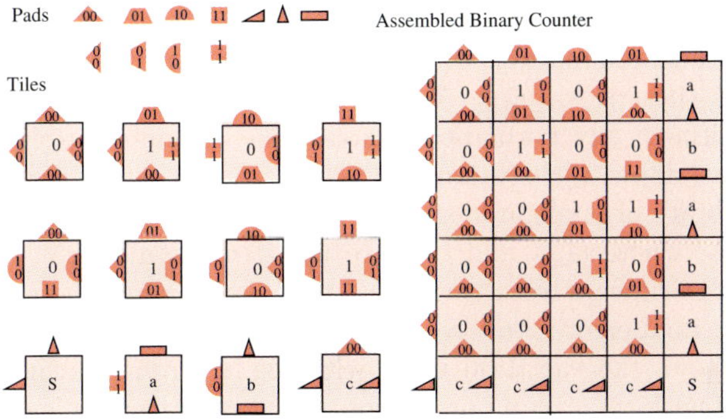

Fig. 13. The construction of a binary counter using error-resilient assemblies version 1. The pads and the tile set are shown on the left and the assembled binary counter is shown on the right. The pads of strength 2 have black borders while the strength 1 pads are borderless. The seed tile is labeled with S. Tiles a, b and c are the other frame tiles.

For the simulation study, we used the Xgrow simulator of Winfree [25] and simulated the assembly of Sierpinski triangles for the following cases:

- assembly without any error correction,
- assembly using Winfree's 2×2 proof-reading tile set,
- assembly using Winfree's 3×3 proof-reading tile set,
- assembly using our error-resilient scheme version 1, T_1 (construction in Fig. 14),
- assembly using our error-resilient scheme version 2, T_2 (construction not shown).

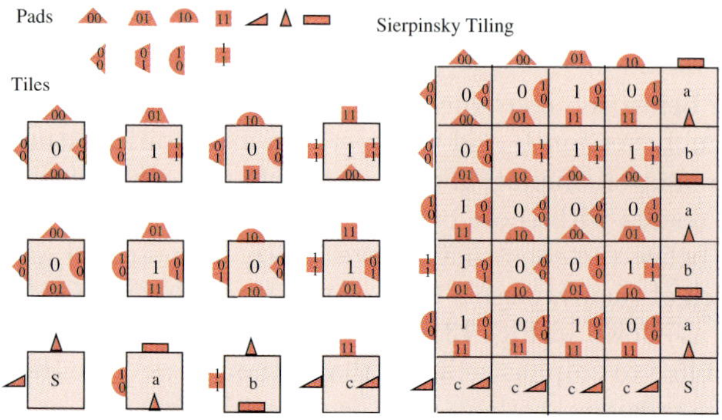

Fig. 14. The construction of a Sierpinski triangle using error-resilient assemblies version 1. The pads and the tile set are shown on the left and the assembled Sierpinski triangle is shown on the right. The pads of strength 2 have black borders while the strength 1 pads are borderless. The seed tile is labeled with S. Tiles a, b, and c are the other frame tiles.

We performed simulations of the assembly process of a target aggregate of 512×512 tiles. A variable N is defined as the largest number of tiles assembled without any permanent error in the assembly in 50% of all test cases. The variations in the value of N are measured as we increase the value of the probability of a single mismatch in pads (ϵ) by changing the values of G_{mc} and G_{se}, where G_{mc} and G_{se} are the free energies [25]. As suggested in [25], the experiments were performed near equilibrim, where $G_{mc} \approx 2G_{se}$, to achieve optimal error rate $\epsilon \approx 2e^{-G_{se}}$.

Fig. 15 shows the variation in N with $\log_e \epsilon$. From the figure it can be seen that the performance of our version 1 (T_1) construction is comparable to Winfree's 2×2 proof-reading tile set construction, while our version 2 (T_2) performs comparably to Winfree's 3×3 proof-reading tile set construction.

6 Discussion

We report a theoretical design that can reduce mismatch errors in algorithmic DNA tiling without increasing the size of the assembled structure. We have proved the correctness of our result using theoretical analysis and computer simulation. Next, we will further test the effectiveness of our construction using wet lab experimental demonstrations. The self-assembled Sierpinski crystals [16] and binary counters [1] provide an amiable platform for experimentally evaluating the effectiveness of our proposed methods. Another candidate system is the "4×4" tile system [29], and we have obtained some preliminary results in assembling binary counter crystals using this system.

There are also some open theoretical problems in our proposed system. First, in the proof of this chapter, we require $\texttt{OP}_1$ to be $\texttt{XOR}$, for concreteness.

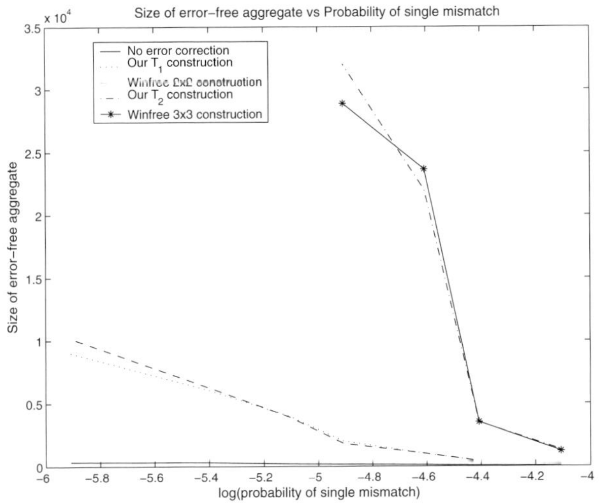

Fig. 15. A graph showing the variation of N versus increasing value of error (probability of single mismatch) ϵ.

However, note that our constructions apply to more general Boolean arrays in which $\texttt{OP}_1$ is an *input sensitive operator*, i.e., the output changes with the change of exactly one input. Second, we note that $\texttt{OP}_1$ and $\texttt{OP}_2$ are both the function `XOR` for the Sierpinski triangle but this is not true for the assembly for a binary counter of N bits, since $\texttt{OP}_2$ is the logical `AND` in that example. It is an open question whether our above error-resilient constructions can be further simplified in the case of special computations, such as the Sierpinski triangle, where the $\texttt{OP}_1$ and $\texttt{OP}_2$ are the same function such as `XOR`. Finally, another open question is to extend the construction into a more general construction such that the error probability can be decreased to ϵ^k for any given k, or alternatively, to prove an upper bound for k.

References

1. R. Barish, P. W.K. Rothemund, and E. Winfree. Algorithmic self-assembly of a binary counter using DNA tiles. 2005. In preparation.
2. B.A. Bondarenko. *Generalized Pascal Triangles and Pyramids, Their Fractals, Graphs and Applications.* The Fibonacci Association, 1993. Translated from Russion and edited by R.C. Bollinger.
3. N. Chelyapov, Y. Brun, M. Gopalkrishnan, D. Reishus, B. Shaw, and L. Adleman. DNA triangles and self-assembled hexagonal tilings. *J. Am. Chem. Soc.*, 126:13924–13925, 2004.
4. H.L. Chen, Q. Cheng, A. Goel, M.D. Huang, and P.M. de Espanes. Invadable self-assembly: Combining robustness with efficiency. In *Proceedings of the 15th annual ACM-SIAM Symposium on Discrete Algorithms (SODA)*, pages 890–899, 2004.

5. H.L. Chen and A. Goel. Error free self-assembly using error prone tiles. In *DNA Based Computers 10*, pages 274–283, 2004.
6. K. Fujibayashi and S. Murata. A method for error suppression for self-assembling DNA tiles. In *DNA Based Computing 10*, pages 284–293, 2004.
7. T.H. LaBean, H. Yan, J. Kopatsch, F. Liu, E. Winfree, J.H. Reif, and N.C. Seeman. The construction, analysis, ligation and self-assembly of DNA triple crossover complexes. *J. Am. Chem. Soc.*, 122:1848–1860, 2000.
8. M.G. Lagoudakis and T.H. LaBean. 2-D DNA self-assembly for satisfiability. In *DNA Based Computers V*, volume 54 of *DIMACS*, pages 141–154. American Mathematical Society, 2000.
9. D. Liu, M.S. Wang, Z.X. Deng, R. Walulu, and C.D. Mao. Tensegrity: Construction of rigid DNA triangles with flexible four-arm DNA junctions. *J. Am. Chem. Soc.*, 126:2324–2325, 2004.
10. D. Liu, S.H. Park, J.H. Reif, and T.H. LaBean. DNA nanotubes self-assembled from triple-crossover tiles as templates for conductive nanowires. *Proc. Natl. Acad. Sci. USA*, 101:717–722, 2004.
11. C. Mao, T.H. LaBean, J.H. Reif, and N.C. Seeman. Logical computation using algorithmic self-assembly of DNA triple-crossover molecules. *Nature*, 407:493–496, 2000.
12. C. Mao, W. Sun, and N.C. Seeman. Designed two-dimensional DNA Holliday junction arrays visualized by atomic force microscopy. *J. Am. Chem. Soc.*, 121:5437–5443, 1999.
13. J.H. Reif. Local parallel biomolecular computation. In H. Rubin and D. H. Wood, editors, *DNA-Based Computers 3*, volume 48 of *DIMACS*, pages 217–254. American Mathematical Society, 1999.
14. P.W.K. Rothemund and E. Winfree. The program-size complexity of self-assembled squares (extended abstract). In *Proceedings of the thirty-second annual ACM symposium on Theory of computing*, pages 459–468. ACM Press, 2000.
15. P.W.K. Rothemund, A. Ekani-Nkodo, N. Papadakis, A. Kumar, D.K. Fygenson, and E. Winfree. Design and characterization of programmable DNA nanotubes. *J. Am. Chem. Soc.*, 126:16344–16353, 2004.
16. P.W.K. Rothemund, N. Papadakis, and E. Winfree. Algorithmic self-assembly of DNA Sierpinski triangles. *PLoS Biology 2 (12)*, 2:e424, 2004.
17. R. Schulman and E. Winfree. Programmable control of nucleation for algorithmic self-assembly. In *DNA Based Computers 10*, LNCS, 2005.
18. N.C. Seeman. DNA in a material world. *Nature*, 421:427–431, 2003.
19. J. von Neumann. Probabilistic logics and the synthesis of reliable organisms from unreliable components. *Autonomous Studies*, pages 43–98, 1956.
20. H. Wang. Proving theorems by pattern recognition II. *Bell Systems Technical Journal*, 40:1–41, 1961.
21. E. Winfree. Complexity of restricted and unrestricted models of molecular computation. In R. J. Lipton and E. B. Baum, editors, *DNA Based Computers 1*, volume 27 of *DIMACS*, pages 187–198. American Mathematical Society, 1996.
22. E. Winfree. On the computational power of DNA annealing and ligation. In R. J. Lipton and E. B. Baum, editors, *DNA Based Computers 1*, volume 27 of *DIMACS*, pages 199–221. American Mathematical Society, 1996.
23. E. Winfree. *Algorithmic Self-Assembly of DNA*. PhD thesis, 1998.
24. E. Winfree. Simulation of computing by self-assembly. Technical Report 1998.22, Caltech, 1998.

25. E. Winfree and R. Bekbolatov. Proofreading tile sets: Error correction for algorithmic self-assembly. In *DNA Based Computers 9*, volume 2943 of *LNCS*, pages 126–144, 2004.
26. E. Winfree, F. Liu, L.A. Wenzler, and N.C. Seeman. Design and self-assembly of two-dimensional DNA crystals. *Nature*, 394(6693):539–544, 1998.
27. E. Winfree, X. Yang, and N.C. Seeman. Universal computation via self-assembly of DNA: Some theory and experiments. In L.F. Landweber and E.B. Baum, editors, *DNA Based Computers II*, volume 44 of *DIMACS*, pages 191–213. American Mathematical Society, 1999.
28. H. Yan, L. Feng, T.H. LaBean, and J.H. Reif. Parallel molecular computation of pair-wise XOR using DNA string tile. *J. Am. Chem. Soc.*, 125(47), 2003.
29. H. Yan, T.H. LaBean, L. Feng, and J.H. Reif. Directed nucleation assembly of DNA tile complexes for barcode patterned DNA lattices. *Proc. Natl. Acad. Sci. USA*, 100(14):8103–8108, 2003.
30. H. Yan, S.H. Park, G. Finkelstein, J.H. Reif, and T.H. LaBean. DNA-templated self-assembly of protein arrays and highly conductive nanowires. *Science*, 301(5641):1882–1884, 2003.

Forbidding–Enforcing Conditions in DNA Self-assembly of Graphs

Giuditta Franco[1] and Nataša Jonoska[2]

[1] Department of Computer Science, University of Verona, Italy.
franco@sci.univr.it
[2] Department of Mathematics, Univ. of South Florida, Tampa, FL 33620, USA
jonoska@math.usf.edu

1 Introduction

Nanoscale guided construction of complex structures is one of the key challenges involving science and technology in the 21st century. This challenge is at the core of the emerging discipline of nanoscience. One of the first synthesized DNA molecule with a structure that deviated from the standard double helix was the stable four-junction molecule (now known as J1) that was designed in the late 1980s in Seeman's laboratory [17]. This molecule is now one of the basic building blocks used for the design and assembly of various different constructions in the rapidly growing field of DNA nanotechnology. It has been used as a basis for a more complex building blocks of double- and triple-cross-over molecules [9, 22], as well as for junction molecules with more than four branches [20]. These armed branched molecules have been employed in the construction of two-dimensional arrays [10], and have been suggested for growing a DNA fractal-like molecule [1], for assembling arbitrary three-dimensional graphs [15], and even for obtaining DNA Borromean rings [12].

The powerful molecular recognition of Watson–Crick complementarity employed in DNA base pairing is also used in various models of biomolecular computing and information processing to guide the assembly of complex DNA structures. The DNA strands have a natural orientation that is maintained by concatenation through phosphodiester bonds, while the Watson–Crick hydrogen bonds cause two strands with opposite orientation to anneal following the base complementarity of adenine $\leftrightarrow$ thymine ($a \leftrightarrow t$) and cytosine $\leftrightarrow$ guanine ($c \leftrightarrow g$). Two- and three-dimensional DNA assemblies have been suggested and demonstrated for information processing and for computing (see, for example, [2, 23]). Experimental demonstrations of some of these ideas have been obtained through the construction of a Sierpinski triangle [14] and by the linear assembly of TX molecules encoding an XOR computation [11]. The use of branched junction molecules for computation through assembling three-dimensional structures was suggested in [7] by a demonstration of how NP-complete problems can be solved by one-step assembly.

This process, in which substructures, driven by their selective affinity, are spontaneously self-ordered into superstructures, is now widely referred to as *self-assembly*. Although there have been some notable successes with the self-assembly process, there is still a lack of consistent methods for constructing complex structures out of a pool of individual molecular components and, in general, understanding the process of self-assembly remains challenging [18]. There have been initial theoretical investigations dealing with the complexity of the self-assembled structures and computational power (see, for example, [6, 13, 19, 21]); however, understanding how the molecular architecture works is a wide-open question. This necessitates much more theoretical and experimental investigation.

In this chapter, we present a theoretical model for the generation of DNA self-assembled forms, considered as a family of graph structures that comply with certain "forbidding" constraints and follow some chemically predetermined "enforcing" conditions. This model is a variant of the model of *forbidding–enforcing systems*, introduced in [5] as a model of chemical processes. We elevate this original model to the construction of three-dimensional structures; in particular, we concentrate here on structures obtained by DNA self-assembly. On the other hand, if we consider it as a systematic way of describing classes of graphs, our model can be considered as a starting point for developing new ways to investigate graphs in classical graph theory.

2 Forbidding–Enforcing Systems

The model of forbidding–enforcing (f-e for short) systems is a nonstandard device to generate formal languages, which is an alternative to the grammar systems of classical formal language theory. It was inspired by chemical processes and has been used to simulate certain DNA-based computations [5], and afterwards it was introduced in the context of membrane computing [3].

The basic idea is to simulate a molecular reaction where "everything that is not forbidden is allowed". This assumes a completely different perspective with from the basic axiom underlying computation by grammars and automata, where "everything that is not allowed is forbidden". In fact, while in a typical formal-language-theory model, a set of rewriting productions establishes how to generate (or recognize) words of a language, in an f-e system, a family of languages is generated by some *enforcing conditions*, which dictate certain evolving rules for the system, and some *forbidding conditions*, given as a group of patterns which cannot occur together at the same time. The enforcing rules ensure that if a certain group of strings is present in the system, then some other strings will *eventually* be present as well.

More formally, given an alphabet Σ, we have the following definitions, as introduced in [5].

Definition 1. *A* forbidding set $\mathcal{F}$ *is a family of finite nonempty subsets of* Σ^+, *and an* enforcing set $\mathcal{E}$ *is a family of ordered pairs* (X, Y), *where* X *and* Y *are finite subsets of* Σ^+ *and* $Y \neq \emptyset$.

We call any element of $\mathcal{F}$ a *forbidder*, and any element of $\mathcal{E}$ an *enforcer*.

Definition 2. *A* forbidding–enforcing system *(f-e system) is a triple* $\Gamma = (\Sigma, \mathcal{F}, \mathcal{E})$, *where* $\mathcal{F}$ *is a forbidding set and* $\mathcal{E}$ *is an enforcing set (over* Σ*).*

As usual, given a language L, we denote by $sub(L)$ the set of all subwords of w for some $w \in L$.

Definition 3. *A language* L *over* $\Sigma^\star$ *is* generated by an f-e system Γ *if* $F \not\subseteq sub(L)$ *for every* $F \in \mathcal{F}$, *and* $X \subseteq L \Rightarrow Y \cap L \neq \emptyset$ *for every* $(X, Y) \in \mathcal{E}$. *The family of all languages generated by* Γ *is indicated by* $\mathcal{L}(\Gamma)$ *and is called an f-e family.*

In order to generate a language, the evolution of an f-e system proceeds according to the "molecular reactions" specified by $\mathcal{E}$ (for every forbidder (X, Y), the presence of all the strings contained in X produces at least one of the strings contained in Y), *but* it is constrained by $\mathcal{F}$, that is, the evolution cannot lead to any group of patterns specified by a forbidder from $\mathcal{F}$. Note that the forbidding set $\mathcal{F}$ contains patterns, for example x and y, that *may not be* in the system or that *may not be simultaneously* in the system; this depends on whether $\{x\}, \{y\} \in \mathcal{F}$ or $\{x, y\} \in \mathcal{F}$, respectively.

Definition 4. *An f-e system* $\Gamma = (\Sigma, \mathcal{F}, \mathcal{E})$ *is* finitary *if, for any finite language* Z, *there is at most a finite number of elements* (Z, Y) *in* $\mathcal{E}$.

In other words, a system is finitary if in one instance, the presence of a finite set of strings in the system enforces the presence of only a finite number of additional strings. Any family of languages that can be specified by an enforcing set can be also specified by a finitary enforcing set [5], and thus there exists a sort of "finitary normal form" of f-e systems.

We conclude the introduction of the basic notions of f-e systems with a simple example. Let $\Sigma = \{a, b\}$, $\mathcal{F} = \{\{aa, bb\}\}$, and $\mathcal{E} = \{(\emptyset, bb)\}$. Then the subsets of $\Sigma^\star \backslash \Sigma^\star aa \Sigma^\star$ containing $\{bb\}$ make up the f-e family of the given system. To see this one can observe that the forbidder is satisfied by subsets from $\Sigma^\star \backslash \Sigma^\star aa \Sigma^\star$ or $\Sigma^\star \backslash \Sigma^\star bb \Sigma^\star$, since at least one of aa or bb cannot appear as a subword in the language. But, as the $\emptyset$ is a subset of any language, the enforcer ensures that bb is in every language of the family. Now $\Sigma^\star \backslash \Sigma^\star bb \Sigma^\star$ excludes bb, so all languages that satisfy the f-e system are subsets of $\Sigma^\star \backslash \Sigma^\star aa \Sigma^\star$ and contain $\{bb\}$.

In what follows, we concentrate on the annealing of DNA strands guided by Watson–Crick complementarity. This will be performed by imposing forbidding-enforcing constraints guided by the physical and chemical constraints of the molecule. Starting from a system containing some given DNA filaments that can be partially annealed, we model the formation of further bonds guided by Watson–Crick complementarity, without destroying or changing any of the bonds that are already present. Each DNA strand is represented as a directed path or a directed cycle (in the case of circular molecules) and the Watson–Crick connections are represented as undirected edges. This idea allows a potentially larger number of "possible" products from the same DNA strands initially present in the pot. Therefore, instead of increasing the number of strings, we model the increase of the annealing bonds among the initial DNA substructures. In this way, we extend the basic idea of f-e systems to graphs and suggest another way to look at DNA self-assembly.

3 A Model for DNA Self-assembly

Regardless of the biochemical and topological properties of the structures seen in various aspects of DNA nanostructures, such forms can be seen as complex structures made of single strands connected to each other by two kinds of "bonds": phosphodiester bonds, i.e., concatenation, and Watson–Crick complementarity.

We describe three-dimensional (3D) DNA forms by means of graphs $G = (V, \mathcal{P}, E, \lambda)$, where V is a set of vertices labeled by elements of $\{a, c, g, t\}^k$; k is a fixed positive integer; $\mathcal{P}$ is a set of directed paths, possibly cycles, on the vertices of V; and E is a partial matching set of undirected edges such that two vertices in V are incident with the same edge only if they have Watson–Crick complementary labels.[3] The labeling function is $\lambda : V \to \{a, c, g, t\}^k$.

The labels of the paths (i.e., the concatenation of the labels of the vertices) represent the given DNA filaments, while the edges of E represent the Watson–Crick bonds generated to form the structure. A simple DNA form described by such a graph is depicted in Fig. 1.

A similar idea was used in [21] for describing a *DNA complex*, on which self-assembly rules were defined. In that case, a DNA complex was considered as a connected directed graph with vertices labeled by symbols from $\{a, t, c, g\}$ and edges from $\{backbone, basepair\}$, with at most one incoming and one outgoing edge of each type at each node. Here we consider strings that have a fixed length k as labels of the vertices, by abstracting from the experimental fact

[3] Given an involution φ on the alphabet $Z = \{a, t, g, c\}$ (a mapping from Z to Z such that φ^2 is equal to the identity mapping), and the usual reverse operation on strings ($rev(a_1 \ldots a_{n-1}a_n) = a_n a_{n-1} \ldots a_1$, where $a_1, \ldots, a_n \in Z$), we call the composition of the reverse operation with φ, extended to a morphic involution on $Z^\star$ *corev*. The order is not relevant, because $\varphi \circ rev = rev \circ \varphi$.

In the case of DNA complementarity, the involution c is the correspondence $a \to t$, $c \to g$, $t \to a$, $g \to c$, and *corev* is the Watson–Crick complementarity on DNA strings, which pairs strings such that the "reverse" of a given string is the image of the other one under φ.

As is customary, $corev(\alpha)$ will be denoted by in this chapter $\overline{\alpha}$, where $\alpha \in Z^\star$.

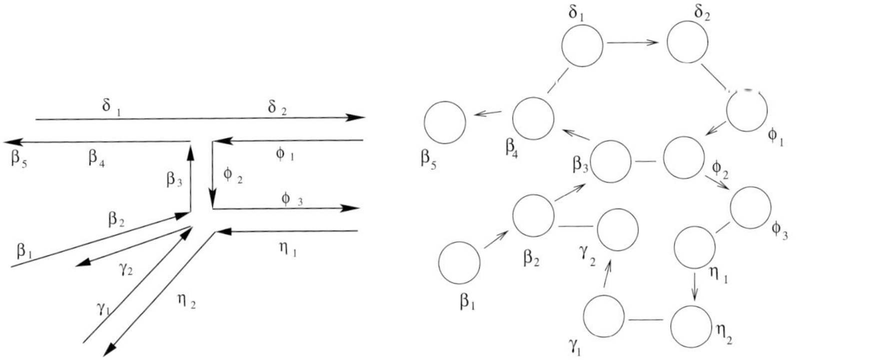

Fig. 1. Example of a self-assembled structure. Here, $\bar{\delta}_1 = \beta_4$, $\bar{\delta}_2 = \phi_1$, $\bar{\beta}_3 = \phi_2$, $\bar{\eta}_1 = \phi_3$, $\bar{\gamma}_2 = \beta_2$, and $\bar{\eta}_2 = \gamma_1$.

that there exists a lower bound on those lengths that will provide Watson–Crick pairing. This bound depends on the temperature, salt concentration, and other parameters of the experiment. It is quite intuitive that, starting from the same filaments the number of possible self-assembled DNA forms increases as the value of k, which in our case represents the (minimum) length of the attached portions, decreases. Here we focus on the graph structures corresponding to self-assembled forms, by fixing the value of k (for example $k = 5$, which is approximately the length of a half-helical turn) and considering the complementarity between two strings of length k.

Another possibility that can appear experimentally but is ignored in this exposition is the overlapping of strings. Consider the ten symbol string $w = actactacta$. For $k = 5$, we can write $w = uv$, where $u = actac$ and $v = tacta$. However there are two occurrences of u as a substring of w, i.e., $w = act \cdot u \cdot ta$. In practice, a string complementary to u can anneal to both of these occurrences. Our model assumes that none of the strings representing DNA strands have such a labeling.

On the other hand, we suppose that the correspondence from vertices to (labeling) strings may not necessarily be injective; in fac,t more than one occurrence of a string may be located along the filaments forming the structure. Therefore we consider a labeling function $\lambda : V \rightarrow \{a, c, g, t\}^k$ that assigns a string from $\{a, t, c, g\}^k$ to each vertex from V. Further, in order to keep the model more realistic, all our graphs are *finite* graphs, where V and $\mathcal{P}$ are (given) finite sets.

The description of a self-assembly structure by means of such a graph simplifies the representation and emphasizes the interrelations between the substructures; for example, a loop (a cycle with only one nondirected edge) corresponds to a hairpin formation [16], and a connected component of the graph corresponds to one DNA structure.

Consider the triple-cross-over molecule TX [9] designed in Seeman's laboratory (see Fig. 2, top). This complex structure is made of six strands, al-

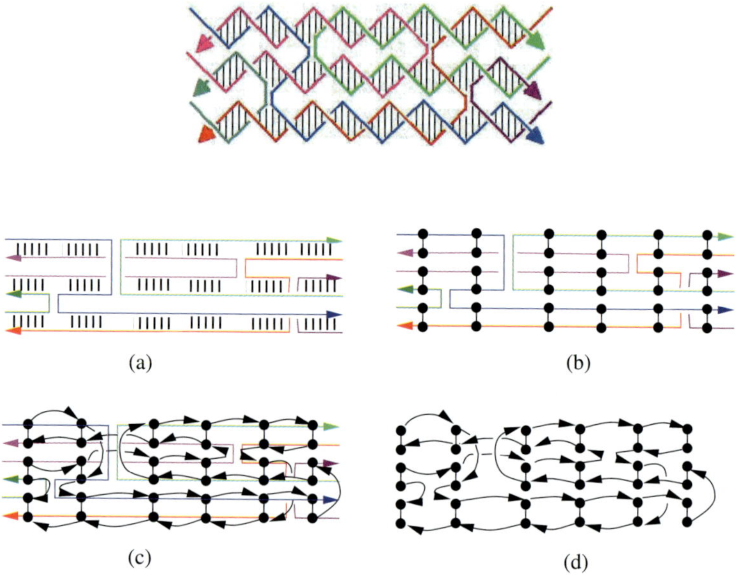

Fig. 2. Triple-cross-over molecule (TX) considered as a self-assembly graph.

though there are examples of similar TX molecules with fewer strands. It can be presented as a graph in the following way. The length of the molecule is about 4.5 helical turns which corresponds to roughly 48 bases. To represent this, we consider 36 vertices, each labeled with a string of length 8. In Fig. 2a, the TX molecule is simplified by ignoring the helical turns, and the Watson–Crick pairing corresponding to eight consecutive nucleotides is identified by a group of five short bars. In Figs 2b–d, the process of obtaining a graph structure corresponding to the TX molecule is presented. The directed edges follow the $5'$–$3'$ direction of the strand, and the undirected edges indicate the sequence of eight base pairs.

theoretically, if size is ignored, any shape can be self-assembled from a few types of initial pieces by spontaneous local bonding. In fact, the Kolmogorov complexity of a shape provides upper and lower bounds on the number of tile types necessary to self-assemble a given shape (at some scale) [19]. Nevertheless, to find strands that generate given structures remains a difficult design problem.

One can consider the converse question of what kind of forms are obtained by adding some given DNA substructures (or filaments) in a pot. In other terms, assume a labeled-graph structure $(V, \mathcal{P}, \lambda, E)$, where V is the set of vertices, $\mathcal{P}$ is the set of paths, λ is the labeling function, and E, a partial matching on the vertices, is given. What are the edges that can be added in the (possibly empty) matching set E such that there exists a DNA structure corresponding to the graph obtained. It is clear that there exist graphs $(V, \mathcal{P}, \lambda, E)$ that do not represent DNA structures; for example, they may take some physical constraints into account. Fig. 3 shows two examples of such structures.

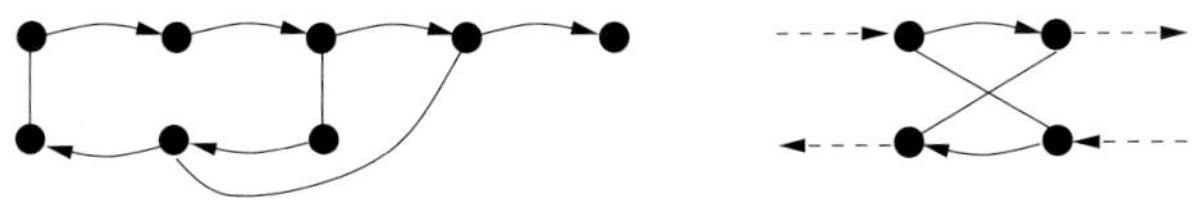

Fig. 3. Examples of graphs that do not correspond to a DNA structure.

Thus, given a collection of directed paths and cycles with vertices labeled by strings, we shall consider a set of *valid* graphs, where "forbidden" structures are not present. In particular, in order to obtain a graph which represents a self-assembled DNA structure, the matching set must respect certain constraints defined by means of a set of forbidden subgraphs which, follow physical and chemical restrictions of the DNA molecule. Such restrictions on the interrelations between DNA strings can be formulated only locally [8]. Moreover, a set of enforcing structures is considered in order to describe the parallelism intrinsic to the nature of self-assembly. This includes the consideration that further pairing of partially annealed molecules is preferred over strands that are far apart.

Molecular self-assembly is an inherently parallel process which begins anywhere it is energetically favored. Here we assume that all thermodynamic conditions necessary for self-assembly are present, such that assembly is obtained wherever it is structurally possible. Another assumption, coming from the chemical structure of DNA, is its *nonflexibility*. For example, for a given k the model forbids formation of double-stranded DNA circular molecules with a length less than nk nucleotides. We can assume that $nk \geq 100$, for example, in which case n would depend on our choice of k.

First, we consider the theoretical model for forbidding–enforcing systems for graphs.

4 Forbidding–Enforcing Systems for Graphs

Consider graphs of the type $G = (V, \mathcal{P}, E)$, where V is a finite set of vertices; $\mathcal{P}$ is a set of oriented paths, possibly cycles, on vertices of V; and E is a partial matching, that is, a set of undirected edges such that any vertex of V is incident with at most one other vertex. We denote this family of graphs with $\mathcal{G}$. If $p \in \mathcal{P}$ we indicate by $A(p)$ the set of arcs included in the path p, and call the set $\cup_{p\in\mathcal{P}} A(p)$ $A(\mathcal{P})$.

Definition 5. *Given a positive integer m, a m-*local g-forbidder *is a graph $(V, \mathcal{P}, E)$ in $\mathcal{G}$ with $|A(p)| < m$ for every $p \in \mathcal{P}$. A* g-enforcer *is an ordered pair (X, Y), where $X = (V, \mathcal{P}, E) \in \mathcal{G}$ and $Y = (V', \mathcal{P}', E') \in \mathcal{G}$ are such that $V = V'$, $\mathcal{P} = \mathcal{P}'$, and $E \subsetneq E'$.*

The constant m is included to ensure that all forbidders act locally, i.e., one needs to concentrate only on paths with not more than m vertices. This constant, in general, may depend on the experimental conditions. In what follows, we assume that m is fixed, and all m-local g-forbidders are referred to simply as g-forbidders.

The set $\mathcal{F}$ of g-forbidders is called the *forbidding set*, and the set $\mathcal{E}$ of g-enforcers is called the *enforcing set* of a family of graphs.

As in the original definition, a forbidding set may be infinite, and the only requirement is that each forbidder is finite [5]. A g-forbidder is finite if it has a finite number of arcs and edges. Moreover, regardless of the presence of the other forbidders, a g-forbidder cannot appear as a subgraph of a graph satisfying that forbidder.

Definition 6. *A* g-f-e system *is a structure* $\Gamma = (V, \mathcal{P}, E, \mathcal{F}, \mathcal{E})$, *where* V *is a finite set of vertices,* $\mathcal{P}$ *is a set of directed paths, possibly cycles, on vertices of* V; E *is a partial matching on* V; $\mathcal{F}$ *is a set of g-forbidders; and* $\mathcal{E}$ *is a set of g-enforcers.*

Given a graph $G = (V, \mathcal{P}, E) \in \mathcal{G}$, we denote by $sub(G)$ the set of all subgraphs $(V_0, \mathcal{P}_0, E_0)$ of G, where $V_0 \subseteq V$, every $p_0 \in \mathcal{P}_0$ is a path on vertices from V_0 such that $p_0 \subseteq p$ for some $p \in \mathcal{P}$, and $E_0 \subseteq E$ is a matching set on V_0. We write $G' \leq G$ for $G' \in sub(G)$. Similarly, $G' < G$ if $G' \leq G$ but $G' \neq G$.

Definition 7. *A graph* $G = (V, \mathcal{P}, E^\star)$ *is* generated by the g-f-e system $\Gamma = (V, \mathcal{P}, E, \mathcal{F}, \mathcal{E})$ *if,* $E \subset E^\star$, $F \notin sub(G^\star)$ *for every* $F \in \mathcal{F}$, *and for every* $(X, Y) \in \mathcal{E}$, *if* $X \in sub(G)$ *then there is a* $Y' \in sub(G)$ *such that* $X < Y' \leq Y$.

The family of all graphs generated by a graph forbidding–enforcing system Γ is indicated by $\mathcal{G}(\Gamma)$. The elements of $\mathcal{G}(\Gamma)$ are called *assembled graphs*.

Similarly to the original definition of forbidding–enforcing systems, the evolution of a g-f-e system proceeds according to the molecular reactions specified through $\mathcal{E}$ by increasing the elements of the matching set E, but does not allow subgraphs that are forbidden by $\mathcal{F}$.

5 Forbidding–Enforcing for DNA Nanostructures

Now we concentrate on models based on of g-f-e systems that simulate the self-assembly process of DNA. In this case the vertices of the graphs are labeled by strings from the alphabet $\{a, g, c, t\}^k$. Hence all graphs belong to the class of graphs $(V, \mathcal{P}, E, \lambda)$ where V, $\mathcal{P}$, and E are the same as in the previous section, and $\lambda : V \to \{a, g, c, t\}^k$ is the labeling of the vertices. All definitions related to g-forbidders, g-enforcers and g-f-e systems are transferred to this class of graphs in a straightforward way. We note that the labeling of the vertices of every subgraph of a graph is preserved.

Consider a g-f-e system $\mathcal{G}(\Gamma)$ where $\Gamma = (V, \mathcal{P}, E, \lambda, \mathcal{F}, \mathcal{E})$ and where the DNA strings associated with the paths $\mathcal{P}$ are given by the (finite number of) initial DNA filaments in the pot, and the set E is given by the Watson Crick bonds present in the initial DNA substructures. We specify a set of forbidders and enforcers that ensures construction of DNA structures.

The forbidding set $\mathcal{F}$ forbids constructions that are "impossible" because of the physical and chemical properties of DNA. We list three g-forbidders that are most straightforward observations and should be included in every g-f-e system that simulates DNA self-assembly:

1. *Proper annealing* (a pair of vertices is matched only if they have complementary labels):

$$F_0 = (V = \{v_1, v_2\}, \mathcal{P} = \emptyset, E = \{e = \{v_1, v_2\}\}, \lambda(v_1) = \alpha, \lambda(v_2) \neq \overline{\alpha})$$

2. *Hairpin constraint* (a strand with a string $\alpha\bar{\alpha}$ without any distance between α and $\bar{\alpha}$ cannot form a hairpin):

$$F_1 = (V = \{v_1, v_2\}, \mathcal{P} = \{p = (v_1, v_2)\}, E = \{e = \{v_1, v_2\}\}, \\ \lambda(v_1) = \alpha, \lambda(v_2) = \overline{\alpha})$$

   To simplify the notation, we describe the forbidders just by listing the labels of the vertices in the paths and the set E. This assumes that all vertices appearing in the listed paths are distinct. Hence, the above forbidder is written $F_1 = (\{\alpha\bar{\alpha}\}, \{\{\alpha, \bar{\alpha}\}\})$.
3. *Noncrossing, i.e., orientation-preserving constraint*:

$$F_2 = (\{\alpha_1\alpha_2, \bar{\alpha}_1\bar{\alpha}_2\}, \{\{\alpha_1, \bar{\alpha}_1\}, \{\alpha_2, \bar{\alpha}_2\}\}).$$

The forbidder F_1 says that a positive length (between the attached portions) is necessary to allow a strand to turn back and attach to itself [16]. The forbidder F_2 avoids physically impossible situations such as those shown in Fig. 3. Note that both structures shown in Fig. 3 are forbidden by F_2.

Owing to the experimental conditions, the purpose of the design, and the length of the labeling strings the complete set of g-forbidders for the self-assembly system may include additional structures, for example the one shown in Fig. 4.

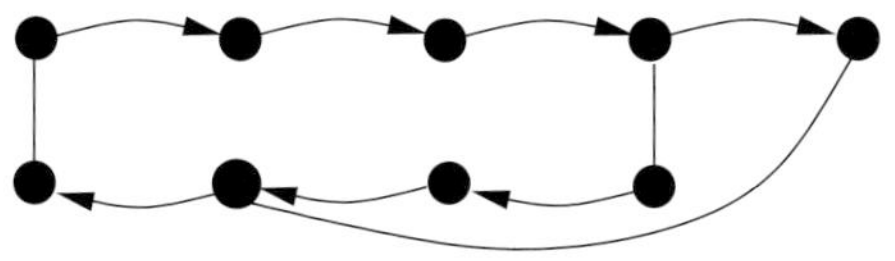

Fig. 4. A forbidden subgraph corresponding to some context constraint.

The basic set $\mathcal{E}$ contains the g-enforcers listed below. From experimental evidence [26], it is clear that DNA strands prefer pairing with complete complements. Also, all of the DNA nanostructures are obtained by using "stick-end" cohesion. Hence, we have the following enforcers.

1. *Annealing*: $E_0 = (X, Y)$ where

$$X = (V = \{v_1, v_2\}, \emptyset, \emptyset, \lambda(v_1) = \overline{\lambda(v_2)})$$

   and

$$Y = (V = \{v_1, v_2\}, \emptyset, \{\{v_1, v_2\}\}, \lambda(v_1) = \overline{\lambda(v_2)}).$$

   This enforcer can be seen as a brute-force enforcer that ensures annealing of complementary edges. If left without any changes (for example, requiring that the vertices v_1 and v_2 belong to paths with certain lengths), E_0 will ensure that all structures in the g-f-e system have all possible complementary vertices connected.
2. *One-side context rules, complete complements* (see Fig. 5):
   $E_1 = (X, Y)$, where (in the simplified notation)

$$X = (\{\alpha\beta, \bar{\beta}\bar{\alpha}\}, \{\{\alpha, \bar{\alpha}\}\}), \quad Y = (\{\alpha\beta, \bar{\beta}\bar{\alpha}\}, \{\{\alpha, \bar{\alpha}\}, \{\beta, \bar{\beta}\}\}),$$

   and $E_1' = (X, Y)$ is essentially the same as E_1 except, that the initial hybridization has occurred at the other side of the molecule:

$$X = (\{\alpha\beta, \bar{\beta}\bar{\alpha}\}, \{\{\beta, \bar{\beta}\}\}), \quad Y = (\{\alpha\beta, \bar{\beta}\bar{\alpha}\}, \{\{\alpha, \bar{\alpha}\}, \{\beta, \bar{\beta}\}\}).$$

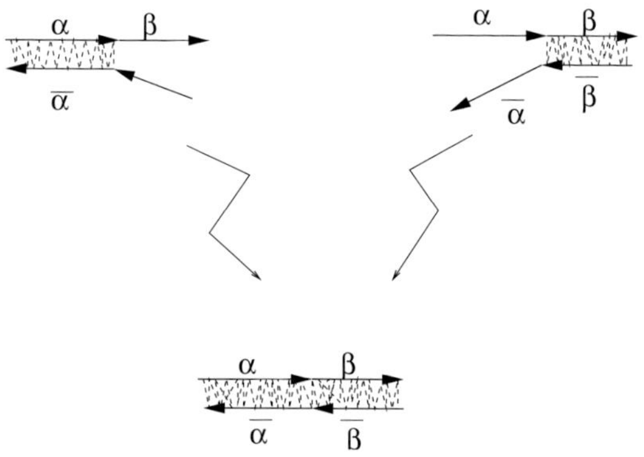

Fig. 5. One-side context enforcing rule allowed by E_1 and E_1'.

The enforcers E_1 and E_1' ensure that full complements of the strands are preferred. This is in accordance with the experimental findings [26], and this fact is a basis for several DNA-based molecular devices [24, 26].

3. *Sticky-end cohesion* (see Fig. 6a): $E_2 = (X, Y)$ where (in the simplified notation)

$$X = (\{\alpha\beta, \bar{\gamma}\bar{\beta}\}, \{\{\alpha, \bar{\alpha}\}, \{\gamma, \bar{\gamma}\}\}),\ Y = (\{\alpha\beta, \bar{\gamma}\bar{\beta}\}, \{\{\alpha, \bar{\alpha}\}, \{\gamma, \bar{\gamma}\}, \{\beta, \bar{\beta}\}\}).$$

4. *Joining* (see Fig. 6b): $E_3 = (X, Y)$ where (in the simplified notation)

$$X = (\{\alpha\beta\gamma\delta, \bar{\delta}\bar{\gamma}, \bar{\beta}\bar{\alpha}\}, \emptyset)$$

$$Y = (\{\alpha\beta\gamma\delta, \bar{\delta}\bar{\gamma}, \bar{\beta}\bar{\alpha}\}, \{\{\alpha, \bar{\alpha}\}, \{\beta, \bar{\beta}\}, \{\gamma, \bar{\gamma}\}, \{\delta, \bar{\delta}\}\})$$

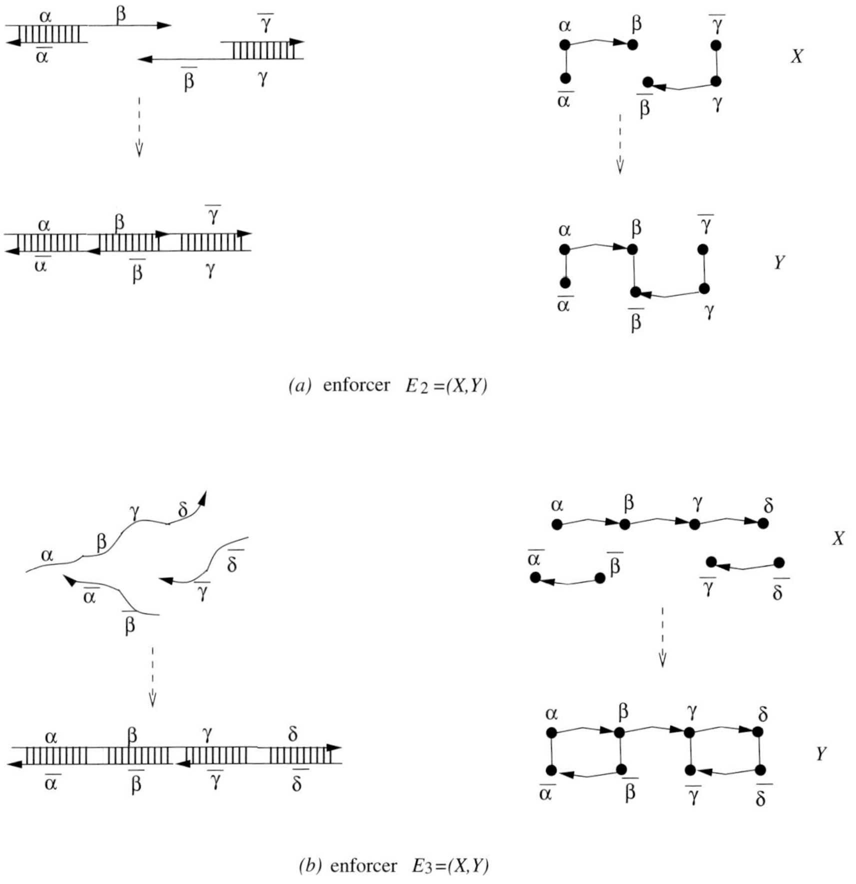

Fig. 6. The enforcing rules for sticky-end cohesion (**a**) and for "gluing" two molecules with a complement to both (**b**).

The enforcers that ensure sticky-end cohesion are depicted in Fig. 6. The left portion of the figure shows the partial annealing of the molecules, and the right portion depicts the graphs corresponding to the two cases. Note that the enforcer E_3 adds one, two, three, or four new undirected edges. If only one edge is added, there is no guarantee that full annealing will happen, but, because of the enforcers E_1 and E_1', there will be at least one more edge added.

As in the case of forbidders, additional enforcers may be added, in accordance with the experimental conditions and the initial designs of the molecules. However, we believe that the above set of enforcers should be included in every model based on a g-f-e system that describes DNA self-assembly.

We conjecture that the family $\mathcal{G}(\Gamma)$ defined by the g-f-e system described above exhibits all graph structures corresponding to the DNA complexes that can be obtained from the initial substructures (given by $(V, \mathcal{P}, E, \lambda)$) by means of self-assembly. Clearly it contains any DNA construction generated from the initial substructures, but it remains an open problem to show that the g-forbidders and the g-enforcers proposed above guarantee that this family contains *only* those graphs corresponding to possible DNA structures.

6 Conclusion

This chapter suggests new directions in both graph theory and DNA self-assembly. The general problem faced here is the following: given a set $\mathcal{P}$ of paths and cycles, a set $\mathcal{F}$ of forbidden structures, and a set $\mathcal{E}$ of enforced structures, what are the graphs included in the set $\mathcal{G}(\Gamma)$ for $\Gamma = (V, \mathcal{P}, E, \lambda, \mathcal{F}, \mathcal{E})$? The model presented focuses in particular on DNA self-assembly and the set of structures obtained through this process. However, the idea of graph forbidding–enforcing systems can certainly be extended to other self-assembly processes in nature, as well as to the pure theoretical methods used to study the mathematical properties of graphs. In the case of DNA self-assembly, the evolution process is described in a very natural way as an increase in the cardinality of the matching set between vertices with complementary labels. For other types of applications, the concept of g-f-e systems may need to be adjusted in a different way that will be more suitable for simulating the evolution in those particular processes.

Taking into account the fact that the labels of the vertices are strings over a finite alphabet, one can consider theoretical questions in the context of formal language theory. It may be interesting to investigate the classes of graphs generated by a g-f-e system where the labels of V belong to a given language taken from one of the Chomsky classes. On the other hand, considering finite languages and investigating how the structure of generated graphs depends on the g-f-e system could be useful in the study of cellular processes, where, for example, the function of signal transduction nets is fairly well understood.

Acknowledgments. Many ideas presented in this chapter are a result of numerous discussions with Nadrian C. Seeman. This work was done during the visit of G. Franco to the University of South Florida in the Spring of 2005 and is in part supported by the grants EIA-0086015, CCF-0523928, and CCF-0432009 from the National Science Foundation, USA.

References

1. A. Carbone, N.C. Seeman, Coding and geometrical shapes in nanostructures: a fractal DNA assembly. *Natural Comput.* **2** (2003) 133–151.
2. A. Carbone, N.C. Seeman, Circuits and programmable self-assembling DNA structures. *Proc. Nat. Acad. Sci.* USA **99** (2002) 12577–12582.
3. M. Cavaliere, N. Jonoska, Forbidding and enforcing in membrane computing. *Natural Comput.* **2**(3) (2003) 215–228.
4. J.H. Chen, N.C. Seeman, Synthesis from DNA of a molecule with the connectivity of a cube. *Nature* **350** (1991) 631–633.
5. A. Ehrenfeucht, G. Rozenberg, Forbidding-enforcing systems, *Theoretical Comp. Sci.* **292**(3) (2003) 611–638.
6. N. Jonoska, G.L. McColm, A computational model for self-assembling flexible tiles, *Unconventional Computing*, (C. Calude et al. eds.), Lecture Notes in Computer Science No. 3699, Springer, Berlin, Heidelberg (2005) 142–156.
7. N. Jonoska, P. Sa-Ardyen, N.C. Seeman, Computation by self-assembly of DNA graphs. *Genet. Program. Evolv. Mach.* **4** (2003) 123–137.
8. N. Jonoska, M. Saito, Boundary components of thickened graphs. in N. Jonoska and N.C. Seeman (eds.) *DNA Computing*, Lecture Notes in Computer Science No. 2340 (2002) 70–81.
9. T.H. LaBean, H. Yan, J. Kopatsch, F. Liu, E. Winfree, J.H. Reif, N.C. Seeman, Construction, analysis, ligation, and self-assembly of DNA triple crossover complexes. *J. Am. Chem. Soc.* **122**(9) (2000) 1848–1860.
10. C. Mao, W. Sun, N.C. Seeman, Designed two-dimensional Holliday junction arrays visualised by atomic force microscopy. *J. Am. Chem. Soc.* **121** (1999) 5437–5443.
11. C. Mao, T. LaBean, J.H. Reif, N.C. Seeman, Logical computation using algorithmic self-assembly of DNA triple crossover molecules. *Nature* **407** (2000) 493–496.
12. C. Mao, W. Sun, N.C. Seeman, Assembly of Borromean rings from DNA. *Nature* **386** (1997) 137–138.
13. J.H. Reif, S. Sahu, P. Yin, Complexity of graph self-assembly in accretive systems and self-destructive systems. *Preliminary Proceedings of 11th International Meeting on DNA Computing*, N. Pierce, A. Carbone (eds.) London, Ontario, Canada, June 6-9 (2005) 101–112.
14. P. Rothemund, N. Papadakis, E. Winfree, Algorithmic self-assembly of DNA Sierpinski triangles. *PLoS Biology* **2**(12) (2004) available at http://biology.plosjournals.org/.
15. P. Sa-Ardyen, N. Jonoska. N.C. Seeman, Self-assembly of graphs represented by DNA Helix Axis Topology. *J. Am. Chem. Soc.* **126**(21) (2004) 6648–6657.

16. K. Sakamoto, H. Gouzu, K. Komiya, D. Kiga, S. Yokoyama, T. Yokomori, M. Hagiya, Molecular computation by DNA hairpin formation. *Science* **288** (2000) 1223–1226.
17. N.C. Seeman et al., Gel electrophoretic analysis of DNA branched junctions. Electrophoresis **10** (1989) 345–354.
18. R.F. Service, How far can we push chemical self-assembly, *Science* **309** (special issue: *What don't we know?*) July, 2005, p. 95.
19. D. Soloveichik, E. Winfree, Complexity of self-assembled shapes. preprint available at http://www.dna.caltech.edu/Papers as file SAshapes_arxiv.pdf.
20. Y. Wang, J.E. Mueller, B. Kemper, N.C. Seeman, The assembly and characterization of 5-arm and 6-arm DNA junctions. *Biochemistry* **30** (1991) 5667–5674.
21. E. Winfree, Algorithmic Self-Assembly of DNA. Ph.D. thesis, Cal.Tech. Pasadena CA, 1998.
22. E. Winfree, F. Liu, L. Wenzler, N. C. Seeman, Design of self-assembly of two-dimensional crystals. *Nature* **494** (1998) 539–544.
23. E. Winfree, X. Yang, N.C. Seeman, Universal computation via self-assembly of DNA: some theory and experiments. in L. Landweber, E. Baum (eds.), *DNA based computers II* AMS DIMACS series **44** (1998) 191–214.
24. H. Yan, X. Zhang, Z. Shen and N.C. Seeman, A robust DNA mechanical device controlled by hybridization topology. *Nature* **415** (2002) 62–65.
25. Y. Zhang, N.C. Seeman, The construction of a DNA truncated octahedron. *J. Am. Chem. Soc.* **160** (1994) 1661–1669.
26. B. Yurke, A.J. Turberfield, A.P. Mills, F.C. Simmel Jr., A DNA fueled molecular machine made of DNA. *Nature* **406** (2000) 605–608.

Part II

Codes for DNA Nanotechnology

Finding MFE Structures Formed by Nucleic Acid Strands in a Combinatorial Set

Mirela Andronescu and Anne Condon

Department of Computer Science, University of British Columbia, Vancouver, BC V6T1Z4, Canada
andrones,condon@cs.ubc.ca

1 Introduction

When designing sets of DNA strands for biomolecular computations, it is often desirable to have a "structure free" combinatorial set of strands, that is, a set of long strands which do not form any secondary structure, obtained by concatenating short strands in the designed set.

The ability to computationally predict the combination in a combinatorial set of strands with lowest minumum free energy (MFE) secondary structure is also useful in the design of strands for directed mutagenesis and SELEX experiments [3] – biochemical analyses of a library of nucleic acid sequences – to determine whether simple mutations of the sequences have desired properties. The input sequence sets can be represented as strings of characters (DNA or RNA nucleotides) and "wild cards", which can code for several different characters using IUPAC (International Union of Pure and Applied Chemistry) format or another format. In the case of SELEX, it is useful to be able to predict not only the combination whose minimum energy structure has lowest energy, but also other combinations with minimum energy structures of low value; this is the problem we address in this chapter.

In earlier work, we described an algorithm, CombFold, that calculates which concatenated long strand in a combinatorial set forms the minimum free energy secondary structure with the lowest energy [2]. In this work, we extend that algorithm to output k secondary structures with the lowest minimum free energies, where k is specified by the user.

We use the following definitions and notation throughout.

- Let *word* denote an RNA or DNA sequence $w = v_1 v_2 \dots v_l$, where $v_i \in \{A, C, G, U\}$ for RNA and $v_i \in \{A, C, G, T\}$ for DNA. The orientation of the strand is from $5'$ to $3'$, unless otherwise stated. For example, `ACGCUAGGCA` is an RNA word of length 10.
- Let *set* denote a set of g *words* of the same length l. Formally we use the notation $S = \{w_1, w_2, \dots,\ w_g \mid length(w_i) = length(w_j), \forall i, j \in$

$\{1, \ldots, g\}, i \neq j\}$. The following *set* (displayed as a column of words) is formed of four words of length 5:

```
AUACG
UAGCG
GCCGA
CUGCG
```

The word order in a set does not matter, but for convenience later, we assume that the words in S are indexed and can be ranked by their index.

- Let *Input-Set* denote a sequence of s *sets*, $IS = S_1, S_2, \ldots, S_s$. For example, the following is an *Input-Set* of five sets:

UAGCGA	CAGCGUAAUAU	AUGCG	AUAGCGGUA	AUCG
AUAGAU	AGAUGCGCGGU		GAGCGCAAG	CUGC
	UAGGCUAGCGU			GCGA

Note that the number of words in each *set* can differ, as can the length of the words across *sets*.
An *Input-Set* can also be written in terms of *words* rather than *sets*: $IS = \{w_{ij}, 1 \leq i \leq s, 1 \leq j \leq g_i, \forall i, 1 \leq i \leq s\}$, where g_i is the number of *words* in the *set* i.

$$\begin{array}{cccc} w_{11} & w_{21} & \ldots & w_{s1} \\ w_{12} & w_{22} & \ldots & w_{s2} \\ \vdots & \vdots & \vdots & \vdots \\ w_{1g_1} & & & \\ & \vdots & & w_{sg_s} \\ & w_{2g_2} & & \end{array}$$

Thus, an *Input-Set* IS is characterized by s sets, where each set S_i has g_i words, of length l_i. In what follows, when IS is fixed, we consider all its characteristics: s, w_{ij}, g_i, l_i, $\forall i, j, 1 \leq i \leq s, 1 \leq j \leq g_i$, to be known.

- Let *Combination* denote an RNA/DNA sequence, formed by concatenating one word w_{ij} of each set S_i from IS, starting at S_1 and finishing at S_s. For example $C = w_{11}w_{21} \ldots w_{s1}$ is a combination formed by concatenating the first word of each set together. Generally, a *combination* is of the form $C = w_{1b_1}w_{2b_2} \ldots w_{sb_s}$, where $1 \leq b_i \leq g_i$. Here, b_i denotes the word rank within the set S_i. A *combination* has the length $n = \sum_{i=1}^{s} l_i$. If we think of a *combination* as a sequence of nucleotides rather than a concatenation of words, we can denote it as $C = c_1c_2 \ldots c_n$, with $c_i \in \{A, C, G, U\}$ for RNA.
- Given an *Input-Set* IS, the set of all possible *combinations* forms the *Combinatorial-Set*: $CS = \{w_{1b_1}w_{2b_2} \ldots w_{sb_s} \mid 1 \leq b_i \leq g_i\}$. Note that

all *combinations* have the same length: $n = \sum_{i=1}^{s} l_i$ and that CS has $g_1 \times g_2 \times \cdots \times g_s$ elements. If $g_i > 1, \forall i$, then the number of elements in CS is exponential in s.

The *optimal MFE combination problem* is: given an *RNA Input-Set* IS and a thermodynamic model M, predict which *combination*, out of all elements of the *Combinatorial-Set* CS formed from IS, folds to a pseudoknot-free secondary structure with the lowest minimum free energy.

An extension of the *optimal MFE combination problem* is to find the k best MFE combinations, rather than the optimal one only. The *k-suboptimal MFE combinations problem* is: given an *RNA Input-Set* IS and a thermodynamic model M, predict which k different *combinations*, out of all elements of the *Combinatorial-Set* CS formed from IS, fold to pseudoknot-free secondary structures with the lowest minimum free energies.

In this chapter, we build on earlier work [2] to develop an algorithm for the k-suboptimal MFE combination problem. In Section 2, we first review our dynamic programming algorithm which runs in polynomial time, for solving the *optimal MFE combination problem.* Then, in Section 3, we present our algorithm for the *k-suboptimal MFE combinations problem.* We provide a theoretical and empirical analysis of the optimal and k-suboptimal MFE combinations problems in Section 4 and show that both run in polynomial time.

2 Review of Algorithm for the Optimal MFE Combination Problem

Our *CombFold* algorithm [2] is based on the classical algorithm of Zuker and Stiegler [4] for finding the minimum free energy secondary structure of a single RNA strand.

One method to solve the optimal MFE combination problem is to create all possible *combinations* and then to run the Zuker–Stiegler algorithm on each of them. However, depending on the characteristics of the *Input-Set*, the number of *combinations* may be very big. If $g_i = g > 1, \forall i$, then there are g^s *combinations.* Since the Zuker–Stiegler algorithm runs in $\Theta(n^3)$ time, where n is the length of the combinations, this approach has running time complexity that is $\Theta(g^s n^3)$. More generally, the number of combinations is exponential in the number of sets which have at least two words. We have implemented this exhaustive search approach under the name of *ExhaustS*, which will be discussed in Section 4.

To avoid this exponential running time, we extended the Zuker–Stiegler algorithm. In the description that follows, we use indices i and j for the nucleotide positions (i.e. columns in Table 1) in a *combination* C. We use $s(i)$ and $s(j)$ to denote the sets in which c_i and c_j are positioned, respectively. We say that c_i and c_j belong to, or are in, the sets $s(i)$ and $s(j)$, respectively.

Table 1. Example of a combinatorial set of short RNA sequences.

		123...	....i......	.....	......j..	...n
b_j	1	UAGCGA	CAGCGUAAUAU	AUGCG	AUAGCGGUA	AUCG
b_i	2	AUAGAU	AGAUGCGCGGU		GAGCGCAAG	CUGC
	3		UAGGCUAGCGU			GCGA

We use b_i and b_j to denote the indices (i.e. the rows in Table 1) of the words containing c_i and c_j within the sets $s(i)$ and $s(j)$. Given a set S, $g(S)$ returns the number of words in S. Hence, b_i can take $g(s(i))$ values. When the *Input-Set* IS and b_i are given, we let the base c_i at position i of a combination that is in column i and row b_i be given by the function $c_i = Nucleotide(IS, b_i, i)$. Table 1 shows an example of the nucleotides c_i and c_j.

Notation for Substructure Free Energy Values

We use the following notation to denote free energy values of various substructures; in our implementation, the values are stored in four-dimensional arrays.

- $W'(j)$ is the lowest minimum free energy of a structure formed from the first j nucleotides $c_1c_2 \ldots c_j$ of a *combination*. Consequently, $W'(n)$ contains the lowest minimum free energy of any structure formed by any combination in the *Combinatorial-Set* corresponding to the *Input-Set* IS.
- $W^c(b_j, j)$ is the lowest minimum free energy of a structure formed from the first j nucleotides of a *combination* in which b_j is the word index of the set $s(j)$.
- $V^c(b_i, b_j, i, j)$ is the lowest minimum free energy of a structure formed from a *combination* fragment $c_i \ldots c_j$ starting at i and ending at j, and with fixed word indices b_i and b_j, assuming that $(c_i.c_j)$ is a base pair.
- $H^c(b_i, b_j, i, j)$ is the lowest free energy of a *combination* fragment $c_i \ldots c_j$ in which b_i and b_j are fixed, assuming that $(c_i.c_j)$ closes a hairpin loop.
- $S^c(b_i, b_j, i, j)$ is the lowest free energy of a *combination* fragment $c_i \ldots c_j$ in which b_i and b_j are fixed, assuming that $(c_i.c_j)$ closes a stacked loop.
- $VBI^c(b_i, b_j, i, j)$ is the lowest minimum free energy of the *combination* $c_i \ldots c_j$ in which b_i and b_j are fixed, assuming that $(c_i.c_j)$ closes an internal loop.
- $VM^c(b_i, b_j, i, j)$ is the lowest minimum free energy of a *combination* fragment $c_i \ldots c_j$ in which b_i and b_j are fixed, assuming that $(c_i.c_j)$ closes a multi-branched loop.
- $WM^c(b_i, b_j, i, j)$ is the lowest minimum free energy value of a *combination* fragment $c_i \ldots c_j$ that forms part of a multi-branched loop, and is used to calculate VM^c values.

Recurrence Relations

The array free energy values are calculated using several recurrence relations. In describing these here, we omit for clarity the calculations involving dangling ends and terminal AU penalties [1]. First, $W'(j)$ is the minimum of the values $W^c(b_j, j)$, over all possible b_j:

$$W'(j) = min_{b_j} W^c(b_j, j).$$

Here,

$$W^c(b_j, j) = \min \begin{cases} \min_{b_{j-1} \in X(\{b_j, j\}, \{j-1\})} W^c(b_{j-1}, j-1) \\ \min_{1 \leq i < j; b_{i-1}, b_i \in X(\{b_j, j\}, \{i-1, i\})} \\ \quad (V^c(b_i, b_j, i, j) + W^c(b_{i-1}, i-1)) \end{cases}$$

where X is a function which returns the feasible range of words for all of the needed (unknown) indexes. For the first line, the word corresponding to $j-1$ depends on the sets to which j and $j-1$ belong, and on b_j:

$$X(\{b_j, j\}, \{j-1\}) = \begin{cases} \{b_j\} & , \text{ if } s(j-1) = s(j) \\ \{1, \ldots, g(s(j-1))\}, & \text{ if } s(j-1) \neq s(j) \end{cases}$$

For the second line of the recurrence for $W^c(b_j, j)$, there are two word indices, b_{i-1} and b_i, that we have to find the ranges for:

$$X(\{b_j, j\}, \{i-1, i\}) =$$
$$\begin{cases} b_j, b_j & , \text{ if } s(i-1) = s(i) = s(j) \\ \{1, \ldots, g(s(i-1))\}, b_j & , \text{ if } s(i-1) \neq s(i) = s(j) \\ \{1, \ldots, g(s(i-1))\}, b_{i-1} & , \text{ if } s(i-1) = s(i) \neq s(j) \\ \{1, \ldots, g(s(i-1))\}, \{1, \ldots, g(s(i))\}, & \text{ if } s(i-1) \neq s(i) \neq s(j) \end{cases}$$

In the first two lines of the equation for W^c above, the feasible values for b_{j-1} (first line), and b_{i-1}, b_i (second line), depend on one other index: j, and its corresponding b_j. However, in a more general case, there are p indexes with known b's, and q indexes with unknown b's, for which we want to find the feasible ranges. The number of *if* lines needed to specify the function X in the general case will be 2^{p+q-1}. Since we are using the nearest neighbour thermodynamic model, the highest values for p and q are $p = 4$ and $q = 4$ in the case of internal loops, and $p = 2$ and $q = 6$ in the case of multi-branched loops, yielding $2^7 = 128$ *if* lines. Instead of enumerating all of these lines in our code, we developed an algorithm to compute the ranges for unknown b's, for arbitrary values of p and q. This procedure is described next.

The function X calculates the ranges for the unknown b's, for any number of known and unknown indexes. Procedure 1 gives the pseudocode for the X procedure. The input is comprised of two groups: the first group contains

Compute X Procedure

input: group of p indexes with known b's $\{b_{i_1} \dots b_{i_p}, i_1 \dots i_p\}$,
group of q indexes with unknown b's $\{j_1 \dots j_q\}$;
output: q groups $B_{j_1} \dots B_{j_q}$ corresponding to $\{j_1 \dots j_q\}$;

```
procedure Compute X
   order the indexes i's and j's;
   identify the sets S_1 ... S_m to which i's and j's belong;
   for (S = S_1 to S_m)
      if (there exists i_k in set S)
         foreach (j_u in set S)
            B_{j_u} = {b_{i_k}};
         end foreach;
      else
         j_v ← the smallest j in S;
         B_{j_v} = {1, ..., g(S)};
         foreach (j_u in set S, with j_u ≠ j_v)
            B_{j_u} = {b_{j_v}};
         end foreach;
      end if;
   end for;
   return B_{j_1}, ..., B_{j_q};
end procedure X.
```

Procedure 1: Pseudocode for the X procedure. Details are described in the text.

the known b's and the known indexes: $\{b_{i_1} \dots b_{i_p}, i_1 \dots i_p\}$. The input has the property that if $s(i_j) = s(i_{j+1})$ then $b_{i_j} = b_{i_{j+1}}$. The known b's help to determine the ranges of the unknown b's. The second group contains the indexes of the unknown b's, $\{j_1 \dots j_q\}$. First, we need to order all the values $i_1 \dots i_p, j_1 \dots j_q$. This is necessary for the second step, which identifies the sets corresponding to each index. The two extreme situations are: (1) all indexes are in the same set, and thus there will be only one possible configuration for the unknown b's; (2) all indexes are in different sets, hence there will be $g(s(j_1)) \times \dots \times g(s(j_q))$ possible values for the unknown b's.

Once we have identified the sets, for each set S, first we check whether there exists the index of a known b in this set. If so, then all the j's in S will have the corresponding, unknown, b's equal to the known b. No other option is available for these unknown b's, since the value for the known b is fixed. If no known b exists in S, then all the unknown b's in S will be in the range $\{1, \dots, g(S)\}$, with the constraint that they will have the same values, being in the same set. In other words, we can give a value to the b of the smallest index in S, and all the other b's in S will have the same value. The function X will return a group of values for the needed unknown b's.

An example of a particular situation, with the groups $\{b_i, b_j, i, j\}$ and $\{i+1, i+2, j-2, j-1\}$ as input, is presented in Table 2, where $s(i) \neq s(i+1) = s(i+2) \neq s(j-2) \neq s(j-1) = s(j)$. In this case, b_{i+1} will take values in the range $\{1, \ldots, g(s(i+1))\}$, b_{i+2} will take the value that b_{i+1} takes, b_{j-2} will be in the range $\{1, \ldots, g(s(j-2))\}$, and b_{j-1} equals b_j. Hence, for this particular situation, there will be $g(s(i+1)) \times g(s(j-2))$ terms over which to minimize.

Table 2. Example of choices for b values for a particular situation. The known b's are in bold. The vertical lines signify that the index to the left is in a different set from the index to the right.

i	$i+1$	$i+2$	$j-2$	$j-1$	j
$\mathbf{b_i}$	1	b_{i+1}	1	b_j	$\mathbf{b_j}$
	$\vdots$		$\vdots$		
	$g(s(i+1))$		$g(s(j-2))$		

Using function X to decide which are the possible values for each word, the remaining recurrence relations for *CombFold* are a logical extension of the corresponding recurrence relations for the Zuker–Stiegler algorithm. The recurrences use free energy values for hairpins, stacked pairs, and interior loops which we denote by $\Delta G\text{-}H^c(IS, b_i, b_j, i, j)$, $\Delta G\text{-}S^c(IS, b_i, b_j, b_{i+1}, b_{j-1}, i, j)$, and $\Delta G\text{-}I^c(IS, b_i, b_j, b_{i'}, b_{j'}, i, j, i', j')$, respectively.

The relations for V^c and H^c are straightforward:

$$V^c(b_i, b_j, i, j) = \begin{cases} +\infty & \text{for } i \geq j \\ \min(H^c(b_i, b_j, i, j), S^c(b_i, b_j, i, j), & \\ \quad VBI^c(b_i, b_j, i, j), VM^c(b_i, b_j, i, j)) & \text{for } i < j \end{cases}$$

$$H^c(b_i, b_j, i, j) = \Delta G\text{-}H^c(IS, b_i, b_j, i, j).$$

We omit the details of the calculation of hairpin free energies; the interested reader can find these in the M.Sc. thesis of Andronescu [1]. For the calculation of stacked loops, finding b_{i+1} and b_{j-1} is imposed again by the nearest neighbour model itself.

$$S^c(b_i, b_j, i, j) = \min_{b_{i+1}, b_{j-1} \in X(\{b_i, b_j, i, j\}, \{i+1, j-1\})} (\Delta G\text{-}S^c(IS, b_i, b_j, b_{i+1}, b_{j-1}, i, j) + V^c(b_{i+1}, b_{j-1}, i+1, j-1)).$$

The internal loop free energy calculation is a minimization over i' and j', i.e. the closing pair of the internal loop. Once i' and j' fixed, we calculate the free energy value for each possible $b_{i'}$ and $b_{j'}$:

$$VBI^c(b_i, b_j, i, j) = \min_{i<i'<j'<j} (\min_{b_{i'}, b_{j'} \in X(\{b_i, b_j, i, j\}, \{i', j'\})} (\Delta G\text{-}I^c(IS, b_i, b_j, b_{i'}, b_{j'}, i, j, i', j') + V^c(b_{i'}, b_{j'}, i', j'))).$$

The free energy for multi-loops adds the minimization over the necessary b's as well. The equations for WM^c and VM^c follow, where $\mathcal{M}_a, \mathcal{M}_b$, and $\mathcal{M}_c$ are penalties for multi-loops, branches, and unpaired bases that determine the standard multi-loop energy function. For $i < j$,

$$WM^c(b_i, b_j, i, j) = \min \begin{cases} V^c(b_i, b_j, i, j) + \mathcal{M}_b, \\ \min_{b_{i+1} \in X(\{b_i, b_j, i, j\}, \{i+1\})} (WM^c(b_{i+1}, b_j, i+1, j) + \mathcal{M}_c), \\ \min_{b_{j-1} \in X(\{b_i, b_j, i, j\}, \{j-1\})} (WM^c(b_i, b_{j-1}, i, j-1) + \mathcal{M}_c), \\ \min_{i \le h < j; b_h, b_{h+1} \in X(\{b_i, b_j, i, j\}, \{h, h+1\})} (WM^c(b_i, b_h, i, h) + \\ \quad WM^c(b_{h+1}, b_j, h+1, j)) \end{cases}$$

$$VM^c(b_i, b_j, i, j) = \mathcal{M}_a + \min_{i<h<j-1; b_{i+1}, b_h, b_{h+1}, b_{j-1} \in X(\{b_i, b_j, i, j\}, \{i+1, h, h+1, j-1\})} (WM^c(b_{i+1}, b_h, i+1, h) + WM^c(b_{h+1}, b_{j-1}, h+1, j-1)).$$

In the implementation of our software *CombFold v1.0*, we did not implement the equation for VM^c as described above. This equation contains the sum of two WM^c terms in order to make sure that the multi-loop obtained has at least three branches (including the closing one), at the cost of increased complexity, i.e. n^3 instead of n^2 for computing VM^c (see also Section 4). In our implementation, $VM^c = \mathcal{M}_a + WM^c(b_{i+1}, b_{j-1}, i+1, j-1)$ where $b_{i+1}, b_{j-1} \in X(\{b_i, b_j, i, j\}, \{i+1, j-1\})$, while we used a mechanism to make sure that the predicted multi-loops have at least three branches. We believe that this does not involve significantly different predictions, and we plan to implement the more accurate formula above in the next version of *CombFold*.

3 An Algorithm for the k-Suboptimal MFE Combinations Problem

The algorithm for the *optimal MFE combination problem*, just described in the previous section, returns only the combination which has the smallest MFE. We next describe how the algorithm can be extended to return the k combinations that have the lowest MFE.

Suppose that the *Input-Set* IS contains s sets S_i, each having g_i words. We will add the superscript "(1)" to the notation of our sets to denote that first we are looking for the optimal combinations. The superscripts for the next combinations will be "(2)" and so on. Thus, $IS^{(1)} = \{S_1^{(1)}, S_2^{(1)}, \ldots, S_s^{(1)}\}$ will be associated with the *Combinatorial-Set* $CS^{(1)}$.

First, we find the optimal MFE combination using the method described in the previous section. Let the combination $C^{(1)} = w_{1C_1}^{(1)} w_{2C_2}^{(1)} \cdots w_{sC_s}^{(1)}$ denote the optimal MFE combination, where C_i denotes the index of the word in the set $S_i^{(1)}$, which belongs to the optimal combination. The *Input-Set* $IS^{(1)}$ contains all the possible combinations of the original set IS. To find the next best combinations, first we partition the set $IS^{(1)}$ into s sets which do not contain $C^{(1)}$:

$$
\begin{array}{llllll}
IS^{(2)1} = \{ & S_1^{(1)} - \{w_{1C_1}^{(1)}\}, & S_2^{(1)}, & \ldots & , S_s^{(1)} & \} \\
IS^{(2)2} = \{ & \{w_{1C_1}^{(1)}\}, & S_2^{(1)} - \{w_{2C_2}^{(1)}\}, & \ldots & , S_s^{(1)} & \} \\
\vdots & & & & & \\
IS^{(2)s} = \{ & \{w_{1C_1}^{(1)}\}, & \{w_{2C_2}^{(1)}\}, & \ldots & , S_s^{(1)} - \{w_{sC_s}^{(1)}\} & \}.
\end{array}
$$

For convenience later, we denote the newly created sets with $S_i^{(2)j}$, where $1 \leq i, j \leq s$, i denotes the set index within the *Input-Set*, as in the previous notations, and j denotes the index of the newly created *Input-Set*:

$$
\begin{aligned}
IS^{(2)1} &= \{S_1^{(2)1}, S_2^{(2)1}, \ldots, S_s^{(2)1}\} \\
IS^{(2)2} &= \{S_1^{(2)2}, S_2^{(2)2}, \ldots, S_s^{(2)2}\} \\
&\vdots \\
IS^{(2)s} &= \{S_1^{(2)s}, S_2^{(2)s}, \ldots, S_s^{(2)s}\}.
\end{aligned}
$$

The *Input-Sets* $IS^{(2)1}, IS^{(2)2}, \ldots, IS^{(2)s}$ have the following properties:

- $C^{(1)} \notin CS^{(2)m}, \forall m, 1 \leq m \leq s;$
- $CS^{(2)m} \cap CS^{(2)m'} = \emptyset, \forall m, m', 1 \leq m, m' \leq s, m \neq m';$
- $\{C^{(1)}\} \cup CS^{(2)1} \cup \cdots \cup CS^{(2)s} = CS^{(1)};$

where $CS^{(i)j}$ denotes the *Combinatorial-Set* associated with the *Input-Set* $IS^{(i)j}$. In other words, (1) the combination $C^{(1)}$ is not included in any of the new *Input-Sets* created by the partitioning process, (2) the new input sets do not have any combinations in common, and (3) the whole space of combinations in $CS^{(1)}$ is covered by the new input sets plus the optimal combination found. This leads to finding the optimal combinations for each of $IS^{(2)1}, IS^{(2)2}, \ldots, IS^{(2)s}$, followed by choosing the one with the smallest MFE. Thus, the free energy of the second best combination, i.e. the combination with the second lowest MFE, will be $\Delta G^{(2)} = \min(\Delta G^{(2)1}, \Delta G^{(2)2}, \ldots \Delta G^{(2)s})$,

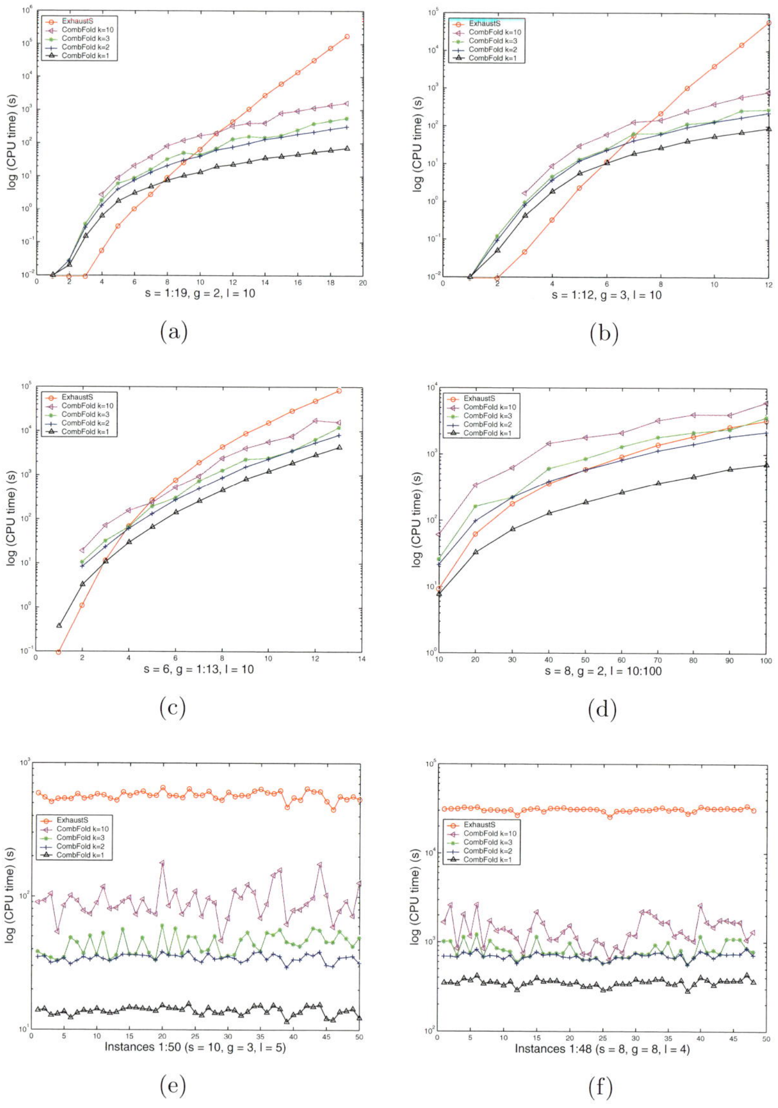

Fig. 2. Performance of *CombFold* with $k = 1, 2, 3, 10$ and *ExhaustS*, on sets with different characteristics: (**a**) 19 instances with s ranging from 1 to 19, and the same $g = 2$ and $l = 10$; (**b**) 12 instances with s ranging from 1 to 12, and the same $g = 3$ and $l = 10$; (**c**) 13 instances with g ranging from 1 to 13, and the same $s = 6$ and $l = 10$; (**d**) 10 instances with l ranging from 10 to 100, and the same $s = 8$ and $g = 2$; (**e**) 50 instances with the same characteristics: $s = 10, g = 3, l = 5$; (**f**) 48 instances with the same characteristics: $s = 8, g = 8, l = 4$.

$k = 1$, and 5–6 times slower than *CombFold* with $k = 10$. For graph (f), where the number of combinations is $8^8 \approx 16.8 \cdot 10^6$ of length 32, the exponential algorithm is substantially slower, being about two orders of magnitude slower than *CombFold*($k = 1$), and more than one order of magnitude slower than *CombFold*($k = 10$).

5 Conclusions

We presented here an algorithm that, given a combinatorial set and parameter k, predicts the k secondary structures with lowest minimum free energies in the combinatorial set. When the number of words in each set of the overall input-set is considered to be a constant, our algorithm runs in $O(skn^3)$ time.

In our algorithms, given a combination C, we look at the minimum free energy structure only. Extensions of these problems would be to find suboptimal structures (i.e. whose free energy is greater than the MFE), or to consider pseudoknots. Another problem for future work would be to find an algorithm with better running time, for example $O(n^3 + k)$.

References

1. M. Andronescu, *Algorithms for predicting the secondary structure of pairs and combinatorial sets of nucleic acid strands*, M.Sc. Thesis, U. British Columbia, November 2004.
   http://www.cs.ubc.ca/grads/resources/thesis/Nov03/Mirela_Andronescu.pdf
2. M. Andronescu, D. Dees, L. Slaybaugh, Y. Zhao, B. Cohen, A. Condon, and S. Skiena, Algorithms for testing that sets of DNA words concatenate without secondary structure, *DNA-Based Computers*, M. Hagiya, A. Ohuchi, eds., Lecture Notes in Computer Science 2568, Springer (2003) 182–195. (Revised version appeared in Natural Computing, 2(4):391–415, 2003.)
3. J.V. Ponomarenko, G.V. Orlova, A.S. Frolov, M.S. Gelfand and M.P. Ponomarenko, SELEX_DB: a database on in vitro selected oligomers adapted for recognizing natural sites and for analyzing both SNPs and site-directed mutagenesis data, *Nucl. Acids. Res.* (2002) 30(1): 195–199.
4. M. Zuker and P. Stiegler, Optimal computer folding of large RNA sequences using thermodynamics and auxiliary information, *Nucl. Acids. Res.* (1981) 9: 133–148.

Involution Solid Codes

Lila Kari and Kalpana Mahalingam

University of Western Ontario, Department of Computer Science,
London, ON N6A5B7, Canada
lila,kalpana@csd.uwo.ca

1 Introduction

DNA sequences consist of four nucleotide bases A, G, C, T (adenine, guanine, cytosine and thymine) and are joined together by phosphodiester bonds. A single strand of DNA, i.e. a chain of nucleotides, also has a "beginning" (usually denoted by $5'$) and an "end" (denoted by $3'$), and so the molecule is oriented. By the well-known Watson–Crick complementarity, A is complementary to T and C is complementary to G. The double-helix DNA strands are formed by a sequence and its complement binding together. The complementary strand is obtained by replacing the base nucleotide with its complement and reversing its direction.

Besides these "perfect" bonds, in practice certain strands can bind to others which are not their exact complements, hence rendering them useless for subsequent computation. Several attempts have been made to address this issue and many authors have proposed various solutions. A common approach has been to use the Hamming distance [1, 5, 6, 7, 26]. Experimental separation of strands with "good" sequences that avoid intermolecular cross-hybridization was reported in [3, 4].

In [11], Kari et al. introduce a theoretical approach to the problem of designing code words. Theoretical properties of languages that avoid certain undesirable hybridizations were discussed in [13, 18, 19, 25]. Based on these ideas and code-theoretic properties, a computer program for generating code words is being developed [12, 22]. Another algorithm based on backtracking, for generating such code words has also been developed by Li [24]. In [21] the authors have introduced a property of a language and showed that the properties discussed in [13, 19, 25] are its special cases. In [23] the author used the notion of partial words with holes for the design of DNA strands. In this chapter we follow the approach introduced in [11].

Every biomolecular protocol involving DNA or RNA generates molecules whose sequences of nucleotides form a language over the four-letter alphabet

$\Delta = \{A, G, C, T\}$. The Watson–Crick complementarity of the nucleotides defines a natural involution mapping θ, $A \mapsto T$ and $G \mapsto C$ which is an antimorphism of Δ^*. Undesirable Watson–Crick bonds (undesirable hybridizations) can be avoided if the language satisfies certain coding properties. In this paper we concentrate on θ-overlap free and θ-solid codes.

We start the chapter with definitions of coding properties that avoid intermolecular cross-hybridizations. The notions of θ-prefix and θ-suffix languages have been defined in [19] under the names of θ-p-compliant and θ-s-compliant, respectively. Here we consider sets of code words where the Watson–Crick complement of a word does not overlap with any other word. Hence, we have two additional coding properties that leads to the notion of θ-overlap-free code and θ-solid code. We make several observations about the closure properties of such languages. In particular, we concentrate on properties of languages that are preserved by union and concatenation. Also, we show that if a set of DNA strands has "good" coding properties that are preserved under concatenation, then the same properties will be preserved under arbitrary ligation of the strands. Section 3 investigates properties of θ-overlap-free codes. Section 4 investigates the properties of θ-solid codes.

2 Definitions

An alphabet Σ is a finite non-empty set of symbols. A word u over Σ is a finite sequence of symbols in Σ. We denote by Σ^* the set of all words over Σ, including the empty word 1 and, by Σ^+, the set of all nonempty words over Σ. We note that with the concatenation operation on words, Σ^* is the free monoid and Σ^+ is the free semigroup generated by Σ. For a word $w \in \Sigma^*$, the length of w is the number of non empty symbols in w and is denoted by $|w|$. Throughout the rest of the chapter, we concentrate on finite sets $X \subseteq \Sigma^*$ that are *codes*, i.e., every word in X^+ can be written uniquely as a product of words in X. For the background on codes we refer the reader to [2, 16, 27]. For a language $X \subseteq \Sigma^*$, let

$$\begin{aligned} \mathrm{PPref}(X) &= \{u \mid \exists v \in \Sigma^+, uv \in X\} \\ \mathrm{PSuff}(X) &= \{u \mid \exists v \in \Sigma^+, vu \in X\} \\ \mathrm{PSub}(X) &= \{u \mid \exists v_1, v_2 \in \Sigma^*, v_1 v_2 \neq 1, v_1 u v_2 \in X\}. \end{aligned}$$

We recall the definitions initiated in [11, 19] and used in [12, 18].

An involution $\theta : \Sigma \to \Sigma$ of a set Σ is a mapping such that θ^2 equals the identity mapping, $\theta(\theta(x)) = x$, $\forall x \in \Sigma$.

Definition 1. *Let $\theta : \Sigma^* \to \Sigma^*$ be a morphic or antimorphic involution and $X \subseteq \Sigma^+$.*

1. *The set X is called* θ-infix *if $\Sigma^*\theta(X)\Sigma^+ \cap X = \emptyset$ and $\Sigma^+\theta(X)\Sigma^* \cap X = \emptyset$.*
2. *The set X is called* θ-comma-free *if $X^2 \cap \Sigma^+\theta(X)\Sigma^+ = \emptyset$.*

3. *The set X is called* θ-prefix *if* $X \cap \theta(X)\Sigma^+ = \emptyset$.
4. *The set X is called* θ-suffix *if* $X \cap \Sigma^+\theta(X) = \emptyset$.
5. *The set X is called* θ-sticky-free *if for all* $w \in \Sigma^+$, $x, y \in \Sigma^*$, $wx, y\theta(w) \in X$ *then* $xy = 1$.
6. *The set X is called* θ-overhang-free *if for all* $w \in \Sigma^+$, $x, y \in \Sigma^*$, $wx, \theta(w)y \in X$ *or* $xw, y\theta(w) \in X$ *then* $xy = 1$.
7. *The set X is called* strictly θ *if* $X \cap \theta(X) = \emptyset$.

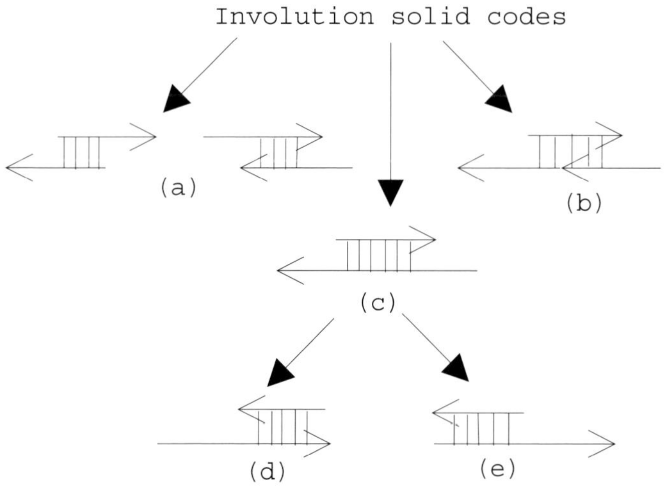

Fig. 1. Various types of intermolecular hybridizations. (**a**) (θ-overlap-free) The prefix or suffix of a code word is a suffix or prefix respectively of a complement of another code word; (**b**) (θ-comma-free) a code word is a reverse complement of a subword of a concatenation of two other code words; (**c**) (θ-infix) one code word is a reverse complement of a subword of another code word; (**d**) (θ-suffix) one code word is a reverse complement of a suffix of another code word; (**e**) (θ-prefix) one code word is a reverse complement of a prefix of another code word. The $3'$ end is indicated with an arrow.

Solid codes were introduced in [28] in the context of the study of disjunctive domains. Certain combinatorial and closure properties of solid codes were discussed in [15]. Properties of maximal solid codes were discussed in [17]. We now recall the definition of solid codes used in [17] which was defined using a characterization given in [15].

Definition 2. *A set* $X \subseteq \Sigma^+$ *is a* solid code *if*

1. *X is an infix code*
2. $\mathrm{PPref}(X) \cap \mathrm{PSuff}(X) = \emptyset$.

27. H.J. Shyr, *Free Monoids and Languages*, Hon Min Book Company 2001.
28. H.J. Shyr, S.S. Yu, Solid codes and disjunctive domains, *Semigroup Forum*, **41** (1990): 23–37.

Test Tube Selection of Large Independent Sets of DNA Oligonucleotides

Russell Deaton[1], Junghuei Chen[2], Jin-Woo Kim[3], Max H. Garzon[4], and David H. Wood[5]

[1] Computer Science and Engineering
University of Arkansas, Fayetteville, AR 72701, USA
rdeaton@uark.edu
[2] Chemistry and Biochemistry
University of Delaware, Newark, DE 19716, USA
junghuei@udel.edu
[3] Biological Engineering
University of Arkansas, Fayetteville, AR 72701, USA
jwkim@uark.edu
[4] Computer Science
University of Memphis, Memphis, TN 38138, USA
mgarzon@memphis.edu
[5] Computer and Information Sciences
University of Delaware, Newark, DE 19716, USA
wood@mail.eecis.udel.edu

1 Introduction

Duplex formation between DNA oligonucleotides is an important technique for assembly of nanostructures [25, 23]. In addition, the reaction itself is capable of performing information processing and computation [1]. Matching of sequence templates during Watson–Crick duplex formation can implement a search among a large number of possibilities for information or potential solutions to computational problems.

In a typical application, DNA oligonucleotide sequences are designed to form specific configurations of duplexes. Through random Brownian motion, molecular structures are assembled through the weak molecular interactions holding the duplexes together, i.e. hydrogen bonding and base stacking. With no explicit external control, this process is referred to as self-assembly.

Of course, it is preferred if the oligonucleotides assemble exactly as designed, but, as pointed out by Seeman [26], control of DNA reactions and structures is a difficult task. In nanotechnology, the effort to control the duplex-forming reactions has been termed DNA word design [15, 11]. It involves the problem of selecting sequences for the DNA oligonucleotides such that the desired assemblies are formed, and errors and defects are minimized.

Unplanned duplex formation is called crosshybridization. Formally, the DNA word design problem is defined as follows.

Definition 1 (DNA Word Design (DWD)) *Given a set of DNA oligonucleotides T, an hybridization energy $J_{ij} = J_{ji} \in Z^- \ \forall\ i,j\ \in T$, a positive integer $K \leq |T|$, and a threshold $B \in Z^-$, does T contain a subset $T' \subseteq T$ such that $|T'| \geq K$, and $J_{ij} \geq B\ \forall\ i,j\ \in T'$?*

In the above formulation, the constraint $J_{ij} \geq B$ reflects a threshold value for the hybridization energy, or an all-or-nothing hybridization model [4]. The set T could be composed of individual oligonucleotides, or pairs of Watson–Crick complements. Let the elements of the set, whether sequences or pairs of Watson–Crick complements, be the nodes of a graph, and connect the nodes if they could from a stable duplex. In other words, an independent set is wanted. A restriction of the DWD problem is equivalent to finding the existence of a given-sized independent set of nodes in a graph [13], which is known to be NP-complete [16]. Thus, it is difficult to find optimal sets of good word designs by conventional computer-aided methods, using both combinatorial and thermodynamic models for duplex formation [20, 17].

Therefore, the work described here has focused on manufacturing good sets of independent DNA oligonucleotides in the test tube with in vitro selection. Research on the development of the protocol and on the characterization of the selection products is reviewed. Some of this work has been published elsewhere [14, 3, 13, 8, 9], and details can be found there. The intent in this paper is to collect these efforts into a comprehensive presentation and review. Furthermore, results are summarized that characterize the protocol and its product, and that demonstrate the independence of the selected oligonucleotides. In addition, experiments and results are described that estimate the number of distinct sequences in the product. Finally, use of the sets in applications is discussed.

2 Methods and Materials for the Selection Protocol

An outline of the protocol is shown in Fig. 1. Synthesized strands are purchased from a manufacturer. These strands have regions of known sequence at either end for primer extension. A region of random sequence, which will become the independent sequences, is in the middle. The temperature is controlled so that only independent oligonucleotides are selected. After a rapid quenching step that freezes pairs of oligonucleotides into mismatched configurations, primer extension is done at a temperature that melts duplexes that have a high degree of mismatch, but not duplexes that are closer to being Watson–Crick complements. Therefore, the protocol selectively copies oligonucleotides that are present in mismatched duplex configurations, and thus have a lower thermal stability.

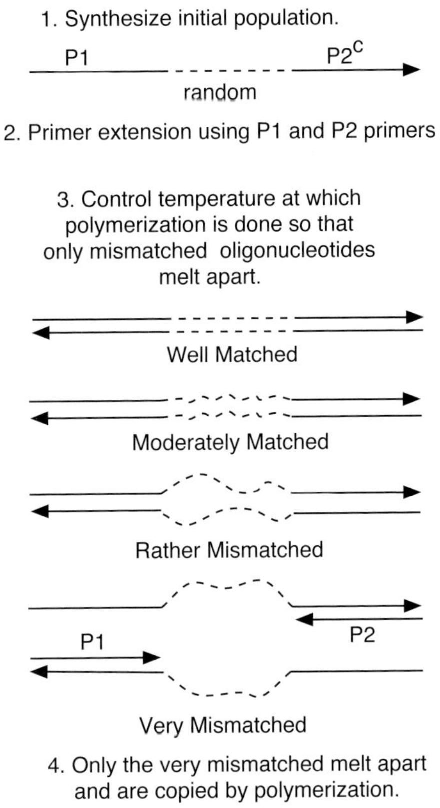

Fig. 1. Protocol to select maximally mismatched oligonucleotides.

Oligonucleotides were purchased from Integrated DNA Technologies. They were purified from denaturing polyacrylamide gels after synthesis. Primer P1 was labeled ^{32}P at the 5′-end using T4 DNA kinase and [γ-^{32}P] ATP. For polymerization, appropriate amounts of ^{32}P-labeled primer P1, primer P2, and double-stranded DNA were incubated in a PCR buffer of 50 mM KCl, 10 mM Tris-HCl, 0.1% Triton X-100, 2.5 mM $MgCl_2$, 0.4 mM 4 dNTP, and 3 U Taq DNA polymerase in a total volume of 10μl at the chosen temperature (43°C). The reaction mixture was incubated for 60 minutes.

3 Gel Characterization

As an initial step, appropriate reaction conditions had to be determined [14]. In addition, the ability of the protocol to preferentially select maximally mismatched oligonucleotides over those closer to being Watson–Crick complements had to be verified [14]. DNA sequence templates with different degrees of mismatching were designed and input into the in vitro selection. The templates included a perfect Watson–Crick complement (T1), a completely mismatched template (T4), and two which contained different types of mismatches in a Watson–Crick duplex, namely two isolated mismatches (T2) and a region of three contiguous mismatches (T3), respectively.

Primer extension was done as detailed in Sec. 2. The Primer extension products were loaded onto 12% (w/v) denaturing polyacrylamide gel (8M urea) with 1X TBE buffer. The characterization gel was run for 1 hour at 60°C. The voltage was 400 V, and the results were documented by autoradiography.

The results are shown in Figs. 2 and 3. Attention should be focused on the topmost band (60-mer) of extension products. In Fig. 2, the maximally mismatched template (T4) was amplified preferentially over the other templates. The degree of amplification reflected the thermal stability of the template, with the perfectly matched template (T1) being the most stable, followed in decreasing order of stability by the templates with two isolated mismatches (T2), a small loop of mismatches in the middle (T3), and a complete mismatch (T4). From the results shown in Fig. 3, the appropriate reaction temperature for the selection protocol was determined. Amplification of the completely matched template (T1) was compared with that of the completely mismatched template (T4) over a range of temperatures. From 37°C to 43°C, the mismatched template was amplified, while the matched one was not. Thus, 43°C was chosen as the reaction temperature for primer extension in the selection protocol. In addition, these results demonstrate that temperature-controlled primer extension works as expected for the selection of independent oligonucleotides. If a completely mismatched template is tested, it is amplified. If a completely matched template is tested, no amplification takes place.

Next, the progress of the protocol over several rounds and the independence of the selected oligonucleotides were characterized [8]. First, however, the primers on the ends of the library sequences (Fig. 1) had to be removed because they interfered with tests for crosshybridization by trapping duplexes into mismatched configurations. The primers on one end were removed by incorporating a ribonucleotide into the extension product. The primer was then cut off with RNAse. This allowed the sequences of interest in the middle to align properly for duplex formation.

It was of interest to determine what percentage of DNA was selected as iteration of the protocol proceeded. As bad sequences are eliminated, one would expect the percentage to increase over subsequent rounds. Thus, intensities of gel bands of DNA for extended and nonextended products were compared with a densitometer. As the cycles progress, the proportion of selected product increases (Fig. 4). Therefore, sequences are eliminated as the protocol progresses, and the hypothesis is that these are the independent oligonucleotides.

This is supported by another gel, shown in Fig. 5, in which the product after primer extension is compared with product that has been heated to a high temperature (95°C) and then reannealed at room temperature for 5 minutes [3]. In addition, the completely mismatched and matched templates

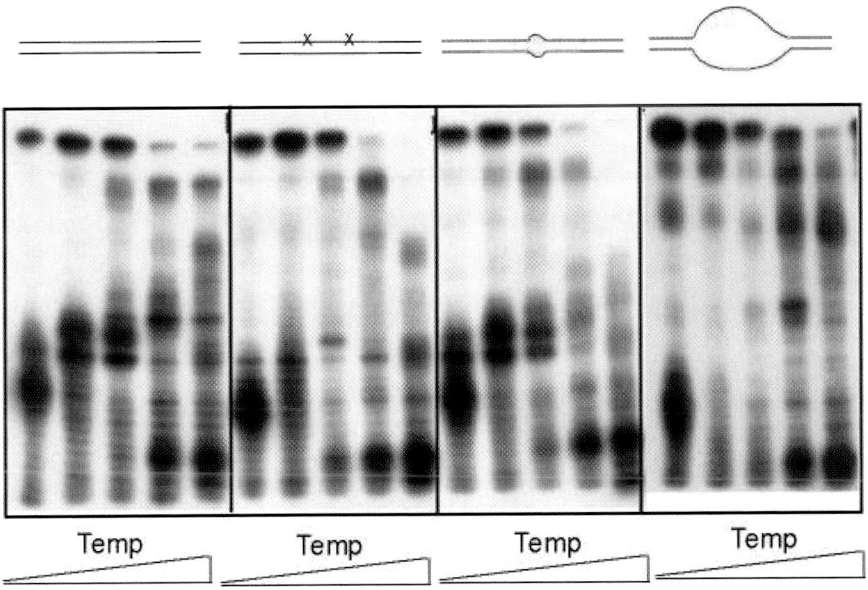

Fig. 2. A denaturing gel, comparing the primer extension products of four different templates, as diagrammed on top of each panel, at various temperatures. The primer was P1. The temperatures were 52°C, 58°C, 64°C, 70°C, and 74°C from left to right in each panel. By focusing on the topmost band in each gel, the degree of full product extension (60-mer) can be observed. At lower temperatures, the maximally mismatched template was preferentially amplified over the perfect Watson–Crick template and the templates with a lower degree of mismatch [14, 3].

were run in separate bands for comparison. In Fig. 5, if attention is focused on the odd-numbered lanes (those with reannealed product), it is seen that the bands move toward the completely mismatched template as the protocol progresses. This supports the contention that independent oligonucleotides are being selected by the protocol.

4 Sample Sequencing of Library Oligonucleotides

In order to check the independence of the protocol product, starting materials (two sequences) and products from one to four cycles of the protocol (four, four, three, and three sequences, respectively) were cloned and sequenced [8]. Using a software tool [13] and the nearest-neighbor model of duplex thermal stability [24], the sequences were checked for bad duplex formation (crosshybridization). The percentage of bad duplexes, excluding hairpins, for each sample is shown in Fig. 6. No crosshybridization was detected in the starting material. Given that potentially there are all possible 20-mers present in the starting material ($\approx 10^{12}$ sequences), it is not surprising that no crosshy-

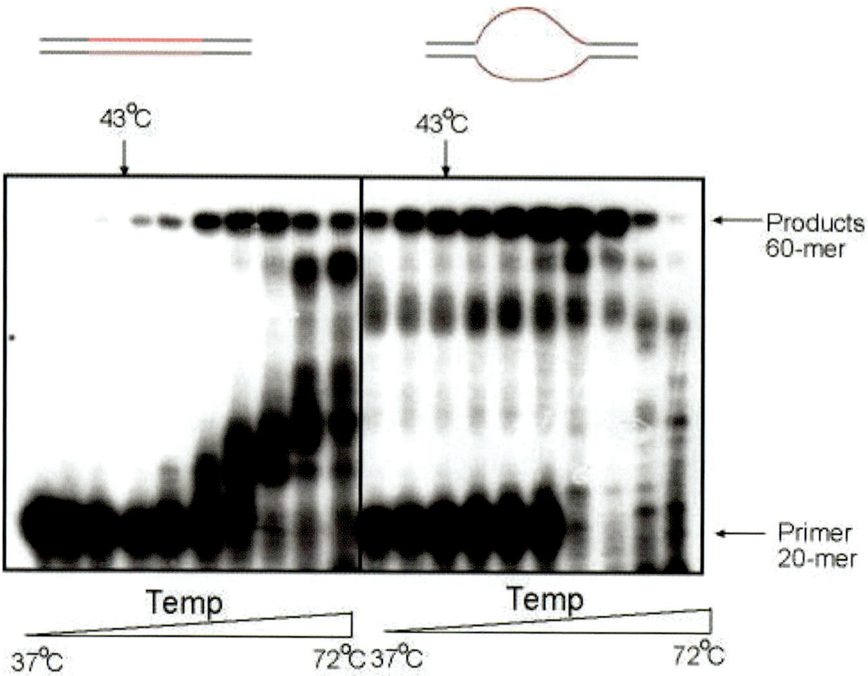

Fig. 3. A denaturing gel, comparing the primer extension products of two templates, perfectly matched and maximally mismatched, as diagrammed on top of each panel, at temperatures ranging from 37°C to 72°C, from left to right (37, 40, 43, 46, 48, 50, 56, 62, 68, 72°C). P1 was the primer. The topmost band of fully extended products (60-mers) shows preferential amplification of the maximally mismatched template over the perfect Watson–Crick template [14, 3].

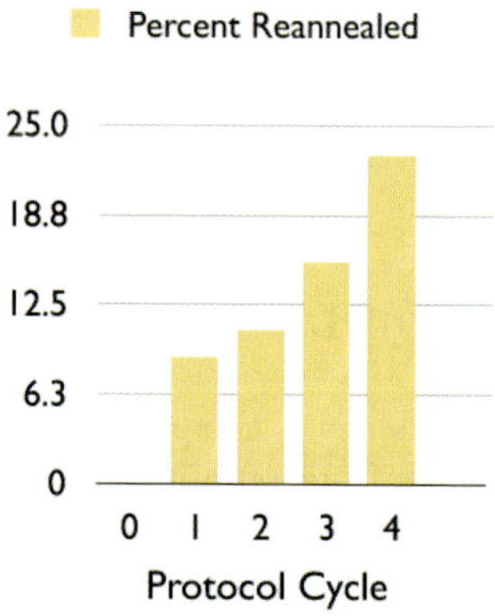

Fig. 4. Percentage of reannealed product versus protocol cycle.

bridization was observed. The sample size was just too small. Once the number

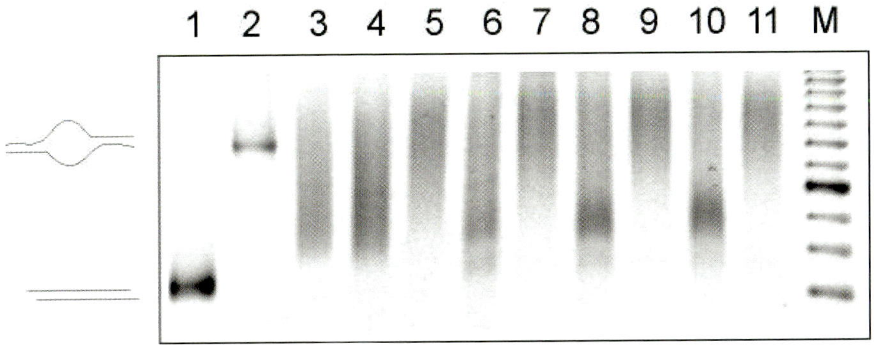

Fig. 5. A stain gel showing results of multiple rounds of the protocol. Lane 1 contains the perfect Watson–Crick template, lane 2 contains the completely mismatched template, and lane 3 contains random starting material with primers. Even-numbered lanes contain protocol product after extension, and odd-numbered lanes contain protocol product after purification and quenching [3].

of sequences was reduced after one and two iterations of the protocol, crosshybridization was observed. No crosshybridization was observed after three and four iterations of the protocol.

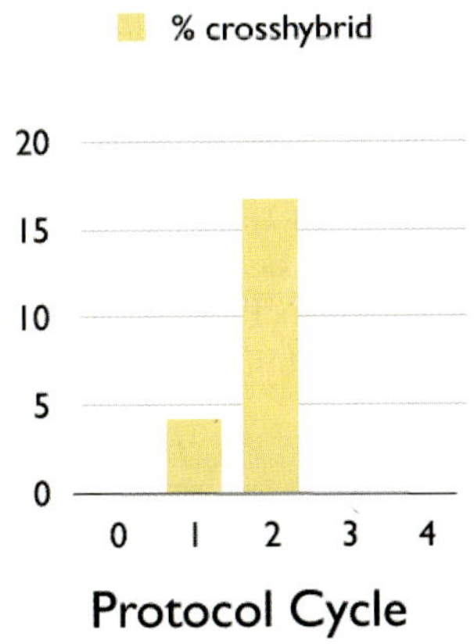

Fig. 6. Percentage of bad duplexes (not counting hairpins) versus protocol cycle.

The sample size of the sequences was too small to state any definitive conclusions. Nevertheless, as the selection protocol was iterated, the independence of the sequences seemed to improve. All cycles had examples of hairpins because the protocol does not explicitly select out self-hybridization. Most of the crosshybridizations, including the self-hybridizations, would have been marginally stable at the selection temperature of 43°C. The exceptions are two hairpins for cycle 1, and three hairpins and one crosshybridization for cycle 2.

5 Spectroscopic Characterization

Given a possible independent set in a test tube, an important issue is to estimate the number of different sequences in the set. When nucleic acids renature, their ultraviolet absorbance, with a maximum around 260 nm, decreases. This effect, called hypochromicity [4], can be used to measure concentrations of nucleic acids. Those sequences that are present in large numbers, i.e. repeated sequences, will renature quickly, while rare or unique sequences will react more slowly. Thus, differences in the kinetics between common and rare sequences can be used to estimate concentrations, or number of sequences, from spectrographic data. This technique is called the C_0t curve technique [6].

The equilibrium rate equation between single-stranded DNA (ssDNA) and double-stranded DNA (dsDNA) follows the stoichiometric equation

$$[\mathrm{ssDNA}_1] + [\mathrm{ssDNA}_2] \underset{k_2}{\overset{k_1}{\rightleftarrows}} [\mathrm{dsDNA}], \tag{1}$$

where the square brackets [ ] denote the molar concentration of nucleotides in either single-stranded or double-stranded in mol/l; k_1 is the forward biomolecular rate constant (renaturation), and k_2 is the reverse rate constant (denaturation). In what follows, the reverse reaction is ignored. Thus, the forward reaction is modeled as a second-order reaction. For renaturation of short DNA, the nucleation can be considered to be the rate-limiting step, after which the duplex zips together rather quickly. The rate constant k_1 can be estimated experimentally as a function of the nucleation rate constant (k_n) as [29]

$$k_1 = \frac{k_\mathrm{n}\sqrt{L_\mathrm{s}}}{X}, \tag{2}$$

where L_s is the length of the shortest strand participating in duplex formation and X is the complexity, which is defined as the number of base pairs in unique sequences [29]. The differential rate equation is

$$\frac{d[\mathrm{dsDNA}]}{dt} = k_1[\mathrm{ssDNA}_1][\mathrm{ssDNA}_2]. \tag{3}$$

Assuming equal starting concentrations of single strands and using conservation of mass, Equation (3) can be rewritten as

$$-\frac{[\mathrm{ssDNA}]}{dt} = k_1[\mathrm{ssDNA}]^2, \tag{4}$$

where [ssDNA] is the total concentration of nucleotides in single strands. The solution is

$$\frac{[\text{ssDNA}_t]}{[\text{ssDNA}_0]} = \frac{1}{1 + k_1[\text{ssDNA}_0]t}, \tag{5}$$

where $[\text{ssDNA}_t]$ is the fraction of single-stranded nucleotides at time t and $[\text{ssDNA}_0]$ is the initial concentration of single-stranded nucleotides. Let C_0 be the total molar concentration of nucleotides in the solution. Assuming that all the single-stranded DNA is initially denatured, $C_0 = [\text{ssDNA}_0]$. If $f_{[\text{ssDNA}]}$ is the fraction of single-stranded DNA at time t, Equation (5) becomes

$$f_{[\text{ssDNA}]} = \frac{1}{1 + k_1 C_0 t}. \tag{6}$$

Equation (6) describes the "C_0t curve". By monitoring the UV absorbance of the renaturation reaction, the sequence complexity of the sample DNA can be estimated. Those sequences present in high concentrations anneal quicker than rarer sequences. The common use of C_0t curves is to estimate the sequence complexity of genomes [6]. Substituting the empirical expression for the forward rate constant (Equation (2)) and using an expression for the nucleotide concentration in terms of the optical density, we obtain the following equation, which was fitted using nonlinear regression to experimental data of UV absorbance versus time. This equation expresses the relationship between the UV absorbance change and the sequence complexity:

$$A(t) = \frac{\triangle A_{\text{total}}}{1 + (A_{\text{ssDNA}} - \triangle A_{\text{total}}) \frac{k_n \sqrt{L_s} t}{X} \left(7.5 \times 10^{-5} \frac{M}{AU_{260\text{nm}}}\right)} + A_{\text{dsDNA}}, \tag{7}$$

where $\triangle A_{\text{total}}$ is the total change in optical density at 260 nm, A_{ssDNA} is the optical density at time $t = 0$, A_{dsDNA} is the optical density at the last reading, k_n is the nucleation rate constant, t is the time (s), L_s is the length of the shortest strand, and X is the complexity (bp).

DNA samples of the selection products were prepared by PCR amplification with DNA primers (Primer #1 (DP#1) = 5′–CAT CGA AGG GGT GTT TTT T–3′; Primer #2 (DP#2) = 5′–TCT TCA TAA GTG ATG CCC G–3′) or DNA primers with ribose tails for C_0t experiments (Primer #1 (RP#1) = 5′–CAT CGA AGG GGT GTT TTT T/3RiboC/–3′; Primer #2 (RP#2) = 5′–TCT TCA TAA GTG ATG CCC G/3RiboU/–3′). The amplified DNA samples obtained with DP#1 and DP#2 were purified using the MiniElute PCR Purification Kit (QIAGEN Inc., Valencia, CA, USA) before the C_0t experiments (final product size = 60 bp). However, the samples amplified with RP#1 and RP#2 for the C_0t experiments were first heated at 90°C for 15 min to sterilize the ribose tails, and then the selection products were extracted from the amplified samples by gel extraction after TBE polyacrylamide gel electrophoresis (PAGE) of the sterilized samples (final product size

6 Potential Advantages and Applications

In vitro manufacturing of independent sets has several advantages. The protocol is simple and cost-effective. Because DNA word design is NP-complete, larger sets can be generated more quickly through the parallelism of the selection protocol than through computer-aided design, though the sets are not guaranteed to be of maximum size. Also, the sets can be manufactured under conditions similar to those in which they will be used. Finally, purchasing random oligonucleotides is less expensive than synthesis of individually designed sequences. Large sets of independent 20-mers for gene deletion in *Saccharomyces cerevisiae* have been reported in [30]. These sequences were found by a BLAST search of genomic DNA, synthesized, and tested against DNA microarrays. In vitro manufacturing has the advantages of cost-effectiveness and optimization of sequences for specific reaction conditions over that technique.

Large libraries of independent oligonucleotides are an enabling resource for DNA-based nanotechnology. For example, fundamental physical limits and increasing costs of fabrication facilities will force alternatives to conventional microelectronics manufacturing to be developed [2]. Seminal work by Seeman [27], Winfree et al. [32], and Mirkin et al. [22] has shown the potential of self-assembly that is directed by sequence-specific, template-matching hybridization reactions among DNA oligonucleotides. Self-assembly of molecular electronic circuits has also been suggested [19]. In self-assembly, weak, local interactions among molecular components spontaneously organize those components into aggregates with properties that range from simple to complex [31]. So far, DNA self-assembly has constructed regular, periodic structures or simple circuits, but more complicated assemblies with greater control of component location are needed to approximate the functionality of microelectronics [5], and to take advantage of the increasing number of nanocomponents [28], from carbon nanotubes and nanoparticles composed of various materials to proteins and antibodies [23]. Libraries of independent oligonucleotides can be used to self-assemble complicated nanostructures with greater control of component location. The in vitro manufacture of NCH libraries provides a large address space of DNA sequences with which to increase both complexity and control. Complexity is increased because nanostructures can be designed for a specific connectivity through a diversity of DNA–nanocomponent conjugates. Control is achieved by assigning unique DNA sequence addresses to specific nanocomponents corresponding to specific locations in the larger structure. Unique sequence addresses might also be exploited to detect and repair specific components, which would increase reliability. Moreover, because the hybridization reactions between library oligonucleotides are not competitive, assembly can occur in parallel. In addition, when compared with synthesis, the manufacture of the library is relatively inexpensive, and thus would make fabrication of nanostructures, with unique addresses for constituent components, more practical. The nanostructures that are enabled by the large DNA libraries will be useful not only for molecular-electronic applications but also for other nanotechnology applications in biology, medicine, and sensors.

To make the libraries useful, the individual oligonucleotides have to be cloned and isolated. The primer sequences can have an embedded restriction site so that the internal sequences can be inserted into a plasmid. These plasmids could be transformed into *E. coli* under conditions that promote one plasmid per cell, and then colonies could be grown. Each clone would, on average, incorporate just one sequence from the starting set. The sequences could then be extracted to build a library of strands, to which nanocomponents could be attached. The restriction site would also allow hybridization of one oligonucleotide to another to form linker molecules for the self-assembly.

It should be possible to assign oligonucleotides from the library to nanocomponents without the sequence being known. Sequencing the DNA is a potential bottleneck. Sequencing of large numbers of oligonucleotides adds cost and time to applications. Thus, it is more efficient and cost-effective to avoid sequencing. An index of which clones were used for specific nanocomponents would be maintained. When a particular component was to be added or addressed, then the index would tell us which oligonucleotide to use.

7 Conclusion

Self-assembly of nanostructures through template-matching hybridization reactions is potentially an important technique in nanotechnology. Given the possibility of errors in hybridization and the difficulty of designing DNA sequences on conventional computers, a viable alternative is to manufacture libraries of oligonucleotides for nanotechnology applications in the test tube. Thus, a protocol has been designed and tested to select mismatched oligonucleotides from a random starting material. Experiments indicate that the selected oligonucleotides are independent, and that there are about 10 000 distinct sequences. Such manufactured libraries are a potential enabling resource for DNA self-assembly in nanotechnology.

References

1. L.M. Adleman, Molecular computation of solutions to combinatorial problems, *Science*, **266** (1994) 1021–1024.
2. A. Bachtold, P. Hadley, T. Nakanishi, C. Dekker, Logic circuits with carbon nanotube transistors, *Science*, **294** (2001) 1317–1320.
3. H. Bi, J. Chen, R. Deaton, M. Garzon, H. Rubin, D.H. Wood, *In Vitro* selection for non-crosshybridizing oligonucleotides for computation, *Natural Comput.*, **2** (2003) 417–426.
4. V.A. Bloomfield, D.M. Crothers, I. Tinoco Jr., Nucleic Acids: Structures, Properties, and Functions, University Science Books, Sausalito, CA, 2000.

5. E. Braun, U. Sivan, DNA-templated electronics, *Nanobiotechnology*, C.M. Niemeyer, C.A. Mirkin eds., Wiley-VCH, Weinheim, (2004) 244–255.
6. R.J. Britten, D.E. Kohne, Repeated sequences in DNA, *Science*, **161** (1968) 529–540.
7. J. Charlton, D. Smith, Estimation of SELEX pool size by measurement of DNA renaturation rates, RNA, **5** (1999) 1326–1332.
8. J. Chen, R. Deaton, M. Garzon, J.W. Kim, D.H. Wood, H. Bi, D. Carpenter, Y.Z. Wang, Characterization of the non-crosshybridizing DNA oligonucleotides manufactured *in vitro*, *DNA Computing: Preliminary Proceedings of the 10th International Workshop on DNA-Based Computers*, G. Mauri, L. Smith eds., Lecture Notes in Computer Science **3384** University of Milano-Bicocca, Milan, (2004) 50–61.
9. J. Chen, R. Deaton, M. Garzon, J.W. Kim, D.H. Wood, H. Bi, D. Carpenter, J.S. Lee, Y.Z. Wang, Sequence complexity of large libraries of DNA oligonucleotides, *DNA Computing: Preliminary Proceedings of the 11th International Workshop on DNA-Based Computers*, N. Pierce, A. Carbone eds., University of Western Ontario, London, Ontario, Canada, 2005.
10. J. Chen, J. Reif, eds., DNA Computing: 9th International Workshop on DNA-Based Computers, Lecture Notes in Computer Science, **2943** Springer, Berlin, Heidelberg, 2004.
11. R. Deaton, M. Garzon, J.A. Rose, D.R. Franceschetti, R.C. Murphy, S.E. Stevens Jr., Reliability and efficiency of a DNA based computation, *Phys. Rev. Lett.*, **80** (1998) 417–420.
12. R. Deaton, J. W. Kim, J. Chen, Design and test of non-crosshybridizing oligonucleotide building blocks for DNA computers and nanostructures, *Appl. Phys. Lett.*, **82** (2003) 1305–1307.
13. R. Deaton, J. Chen, H. Bi, J.A. Rose, A software tool for generating non-crosshybridizing libraries of DNA oligonucleotides, *DNA Computing: 8th International Workshop on DNA-Based Computers*, M. Hagiya, A. Ohuchi eds., Lecture Notes in Computer Science, **2568** Springer, Berlin, Heidelberg, (2003) 252–261.
14. R. Deaton, J. Chen, H. Bi, M. Garzon, H. Rubin, D.H. Wood, A PCR-based protocol for *in vitro* selection of non-crosshybridizing oligonucleotides, DNA Computing: 8th International Workshop on DNA-Based Computers, M. Hagiya , A. Ohuchi eds., Lecture Notes in Computer Science, Springer, Berlin, Heidelberg, **2568** (2003) 196–204.
15. A.G. Frutos, Q. Liu, A.J. Thiel, A.M.W. Sanner, A.E. Condon, L.M. Smith, R.M. Corn, Demonstration of a word design strategy for DNA computing on surfaces, Nucleic Acids Res., **25** (1997) p. 4748.
16. M.R. Garey, D.S. Johnson, Computers and Intractability, Freeman, New York, 1979.
17. M.H. Garzon, R.J. Deaton, Codeword design and information encoding in DNA ensembles, *Natural Comput.*, **3** (2004) 253–292.
18. M. Hagiya, A. Ohuchi, eds., DNA Computing: 8th International Workshop on DNA-Based Computers, Lecture Notes in Computer Science, **2568** Springer, 2003.
19. K. Keren, R.S. Berman, E. Buchstab, U. Sivan, E. Braun, DNA-templated carbon nanotube field-effect transistor, *Science*, **302** (2003) 1380–1382.

20. A. Marathe, A.E. Condon, R.M. Corn, On combinatorial DNA word design, *DNA Based Computers V*, E. Winfree, D.K. Gifford, eds., Providence, RI, DIMACS, American Mathematical Society, (1999) 75–90.
21. G. Mauri, L. Smith, eds., DNA Computing: Preliminary Proceedings of the 10th International Workshop on DNA-Based Computers, Lecture Notes in Computer Science **3384**, Springer, Berlin, Heidelberg, 2005.
22. C. Mirkin, R. L. Letsinger, R. C. Mucic, J. J. Storhoff, A DNA-based method for rationally assembling nanoparticles into macroscopic materials, *Nature*, **382** (1996) 607–609.
23. C.M. Niemeyer, C.A. Mirkin, eds., Nanobiotechnology, Wiley-VCH, Weinheim, 2004.
24. J. SantaLucia Jr., A unified view of polymer, dumbbell, and oligonucleotide DNA nearest-neighbor thermodynamics, *Proc. Natl. Acad. Sci.* USA, **95** (1998) 1460–1465.
25. N.C. Seeman, H. Wang, X. Yang, F. Liu, C. Mao, W. Sun, L. Wenzler, Z. Shen, R. Sha, H. Yan, M.H. Wong, P. Sa-Ardyen, B. Liu, H. Qiu, X. Li, J. Qi, S.M. Du, Y. Zhang, J.E. Mueller, T. Fu, Y. Wang, J. Chen, New motifs in DNA nanotechnology, *Nanotechnology*, **9** (1998) 257–273.
26. N.C. Seeman, H. Wang, B. Liu, J. Qi, X. Li, X. Yang, F. Liu, W. Sun, Z. Shen, R. Sha, C. Mao, Y. Wang, S. Zhang, T. Fu, S. Du, J.E. Mueller, Y. Zhang, J. Chen, The perils of polynucleotides: the experimental gap between the design and assembly of unusual DNA structures, Proceedings of the Second Annual Meeting on DNA Based Computers L.F. Landweber, E.B. Baum eds., **44** Providence, RI, DIMACS, American Mathematical Society, (1998) 191–205.
27. N.C. Seeman, DNA nanostructures for mechanics and computing: nonlinear thinking with life's central molecule, Nanobiotechnology, C.M. Niemeyer, C.A. Mirkin eds., Wiley-VCH, Weinheim, (2004) 308–318.
28. C.S. Thaxton, C.A. Mirkin, DNA–gold nanoparticle conjugates, *Nanobiotechnology*, C.M. Niemeyer, C.A. Mirkin eds., Wiley-VCH, Weinheim, (2004) 288–307.
29. J.G. Wetmur, DNA probes: applications of the principle of nucleic acid hybridization, *Crit. Rev. Biochem. Mol. Bio.*, **26** (1991) 227–259.
30. G.M. Whitesides, M. Boncheva, Characterization of synthetic DNA bar codes in *Saccharomyces cerevisiae* gene-deletion strains, *Proc. Natl. Acad. Sci.* USA, **101** (2004) 11046–11051.
31. G.M. Whitesides, M. Boncheva, Beyond molecules: self-assembly of mesoscopic and macroscopic components, *Proc. Natl. Acad. Sci.* USA, **99** (2002) 4769–4774.
32. E. Winfree, F. Liu, L.A. Wenzler, N.C. Seeman, Design and self-assembly of two-dimensional DNA crystals, *Nature*, **394** (1998) 539–544.

Part III

DNA Nanodevices

DNA-Based Motor Work at Bell Laboratories

Bernard Yurke

Bell Laboratories, 600 Mountain Ave., Murray Hill, NJ 07974, USA
yurke@lucent.com

1 Moore's Law

Many DNA-based nanodevices have been constructed which can execute motion on a nanoscale. By the end of 2004, 19 different forms of such devices have been published, as witnessed by references [1–19]. The number of published DNA-based nanodevices as a function of time is plotted in Fig. 1. One sees that the growth, so far, has been approximately exponential with a doubling time close to 18 months, the doubling time of the growth curve for the component density in the semiconductor industry. This Moore's law indicates a healthy vibrant field. Also plotted as the black rectangles are the cumulative number of DNA-based devices coming out of Ned Seeman's lab. One sees that his are the first two entries in the field [1, 2] and that his lab continues to have a strong presence in this field. In fact, among the most sophisticated DNA-based nanodevices are his PX-JX_2 machine [7] and his translation machine for polymer assembly [13]. For all of us working on DNA-based nanodevices Ned has been a great inspiration.

I will use my space in this volume celebrating Ned's sixtieth birthday as an opportunity to tell a few stories about DNA-based self-assembly work at Bell Laboratories, work that was inspired by Ned's research.

2 Bad Luck

In this section I will relate how Allen Mills, Jr. and I stumbled upon a means of actuating DNA nanostructures using DNA alone. Allen and I had developed a common interest in self-assembly. My own interest in this field sprang from the long tradition of research at Bell Laboratories directed towards making ever smaller transistors and from more resent research at Bell Laboratories on organic electronic devices. Self-assembly at the molecular scale opened up the possibility of making electronic devices much smaller than could be made with photolithographic techniques or electron-beam writing techniques.

I saw it as a means of leapfrogging Moore's law and arriving at the end point of miniaturization. I had collected many of Ned Seeman's papers on the fabrication of DNA-based nanostructures and suggested to Allen that we begin a program on DNA-based self-assembly.

DNA computing, with its promise of massively parallel computation, quickly captured Allen's imagination. We decided to try to implement Boolean logic in DNA. It dawned on us that linking two strands of DNA, A and B, together by a third strand of DNA C that had base sequences complementary to both A and B was a way of implementing an "and" functionality. The strand C could be considered an "and" operator in that the structure A and B linked together is formed only if A and B were both present. We designed some DNA strands to test this idea. Allen had written software that would generate lists of random strings of A, C, G, and T that were ten letters long. The strings were selected to differ from each other by a certain Hamming distance and were selected to minimize self-complementarity. We constructed the base sequences for our strands by concatenating these ten-base sequences into longer sequences. Even though the ten-base strings may only have self-complementary or cross-complementary regions that are three bases long, when joined together into a longer string the longer string could have self-complementary regions that are considerably bigger than three bases. We thought, however, that this would be unlikely. We designed our A, B, and C strands so that the complementary region between A and C and the region between B and C were each 20 nt long.

To test the "and" operation we did a gel shift experiment in which we ran the strands A, B, and C in separate lanes. These served as references for comparison with the bands in lanes in which various combinations of A, B, and C were present. To our surprise, although A, B, and C were the same length, C migrated much faster than A and B. We surmised that C was forming a hairpin. On checking the base sequence we found that C, in fact, had a self-complementary region with an 8 nt overlap. We had been victims of bad luck. However, from the lanes containing A and C, B and C, or A, B, and C it was evident that the hairpin did not prevent the hybridization of C with A or B. We concluded that the A and B strands were able to open up the hairpin. This experiment, performed on April 29, 1997 provided a key "Aha" moment. It made us realize that DNA hybridization could be used to induce structural changes in DNA nanostructures. We were now halfway towards making a molecular motor. What we still needed was a way to remove the DNA strand after it had induced a structural change. It occured to us that some kind of nicking enzyme could do this and there are drawings in the May 6, 1997 entries in my notebook of devices operated this way. One is of a DNA walker.

My interest in molecular motors, at the time, was limited to making one from scratch for which each step of the machine cycle was intentionally designed. I felt that using an enzyme to run part of the machine cycle would be cheating, in this regard, because one does not know how to design such

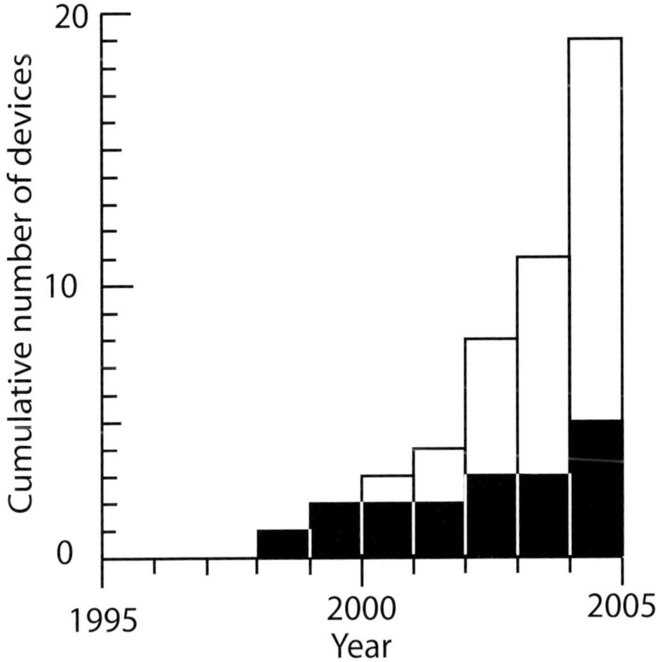

Fig. 1. The cumulative number of DNA-based nanodevices as a function of time. The tops of the open rectangles indicate the total number of DNA-based devices that have been constructed as a function of time. The heights of the black rectangles indicate the cumulative number of DNA-based nanodevices that have come out of Ned Seeman's lab. One sees that Ned produced the first such devices and that his lab continues to be active in the production of DNA-based devices.

enzymes from scratch. So, we did not pursue these schemes. Others have, however, pursued similar kinds of schemes [17, 18]. The DNA-based motor field has been enriched by their work. Eventually, it dawned on us that a DNA strand could be removed through competitive binding. To investigate the effectiveness of this phenomenon, in July, 1997 Allen and I began a series of experiments to explore toehold-mediated strand exchange. By January 15, 1998 we had used toehold mediated-strand exchange to successively open and close a hairpin, as indicated by gel shift experiments. However, we felt something more than a gel shift experiment was needed if one really wanted to claim that one was inducing conformational changes via hybridization. This motivated us to learn how to do fluorescence resonant energy transfer (FRET) experiments. To this end we built a fluorometer powered by a 1 kW Xe arc lamp. This fluorometer was used to take reaction kinetics data in our toehold-mediated strand exchange studies [20]. The 1 kW Xe arc lamp proved inconvenient to work with and was eventually replaced by an argon ion laser.

In the fall of 1998 Andrew Turberfield arrived at Bell Laboratories to spend his sabbatical in my lab. The first task Andrew took upon himself was to rebuild the laser-based fluorometer so that we could make measurements on multiple samples simultaneously. We first tried using fiber optic beamsplitters to distribute the laser light among eight samples simultaneously. This did not work in a satisfactory manner. Guiding light through optical fibers was abandoned in favor of free-space light propagation and conventional beamsplitters were used to distribute the light. This fourth incarnation of the fluorometer worked very well. We then turned our attention to devising and studying catalytic DNA-based systems that could serve as fuel for free-running molecular machines [9]. In December of that year I designed the DNA strands for the DNA tweezers and in January we began performing experiments on the tweezers. The FRET experiments performed as expected and we took this to mean that the tweezers were opening and closing properly. So, we submitted a manuscript for presentation at the DNA 5 conference in Boston, Massachusetts [21]. There was, however, another interpretation of the data that did not occur to us until later, namely, that the fuel strand was bringing donor and quencher dyes together through the formation of dimers, or more generally multimers.

At the beginning of June 1999 Jennifer Neumann, a senior undergraduate at Rutgers University, joined my lab as an intern through Lucent's Summer Research Program. She was given the task of testing DNA tweezer operation via gel shift experiments. Shortly before the DNA 5 conference Jennifer obtained anomalous results that made me begin to suspect that the tweezers were not closing but were forming multimers. At DNA 5 Ned raised the same issue. The tweezers at that time did not have a region of unpaired bases to serve as a flexible hinge. The hinge was, in fact, a nick in double-stranded DNA. Ned pointed out that it was the conventional wisdom that nicked DNA consisted of a linear molecule very similar to double-stranded DNA without a nick. Ned had done crystallography on DNA with sticky ends. So, he knows these sorts of things [22]. Immediately after returning from DNA 5 I ordered DNA strands to make tweezers with single stranded-regions of varying length to serve as a hinge. These were tested by Jennifer and based on her studies we decided to focus our efforts on the study of a tweezer with a hinge 4 nt in length. Jennifer obtained gel shift data indicating that most of the tweezers were opening and closing properly. With gel shift data and new fluorescence-quenching data in hand we wrote up our results and submitted them to *Nature*.

The paper went through three rounds of refereeing. The process was slowed down because I was short-handed. After the summer Andrew had returned to Oxford University and Jennifer was now a graduate student at Rutgers. In the winter of 2000 Friedrich (Fritz) Simmel joined the effort as a postdoctoral student. As a graduate student he had done work in experimental condensed matter physics, studying electron transport through quantum dots. He accepted a postdoctoral position with me because he was interested in en-

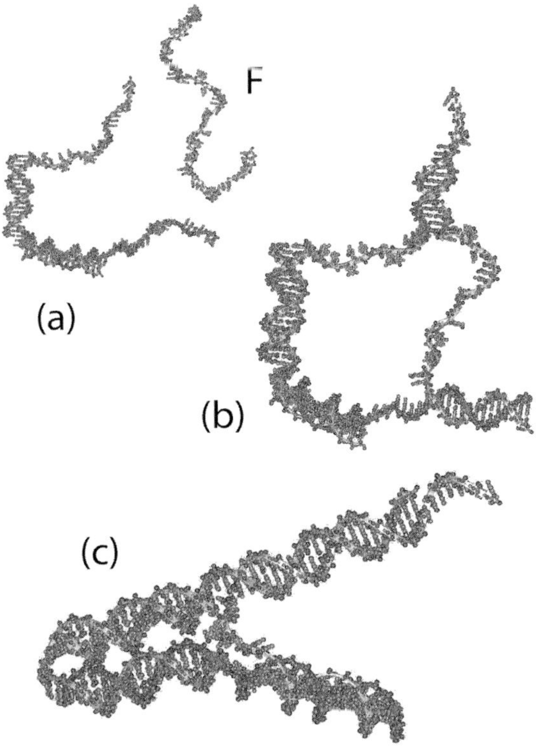

Fig. 2. A sequence showing the closing of the tweezers. (**a**) Depicts the open tweezers. F is the fuel strand. (**b**) Depicts the tweezers as the fuel strand begins to hybridize with the single-stranded extensions off the ends of the tweezers. These single-stranded extensions constitute the motor that pulls the arms together. (**c**) Depicts the tweezers fully closed.

tering a field more closely related to biology. His first task was to perform a series of experiments addressing the concerns of the referees. One issue raised was whether the DNA strands were hybridizing to form some structure other than the tweezers. To address this issue we ordered oligomers that consisted of subsegments of the tweezers' strands. By mixing pairwise combinations of these strands together or by mixing a given segment together with one of the tweezer strands we were able to show via gel shift experiments that hybridization occured only if the two strands happened to have complementary regions. This provided evidence that the tweezers were the only stable structure that the strands could form. To provide evidence, in addition to the gel shift experiments, that the tweezers were opening and closing properly, a variety of experiments were performed for which the FRET signal produced by the closing of the tweezers was different from the FRET signal produced by multimer formation. Fritz's work paid off. The tweezers paper was accepted and scheduled for publication in the August 10 issue of *Nature* [3].

3 Dealing with the Press

Since the paper was likely to receive a lot of press, Lucent had me take a several-hour training course for how to deal with the media. A purpose of the course was to teach us techniques to avoid creating a public relations problem for Lucent. A lot of what was covered was common sense stuff. It was emphasized that we were to stick to the telling of the story of the science we had done. We were not to say anything derogatory about our scientific competitors. We were not to comment on issues outside the domain of science, such as, for example, the state of the telecommunications industry or Lucent's business. We were told to make the story of the science as simple and clear as possible. And we were subjected to practice interviews in front of a video camera. One of the participants in the training sessions I was involved in was Hendrik Schon. Hendrik was the wunderkind of Bell Laboratories. In the few years he had been at Bell Labs he had produced an amazing string of breakthrough papers in *Nature*, *Science*, and the *Journal of Applied Physics*. He was about to announce some more stunning breakthroughs. He mumbled softly through his practice interview not showing a great deal of excitement about his achievements. I thought this was peculiar. At the time, I took it to mean that he was a really humble, low-key, kind of guy. We were all fooled. Hendrik would soon be at the center of accusations of scientific fraud and his collaborators would publish a retraction of all those papers.

At Lucent, interviews by the press are arranged through a media relations person. A media relations person is also present at the interview. Keeping the interview on topic and ensuring that the interview goes smoothly are among the functions of the media relations person. Dan Coulter was the media relations person who handled my interviews. Since my work involved creating nanomachines out of DNA, there was some concern that reporters might raise ethical issues having to do with "tampering with life." Because of this concern Dan Coulter wanted to know where the DNA I was using came from. I told him that it was manufactured in a factory. We designed the base sequences of the DNA we wanted and emailed the sequences to the factory. There the sequences were fed into a machine that had four bottles attached to it labeled A, C, G, and T. The machine would then make DNA strands having those base sequences. Dan then asked me where the stuff inside the bottles came from. I told him that I did not know. I offered him the guess that the nucleotides were extracted from some biological tissue. The scenario I laid out was that DNA was extracted from tissue, broken down into individual nucleotides, chemically separated, chemically modified, purified, and then stuffed into the bottles. That was not a good enough answer for Dan. He demanded that I find out where the stuff in the bottles came from. I called Ned Seeman to see if he could tell me where the nucleotides came from. He did not know, but he shared my opinion that the stuff must be extracted from biological tissues rather than synthesized from scratch. I then called technical support at Glen Research, a major supplier of nucleotides for DNA synthesis. I was put in touch with someone who told me that the nucleotides are produced from salmon sperm, a byproduct of the salmon fishing industry. That made sense to me. Sperm would have a much higher concentration of DNA per unit weight than any other biological tissue. So, it would be a good starting material for an extraction and purification process. I was also told that with the rise of genome sequencing efforts, the demand for nucleotides was threatening to outpace what the salmon fishing industry could supply. Consequently, bacteria culturing was being seriously considered as an alternative source for DNA. When I reported what I had learned, Dan chuckled. I told him: "We now have an answer for where the stuff in the bottles comes from should a reporter ask, but let's not volunteer it to the press."

Nature's press blockade on the contents of the August 10 issue of *Nature* would be lifted during the time I was planning to vacation by accompanying my wife, Jeanne, as she attended her professional conference. Jeanne is a Seventh Day Baptist minister and the gathering was the General Conference of the Seventh Day Baptist denomination. It was being held on the campus of the West Virginia Wesleyan College in Buckhenon. Hence, it was expected that there would be a lot of requests for interviews while I was there. Jeanne and I would be staying in a student dorm without a telephone. Thus, I purchased my first cell phone for the sole purpose of being interviewed by the press. The quality of the cell phone service turned out to be poor. So, it was decided that the live radio interviews would be conducted via an outside pay phone near the dorm were I was staying. Toward the end of my first live interview the reporter asked something like: "I understand that you make your motors out of salmon DNA. What is the ethics of using the substance of life to make machines." In my mind I was thinking: "Thanks for letting the cat out of the bag, Dan!" I went on to try to explain that, apart from the base sequence, all DNA is chemically the same and that the DNA I was using was synthetic with base sequences not related to any biological organism. While I was doing radio interviews via telephone in the wilds of West Virginia, Fritz was doing live interviews at Bell Labs with reporters who wanted video footage.

CNN carried a piece on the DNA tweezers. It went like this: First there came the teaser, an image of salmon swimming and the narrator saying:

> They can navigate thousands of miles to the river of their origin. Now these salmon may guide us to the next generation of computing.

Next, appeared an animation of the DNA tweezers opening and closing, followed by myself pipetting. The first image of the news report depicted a computer chip and a salmon. So, we have fish and chips. Get it? The narration while this image was being displayed was:

> Now this is a breakthrough that could stretch the frontiers of technology and make computers 1,000 times more powerful: Lucent technologies says it has discovered how to use DNA of salmon to carry information like a computer chip. The scientific journal *Nature* comes out with a report tomorrow ...

A discussion of DNA as a carrier of genetic information and its use in forensics followed. Next, returning to the main subject, the narrator said:

> Now Lucent Technologies' Bell Laboratories has discovered another use for it as a motorized machine. Researchers there say they can manipulate DNA in such a way that it may someday help build computer components on a molecular scale with astonishing ability. If it sounds fishy you are partly right. Lucent researchers took DNA from salmon sperm and had it synthesized with a set of instructions. The DNA are shaped like tweezers with two strands connected by a hinge.

Clearly this "salmon" thing had gotten out of hand. Next, a reasonable discussion of the operation of the tweezers was given. That was followed by a video clip of Ned Seeman saying:

> With DNA nanotechnology we are now down to a scale where it really wasn't imaginable a few years ago to be able to actually organize matter as well as we can today.

Next, a video clip was shown of Fritz explaining DNA-based self-assembly. With the airing of this CNN piece, my boss, Dick Slusher, now had an internal public relations problem to deal with. Irate people from all over Lucent where calling him on the phone and asking: "What the %#&@ is Lucent doing research on salmon for?!"

WNBC also carried the story. As a bit of background, shortly after Fritz had arrived in my lab he told me that he had this inner need to shock people. When people in Germany asked him what he would be working on at Lucent he would say: "Remember that X-File episode where that guy gets infected with nanorobots? Well, I'm going to be working on that." Appropriately enough, the WNBC story starts out:

> It's a story straight out of The X-Files, tiny machines injected into the blood stream multiplying rapidly and attacking a man's body with a villain in control using a hand-held remote. Pure science fiction? – Maybe not.

This narration is, of course, accompanied by segments from The X-Files' episode S.R. 819-6×10. A reasonable discussion of how the DNA tweezers work follows. Then a clip appears in which Fritz says:

> Because they are so small you can imagine they could enter a body or cell which carries a disease and lets free a pharmaceutical agent at exactly the point where it is needed to cure the disease.

It is appropriate that Fritz should be the first to devise a DNA nanomachine that could operate on a protein, thrombin, which is involved in blood clotting [12].

What lessons can we learn from this. Perhaps, the first is that it is almost guaranteed that news media will garble the science it is reporting. The second is that the news media will have fun with you at your own expense. But that is OK. Reporting of science is done because people are interested in science. It is the public who ultimately funds our work and we should make the effort to let them know that their money is being well-spent, even if it has to be done through an imperfect medium. My close relatives got to see me on TV because I let them know ahead of time when the news reports would appear. What surprised me was that neighbors and friends who were not alerted caught the programs. Some of them even caught the programs while they were on vacation outside of the United States at the time.

References

1. X. P. Yang, A.V. Vologodskii, B. Lui, B. Kemper, N.C. Seeman, Torsional control of double-stranded DNA branch migration, *Biopolymers* **45**, 69 (1998).
2. C. Mao, W. Sun, Z. Shen, N.C. Seeman, A nanomechanical device based on the B-Z transition of DNA, *Nature* **397**, 144 (1999).
3. B. Yurke, A.J. Turberfield, A.P. Mills, Jr., F.C. Simmel, J.L. Neumann, A DNA-fuelled molecular machine made of DNA, *Nature* **406**, 605 (2000).
4. F.C. Simmel, B. Yurke, Using DNA to construct and power a nanoactuator, *Phys. Rev.* E **63**, 041913 (2001).
5. F.C. Simmel, B. Yurke, A DNA-based molecular device switchable between three distinct mechanical states, *Appl. Phys. Lett.* **80**, 883 (2002).
6. J.J. Li, W. Tan, A single DNA molecule nanomotor, *Nano Lett.* **2**, 315 (2002).
7. H. Yan, X. Zhang, Z. Shen, N.C. Seeman, A robust DNA mechanical device controlled by hybridization topology, *Nature* **415**, 62 (2002).
8. J.C. Mitchell and B. Yurke, DNA Scissors, in *DNA Based Computers VII*, in *DNA computers*, N. Jonoska, N.C. Seeman, eds., LNCS No. 2340 Springer Verlag, Heidelberg, 2002.
9. A.J. Turberfield, J.C. Mitchell, B. Yurke, A.P. Mills, Jr., M.I. Blakey, F.C. Simmel, DNA fuel for free-running nanomachines, *Phys. Rev. Lett.* **90**, 118102 (2003).
10. L. Feng, S.H. Park, J.H. Reif, H. Yan, A two-state DNA lattice switched by DNA nanoactuator, *Angew. Chem. Int. Ed.* **42**, 4342 (2003).
11. P. Alberti and J.L. Mergny, DNA duplex-quadruplex exchange as the basis for a nanomolecular machine, *Proc. Natl. Acad. Sci.* USA **100**, 1569 (2003).
12. W.U. Dittmer, A. Reuter, and F.C. Simmel, A DNA-based machine that can cyclically bind and release thrombin, *Angew. Chem. Int. Ed.* **43**, 3549 (2004).
13. S. Liao and N.C. Seeman, Translation of DNA signals into polymer assembly instructions, *Science* **306**, 2072 (2004).
14. W. B. Sherman N.C. Seeman, A precisely controlled DNA biped walking device, *Nano Letters* **4**, 1203 (2004).

15. J.S. Shin, N.A. Pierce, A synthetic DNA walker for molecular transport, *J. Am. Chem. Soc.* **126**, 10834 (2004).
16. Y. Chen, S.-H. Lee, C. Mao, A DNA nanomachine based on a duplex-triplex transition, *Angew. Chem. Int. Ed.* **43**, 5335 (2004).
17. Y. Chen, M. Wang, C. Mao, An autonomous DNA nanomotor powered by a DNA enzyme, *Angew. Chem. Int. Ed.* **43**, 3554 (2004).
18. P. Yin, H. Yan, X.G. Daniell, A.J. Tuberfield, J.H. Reif, A unidirectional DNA walker that moves autonomously along a track, *Angew. Chem. Int. Ed.* **43**, 4906 (2004).
19. D.C. Lin, B. Yurke, and N.A. Langrana, Mechanical Properties of a reversible, DNA-crosslinked polyacrylamide hydrogel, *J. Biomech. Eng.* **126**, 104 (2004).
20. B. Yurke and A.P. Mills, Jr., Using DNA to power nanostructures, *Genet. Program. Evol. Mach.* **4**, 111 (2003).
21. A.J. Turberfield, B. Yurke, A.P. Mills, Jr., DNA hybridization catalysts and molecular tweezers, in *DNA Based Computers V, DIMACS Series in Discrete Mathematics and Theoretical Computer Science* Vol. 54, E. Winfree, D. K. Gifford, eds., American Mathematical Society, 2000, pp. 171-182.
22. H.X. Qiu, J.C. Dewan, N.C. Seeman, A DNA decamer with sticky end: The crystal structure of d-CGACGATCGT, *J. Mol. Bio.* **267**, 881 (1997).

Nanoscale Molecular Transport by Synthetic DNA Machines*

Jong-Shik Shin[1] and Niles A. Pierce[1,2]

[1] Department of Bioengineering
[2] Department of Applied & Computational Mathematics,
California Institute of Technology, Pasadena, CA 91125, USA
niles@caltech.edu

1 Introduction

Biological systems have evolved motor proteins programmed to perform intracellular transport powered by ATP hydrolysis [23, 14]. Striding along a microtubule with a hand-over-hand gait and a step size of ≈8 nm [26], kinesin is capable of taking ≈100 steps per second, typically negotiating ≈100 steps before falling off the microtubule [4]. Replicating these performance characteristics with a synthetic mimic presents a daunting challenge to molecular engineers pursuing programmable active transport as a means to assembling or probing nanoscale systems.

DNA nanomachines have been demonstrated to perform mechanical switching between stable states in response to external stimuli [1, 3, 9, 11, 17, 19, 25, 28], and recently several processive DNA motors have been successfully designed and constructed. Sherman and Seeman demonstrated a bipedal walker that locomotes in an inchworm fashion, with one foot trailing the other [16]. Inspired by kinesin, the present work describes a bipedal walker that moves by advancing the trailing foot to the lead at each step [18]. Mimicking the periodic structure and directional polarity of a microtubule, we construct a DNA track that allows multiple walkers to haul cargo along the same track with directional specificity. These DNA walkers fail to replicate the autonomous nature of kinesin locomotion because they require the manual sequential addition of auxiliary DNA fuel strands. To demonstrate autonomous locomotion, Yin et al. [27] developed a transport mechanism employing ligation and restriction enzymes powered by ATP, and Tian et al. [21] and Stojanovic and co-workers [20] have developed walkers powered by RNA-cleaving DNAzymes. In principle, autonomous nanomachines can also be powered by hybridization catalysis of metastable DNA molecules [22]. Relative to the speed, versatility and robustness of kinesin, all of these walkers represent initial primitive steps towards a useful framework for achieving programmable active transport with nanoscale resolution.

2 A DNA Walker with the Gait of Kinesin

The walker system has four components (Table 1): a walker (W), a track (T), attachment fuel strands (A), and detachment fuel strands (D). The walker consists of two partially complementary oligonucleotides, with a 20-bp helix joining two single-stranded legs (each 23 bases). The track, constructed of six oligonucleotides, has four protruding single-stranded branches (each 20 bases) separated by 15-bp scaffold helices. Neighboring branches run in opposite directions, so spacing of 1.5 helical turns places all branches on the same side of the track approximately 5 nm apart (Fig. 1).

As shown in (Fig. 1), the walker strides along the track under the external control of A and D strands. An A strand specifically anchors the walker to a branch by forming helices with the corresponding leg (18-bp) and branch (17-bp). Single-stranded hinges adjacent to either end of these helices (underlined in Table 1) provide flexibility for adopting different conformations depending on the fuel species that are present. When both legs are bound to the track, the trailing leg is released using a D strand that nucleates with the perfectly complementary A strand at a 10-base overhang and then undergoes a strand displacement reaction [28] to produce duplex waste and free the walker leg for the next step.

Sequence selection for these system components represents a multi-objective optimization problem requiring the conditional stability of many different secondary structures depending on the subsets of strand species that are considered. Primary sequence design was performed using automated sequence selection software [5] to minimize sequence symmetry [15] while maximizing the probability [7] of adopting a compound secondary structure involving all the strands in the walker system. Subsequent secondary structure prediction [29] for various pairs of strands revealed a small number of undesirable interactions, which were eliminated by slight sequence modifications.

To detect walker locomotion, all four branches are end-labeled with spectrally distinct dyes and the two walker legs are end-labeled with quenchers (Table 1 and Sect. 5) to allow monitoring of the fluorescence changes associated with each dye. Proper monomeric association between the walker and track is demonstrated by examining intermediates during two forward steps using native gel electrophoresis visualized with two different fluorescent scans (Fig. 2). The major band corresponding to fully assembled track (lane 1) is accompanied by minor bands representing partially formed tracks. These partial tracks result from slight discrepancies in stoichiometry between the six

Table 1. DNA sequences for the walker and conveyor systems[a]

Walker strands (W)	
W1[b]	5′-GGCTGGTTTCTGCTCTCTAGTTCGCGAGGTGCAATCTCCTATC-3′-BHQ1
W2	IBRQ-5′-GTCTGGGATGCTGGATACTGAACCTAGAGAGCAGAAACCAGCC-3′
Track strands (T)[c]	
T1	HEX-5′-GCTGTACTTTGGTTACTGAAAGGGAGTGGCTCGGA-3′
T1*	HEX-5′-GCTGTACTTTGGTTACTGAAAGGGAGTGGCTCGGAAGCGTATGTTTGAGA-3′
T2	5′-TCCGAGCCACTCCCTGGACACCATCTACAAACTTGTATGGGACGTAGCGT-3′-Cy5
T3	FAM-5′-TCCACATCGGACTCTGTAATGCAAAGACACGTTACTTGTAGATGGTGTCC-3′
T4	5′-GTAACGTGTCTTTGCTCTCAAACATACGCTCTTCATGGCATTCGTCACCA-3′-Texas Red
T5	5′-AACTCTTAGCCAAGATCGTAAGCGTATGTTTGAGA-3′
T6	5′-TACGATCTTGGCTAAGAGTT-3′
Attachment strands (A)	
A1	5′-AGTAACCAAAGTACAGCACTGCGATAGGAGATTGCACCTCCAATTTACCC-3′
A2	5′-GACTGTTACGGTATCCAGCATCCCAGACGGTCAACGCTACGTCCCATACA-3′
A3[d]	5′-ACAGAGTCCGATGTGGAAGTCAGATAGGAGATTGCACCTCATCATTGTCG-3′
A4[d]	5′-GGATCAGTTAGTATCCAGCATCCCAGACCTAAGTGGTGACGAATGCCATG-3′

[a]Color use is consistent in all graphics. [b]For the conveyor experiments, W1 is biotinylated on the 5′ end to facilitate cargo attachment. [c]For the walker experiments, a track with four branches is constructed from strands T1–T6. For the conveyor experiments, a track with four periodically repeating branches is constructed from strands T1* and T2–T4. [d]Strands not depicted in Figs. 1 and 4

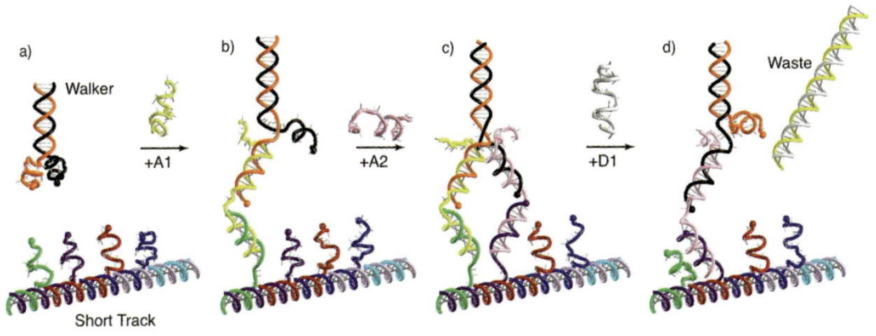

Fig. 1. Schematic of walker locomotion. Colored spheres represent dyes (HEX, green; Cy5, purple; FAM, red; Texas Red, blue) and quenchers (BHQ1, orange; IBRQ, black) for detecting walker movement. The diagrams depict: (**a**) unbound walker, (**b**) walker attached to T1, (**c**) walker attached to T1 and T2, (**d**) walker released from T1 to yield duplex waste

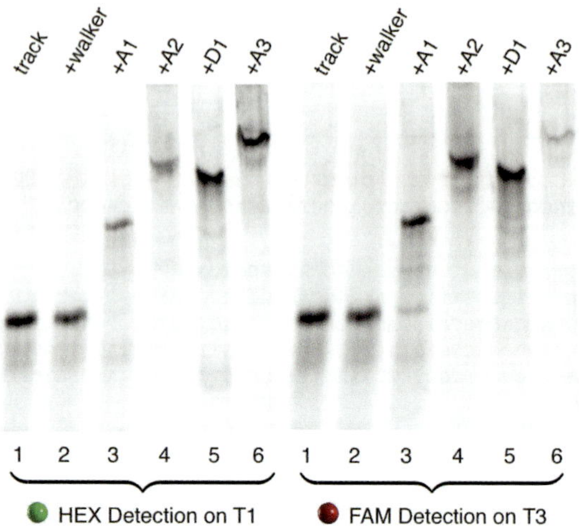

Fig. 2. Native polyacrylamide gel electrophoresis demonstrating specific fuel-mediated association of the walker with the track. This figure is a composite of two images obtained by different fluorescent scans of the same gel, using excitation and emission wavelengths that target either HEX on T1 or FAM on T3 (see Sect. 5)

track species and lead to the observation of minor bands during subsequent stages of the experiment. The track does not exhibit nonspecific interactions with the walker (lane 2). Anchoring of the walker and subsequent translation results in less mobile intermediates (lanes 3–6). The band intensities of these intermediates are consistent with the expected position of the walker for both wavelength scans. For HEX detection, there is a reduction in band intensity for lanes 3 and 4 relative to lanes 1, 2, 5 and 6. For FAM detection, there is a reduction in band intensity for lane 6 relative to lanes 1–5. The absence of empty track during the operation demonstrates the processivity of walker movement (i.e., at least one leg stays bound to the track).

Real-time monitoring of walker movement was carried out by multiplexed fluorescence quenching measurements (Fig. 3). Traversing the track from one end to the other and back, the fluorescent signal from each branch responds specifically to the addition of cognate fuel strands, illustrating high fidelity in controlling walker movement with nanoscale precision.

3 Hauling Molecular Cargo on a DNA Conveyor

A periodic track with the same four branches can be constructed from strands T1* and T2–T4 of Table 1. Multiple walkers can then operate synchronously on the same track to haul tethered molecular cargo with directional specificity under the control of attachment (A) and detachment (D) fuel strands (Fig. 4).

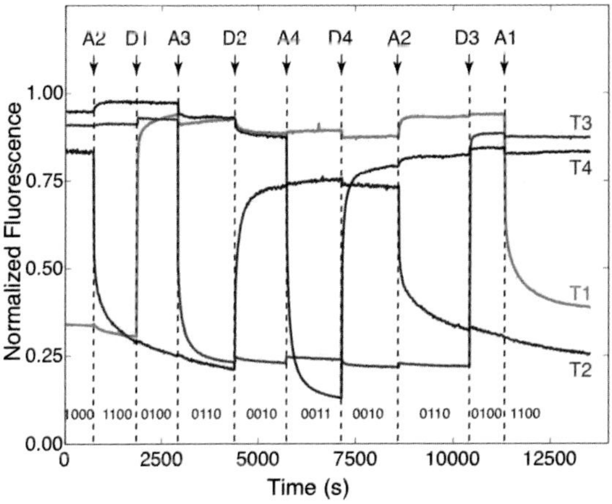

Fig. 3. Real-time multiplexed fluorescence monitoring. The track was preincubated with equimolar walker and A1 for 4 hr at room temperature. Equimolar amounts of A and D fuel strands were successively added from 100× stocks and mixed by rapid pipetting. A different excitation/emission wavelength pair was used to specifically monitor each dye (see Sect. 5). Fluorescence intensities are normalized by the initial track values. The binary digits at the bottom represent the location of the walker on branches T1–T4 (0 for unbound state, 1 for bound state)

To verify the elongated structure of the periodic track, the 3′ end of T1* and the 5′ end of T2 (i.e., consecutive ends imbedded in the track helix) were each modified with a thiol group and coupled to maleimide-functionalized gold nanoparticles 1.4 nm in diameter. Transmission electron microscopy images (TEM) of the gold-labeled track confirm the formation of long tracks with linear connectivity (Fig. 5). It is not possible to discern the anticipated non-constant spacing between gold particles (expected from labeling neighboring as opposed to evenly spaced strands). Several factors may contribute: incomplete labeling efficiency (86%), flexibility in the gold linkers, and flexibility of the track backbone under the low-humidity conditions required for TEM analysis. Consistent with significant backbone flexibility, the linear density of gold particles along the (assumed linear) track axis (0.17 ± 0.04 nm^{-1}) is higher than the estimated values of 0.10 nm^{-1} for B DNA (expected in solution) and 0.13 nm^{-1} for A DNA (expected at low humidity) [12].

Cargo is tethered to the head of the walker by biotinylating the 5′ end of W1 (Table 1), allowing binding of either streptavidin or anti-biotin antibody. The gel mobility of the walker decreases when cargo is bound (Fig. 6), with walkers bound to streptavidin (lane 2) migrating faster than those bound to the heavier antibody (lane 3).

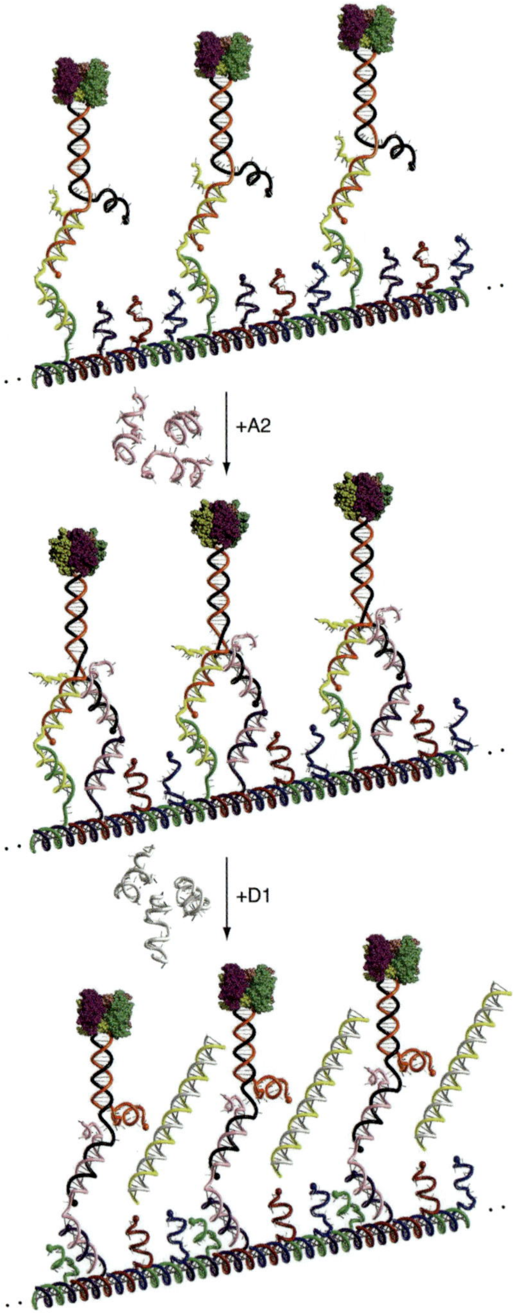

Fig. 4. Operation schematic of a molecular conveyor in which multiple walkers haul streptavidin cargo along a periodic track

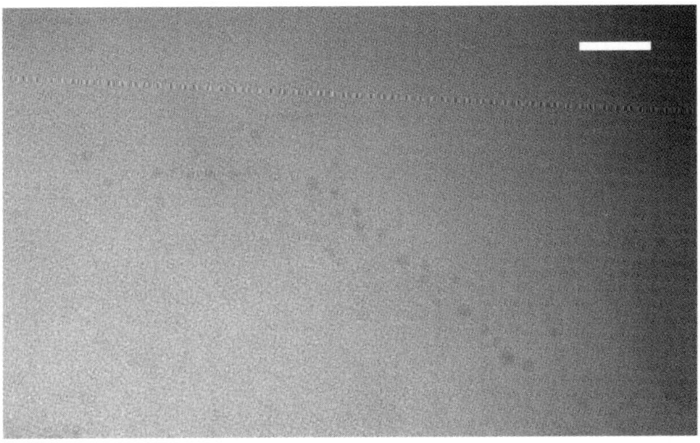

Fig. 5. Transmission electron microscopy image of a gold-labeled conveyor track (scale bar = 20 nm)

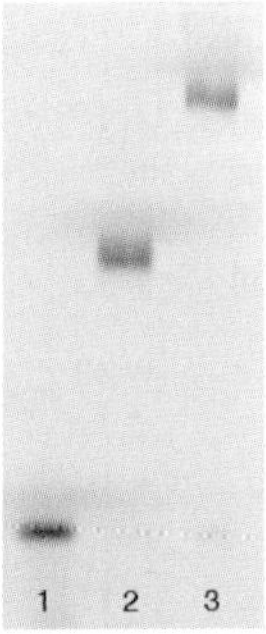

Fig. 6. Native polyacrylamide gel electrophoresis demonstrating cargo binding. When bound to a biotinylated walker, the mobility of 60 kDa streptavidin [8] is greater than that of the 150 kDa anti-biotin antibody [6]. Lane 1: walker. Lane 2: walker bound to streptavidin. Lane 3: walker bound to anti-biotin antibody

Real-time monitoring of organized cargo transport is achieved via multiplexed fluorescence quenching measurements as for the previous walker experiments. Fig. 7 demonstrates cargo transport in both forward and backward modes. After adding walkers to a solution containing pre-assembled periodic tracks, cargo transport is controlled by successive addition of A and D fuel strands.

To investigate the effect of cargo size on attachment and detachment kinetics, apparent rate constants were obtained for the cases of no cargo, streptavidin cargo and anti-biotin antibody cargo. Both attachment and detachment processes exhibit double-exponential kinetics (data not shown), with fast and slow phases that are relatively insensitive to cargo size (Table 2). Decreases in the attachment rate of ≈10–20% exhibit a positive correlation with increasing cargo size but no trend is evident in the ≈15% variation in detachment rates. Overall, the cargo appears to have a mild impact on the kinetics of walker locomotion.

4 Discussion

We have demonstrated a synthetic molecular walker that mimics the bipedal gait of kinesin. The present 5 nm step size is smaller than the 8 nm stride

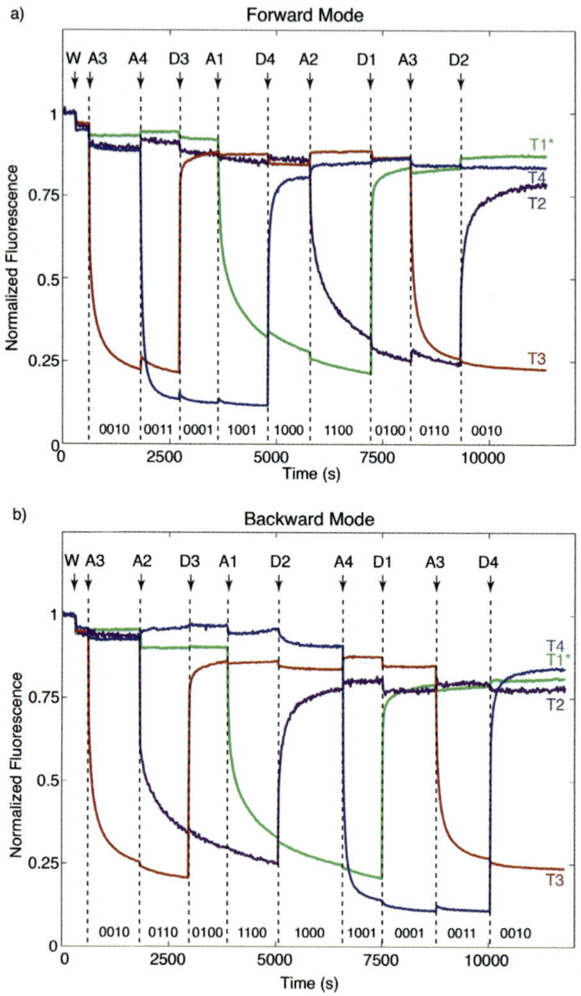

Fig. 7. Real-time monitoring of organized cargo transport a) forwards or b) backwards along the track. After pre-assembling the tracks, equimolar walker and A and D fuel strands were added from 100× stocks and mixed by rapid pipetting. A different excitation/emission wavelength pair was used to specifically monitor each dye (see Sect. 5). Fluorescence intensities are normalized by the initial track values. The binary digits at the bottom represent the location of the walker on branches T1*, T2–T4 (0 for unbound state, 1 for bound state).

Table 2. Apparent rate constants for cargo transport by the molecular conveyor

	Attachment (0010→0011)		Detachment (0011→0010)	
Cargo	fast phase ($\times 10^{-2}s^{-1}$)	slow phase ($\times 10^{-3}s^{-1}$)	fast phase ($\times 10^{-2}s^{-1}$)	slow phase ($\times 10^{-3}s^{-1}$)
None	2.98	5.25	1.75	1.98
Streptavidin	2.70	4.75	1.66	1.94
Anti-biotin antibody	2.36	4.29	2.01	2.18

of kinesin on a microtubule [4], and is tunable by adjusting the design of the track scaffold. Construction of a periodic DNA track allows multiple walkers to function as a molecular conveyor by hauling streptavidin cargo along the track in a prescribed direction. The kinetics of walker locomotion are not dramatically altered by the protein cargo. Generalization to much longer tracks should be achievable by attaching branches to more rigid substrates such as planar [24] or tubular [10, 13] DNA crystals. However, the significant bottleneck of manually introducing fuel strands at each step makes this approach impractical and highlights the importance of developing autonomous walker designs that operate without human intervention.

As a benchmark for walker performance, consider competing with kinesin in a nanomile foot race. Kinesin would make short work of the distance in approximately two seconds. By comparison, the present non-autonomous walker would require one week and one postdoc (ignoring difficulties of yield and exhaustion, respectively). For molecular engineers, there is still much work to do in catching up to the exquisite designs that have evolved in nature.

5 Methods

Oligonucleotide and device preparation. DNA oligonucleotides were synthesized, labeled and purified by Integrated DNA Technologies. DNA stock solutions were prepared in TE buffer (10 mM Tris, 1 mM EDTA, pH 8.0) and concentrations were determined at 260 nm using the molecular extinction coefficients provided by the supplier.

For the walker experiments, the track was prepared by equimolar mixing of the six track strand species (T1–T6) in TSE buffer (10 mM Tris, 150 mM NaCl, 1 mM EDTA, pH 7.5), followed by incubation for 3 hr at 37°C. The walker was prepared by the same procedure using the two walker strand species (W1, W2).

For the conveyor experiments, the periodic track was prepared by equimolar mixing of the four track strand species (T1*, T2–T4) in TSE buffer, followed by heating to 80°C for 5 minutes and then slowly cooling to room temperature over 2 hr. Walkers were assembled by combining equimolar W1 biotinylated at the 5′ end with W2 in TSE buffer for 3 hr at 37°C. Protein cargo was loaded onto the assembled walkers by introducing a 4-fold excess of either streptavidin or anti-biotin antibody (Sigma) and incubating for 1 hr at 37°C.

Native polyacrylamide gel electrophoresis. Fuel-mediated association of the walker and track was analyzed by native PAGE. Initial anchoring of the walker on T1 was achieved by adding equimolar A1 to a 1 μM reaction mixture of track and walker in TSE buffer, followed by 1 hr incubation at 37°C (10 μl reaction volume). Subsequent walker movement was carried out by successively adding different fuel species in equimolar amounts from 100× stocks followed by 1 hr incubation at 37°C. Gel electrophoresis was performed on the entire sample in 1× TBE buffer (89 mM Tris, 2 mM EDTA, 89 mM boric acid, pH 8.0) at 50 V and 4°C with 6.7% acrylamide. The bands were visualized by fluorescent scanning (Molecular Imager FX Pro Plus, Bio-Rad) using 532 nm excitation and a 555 nm cut-off filter to detect HEX on T1, and 488 nm excitation and a 30 nm bandpass filter centered at 530 nm to detect FAM on T3.

For the conveyor system, the binding of protein cargo to the biotinylated walkers (1 μM, 20 μl reaction volume) was verified using native PAGE in 1× TBE buffer at 50 V and 20°C with 5 % acrylamide and Stains-All dye (Sigma).

Multiplexed real-time fluorescence measurement. Fluorescence measurements were carried out with a fluorometer (Photon Technology International) at room temperature. Bandwidths for excitation and emission were set to 4 nm and the working volume for measurements was 100 μl. Fluorescence signals from four different dyes were collected during the same run using the multi-dye mode (Felix32 software, Photon Technology International) to monitor fluorescence intensities at four excitation/emission wavelengths (495/520 nm for FAM, 538/555 nm for HEX, 598/617 nm for Texas Red, and 648/668 nm for Cy5).

For the walker experiments, track (0.5 μM) was preincubated with equimolar walker and A1 in TSE buffer for 4 hr at room temperature. Equimolar amounts of A and D fuel strands were successively added from 100× stocks and mixed by rapid pipetting.

For the conveyor experiments, equimolar amounts of the streptavidin-loaded walker and fuel strands were successively added to the periodic track (track strands at 0.5 μM in TSE buffer) from 100× stocks and mixed by rapid pipetting.

For cargo kinetics studies, equimolar fuel (A4 or D4) was added to 0.5 μM conveyor in TSE buffer (state 0010 or 0011) and the solution was mixed by rapid pipetting. Excitation and exmission wavelengths were 598/617 nm to measure the fluorescence changes of Texas Red. Apparent rate constants were obtained by curve-fitting to a double exponential function.

Independence of fluorescence signals. To monitor walker movement, four fluorescent dyes were selected (FAM, HEX, Texas Red, Cy5 in order of increasing excitation/emission wavelengths) from many possible candidates. To minimize FRET between dyes, branches T1–T4 on the track were labeled in the order (HEX, Cy5, FAM, Texas Red) to double the distance between dyes with adjacent fluorescence spectra. For the periodic track, this property no longer holds where T4 and T1 are proximal. These concerns were particularly relevant because the equilibrium distance between neighboring dyes (5 nm) is typical of the Förster radius for many dye pairs. Excitation and emission wavelengths for each dye were chosen so as to minimize the response of the other three dyes (Fig. 8 and Table 3). Columns 2–5 of Table 3 list fluorescence intensities of the four dye-labeled T strands at the four excitation/emission wavelength pairs employed for multiplexed fluorescence measurements. At each wavelength, one dye fluoresces strongly and the fluorescence of the other dyes is negligible by comparison (1% or less). Column 6 shows that intensities for the 4-dye full track are within 10% of those observed for the dominant single strand dye at each excitation/emission wavelength. In addition to possible FRET between dyes, these differences may also result from changes in the local structural environments of the dyes, which often alter fluorescence emissions (even in the absence of a dye pair).

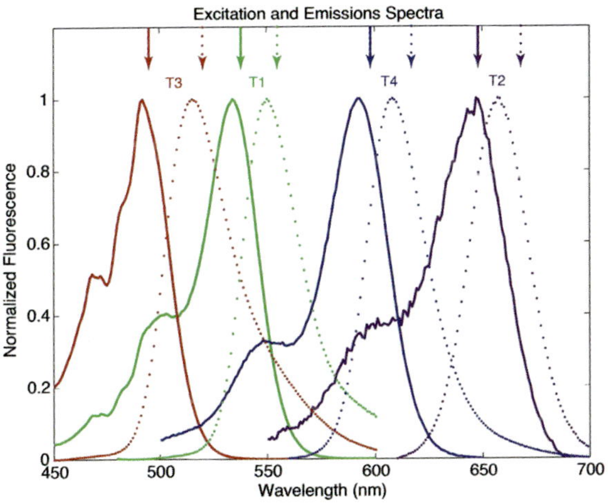

Fig. 8. Excitation (solid lines) and emissions spectra (dotted lines) for the four dye-labeled track strands (HEX-T1, Ex/Em = 470/610 nm; T2-Cy5, Ex/Em = 600/700 nm; FAM-T3, Ex/Em = 440/570 nm; T4-Texas Red, Ex/Em = 550/660 nm). The excitation and emissions wavelengths employed for subsequent kinetics experiments are denoted by the arrows at the top of the plot

Table 3. Fluorescence intensities of dye-labeled track strands at different wavelengths[a,b]

Ex/Em (nm)	HEX-T1	T2-Cy5	FAM-T3	T4-Texas Red	4-dye Track
495/520	23,248 (0.01)	5,148 (0.00)	3,636,790 (1.00)	647 (0.00)	3,290,060 (0.90)
538/555	1,690,730 (1.00)	3,650 (0.00)	4,076 (0.00)	844 (0.00)	1,527,750 (0.90)
598/617	162 (0.00)	1,593 (0.00)	105 (0.00)	741,071 (1.00)	759,613 (1.03)
648/668	24 (0.00)	124,220 (1.00)	22 (0.00)	126 (0.00)	130,448 (1.05)

[a] Fluorescence intensity (counts per second) measured at room temperature with samples (0.5 μM) in TSE buffer. [b] Values in parentheses are normalized intensities at a given wavelength (compare within row)

Transmission electron microscopy. Gold-labeling of oligonucleotides and TEM analysis were performed using standard procedures [2]. Thiol-modified T1* and T2 oligonucleotides (Integrated DNA Technologies) were labeled with nanogold particles by mixing with a 10-fold excess of monomaleimide-nanogold particles (Nanoprobes) in phosphate buffer containing 10% isopropanol at 4°C for 24 hr. The oligonucleotide-gold conjugates were combined and then the resulting complex was purified from excess gold particles on a gel filtration column (Superose 12 in TSE buffer). Labeling efficiency was determined by measuring the optical density of the gold-labeled T1*–T2 complex at 280 and 420 nm using molar extinction coefficients for the gold particles provided by the supplier. The periodic track was assembled by adding the T1*–T2 complex to equimolar T3 and T4. For TEM analysis (Philips EM430 operated at 300 kV), a dilute sample (50 nM) was deposited on copper-grids pretreated with an air plasma and polylysine coating.

Acknowledgment

This work was supported by DARPA and the Air Force Research Laboratory, the Ralph M. Parsons Foundation, the Charles Lee Powell Foundation, and the Caltech Center for Biological Circuit Design.

References

1. P. Alberti and J.-L. Mergny. DNA duplex-quadruplex exhange as the basis for a nanomolecular machine. *Proc. Natl. Acad. Sci. USA*, 100:1569–1573, 2003.
2. A.P. Alivisatos, K.P. Johnsson, X.G. Peng, T.E. Wilson, C.J. Loweth, M.P. Bruchez, and P.G. Schultz. Organization of 'nanocrystal molecules' using DNA. *Nature*, 382(6592):609–611, 1996.
3. J.D. Badjic, V. Balzani, A. Credi, S. Silvi, and J.F. Stoddart. A molecular elevator. *Science*, 303:1845–1849, 2004.
4. S.M. Block. Kinesin: What gives? *Cell*, 93:5–8, 1998.

5. R.M. Dirks and N.A. Pierce. A partition function algorithm for nucleic acid secondary structure including pseudoknots. *J. Comput. Chem.*, 24:1664–1677, 2003.
6. E. Harlow and D. Lane. *Antibodies: A Laboratory Manual.* Cold Spring Harbor Laboratory, Cold Spring Harbor, NY, 1988.
7. I.L. Hofacker, W. Fontana, P.F. Stadler, L.S. Bonhoeffer, M. Tacker, and P. Schuster. Fast folding and comparison of RNA secondary structures. *Chemical Monthly*, 125:167–188, 1994.
8. K. Hofmann, S.W. Wood, C.C. Brinton, J.A. Montibeller, and F.M. Finn. Iminobiotin affinity columns and their application to retrieval of streptavidin. *Proc. Natl. Acad. Sci. USA*, 77(8):4666–4668, 1980.
9. D.A. Leigh, J.K.Y. Wong, F. Dehez, and F. Zerbetto. Unidirectional rotation in a mechanically interlocked molecular rotor. *Nature*, 424:174–179, 2003.
10. D. Liu, S.H. Park, J.H. Reif, and T.H. LaBean. DNA nanotubes self-assembled from triple-crossover tiles as templates for conductive nanowires. *Proc. Natl. Acad. Sci. USA*, 101(3):717–722, 2004.
11. C. Mao, W. Sun, Z. Shen, and N.C. Seeman. A nanomechanical device based on the B-Z transition of DNA. *Nature*, 397(6715):144–146, 1999.
12. D.L. Nelson and M.M. Cox. *Leninger Principles of Biochemistry.* Worth, New York, NY, 2000.
13. P.W.K. Rothemund, A. Ekani-Nkodo, N. Papadakis, A. Kumar, D.K. Fygenseon, and E. Winfree. Design and characterization of programmable DNA nanotubes. *J. Am. Chem. Soc.*, 126(50):16344–16352, 2004.
14. M. Schliwa and G. Woehlke. Molecular motors. *Nature*, 422:759–765, 2003.
15. N.C. Seeman. Nucleic acid junctions and lattices. *J. Theor. Biol.*, 99:237–247, 1982.
16. W.B. Sherman and N.C. Seeman. A precisely controlled DNA biped walking device. *Nano Lett.*, 4(7):1203–1207, 2004.
17. J.-S. Shin and N.A. Pierce. Rewritable memory by controllable nanopatterning of DNA. *Nano Lett.*, 4(5):905–909, 2004.
18. J.-S. Shin and N.A. Pierce. A synthetic DNA walker for molecular transport. *J. Am. Chem. Soc.*, 126:10834–10835, 2004.
19. F.C. Simmel and B. Yurke. A DNA-based molecular device switchable between three distinct mechanical states. *Appl. Phys. Lett.*, 80(5):883–885, 2002.
20. M.N. Stojanovic. Personal communication, 2005.
21. Y. Tian, Y. He, Y. Chen, P. Yin, and C. Mao. A DNAzyme that walks processively and autonomously along a one-dimensional track. *Angew. Chem. Int. Ed.*, 44:2–5, 2005.
22. A.J. Turberfield, J.C. Mitchell, B. Yurke, A.P. Mills, Jr., M.I. Blakey, and F.C. Simmel. DNA fuel for free-running nanomachines. *Phys. Rev. Lett.*, 90(11):118102, 2003.
23. R.D. Vale. The molecular motor toolbox for intracellular transport. *Cell*, 112:467–480, 2003.
24. E. Winfree, F. Liu, L.A. Wenzler, and N.C. Seeman. Design and self-assembly of two-dimensional DNA crystals. *Nature*, 394:539–544, 1998.
25. H. Yan, X. Zhang, Z. Shen, and N.C. Seeman. A robust DNA mechanical device controlled by hybridization topology. *Nature*, 415(6867):62–5, 2002.
26. A. Yildiz, M. Tomishige, R.D. Vale, and P.R. Selvin. Kinesin walks hand-over-hand. *Science*, 303:676–678, 2004.

27. P. Yin, H. Yan, X.G. Daniell, A. J. Turberfield, and J.H. Reif. A unidirectional DNA walker that moves autonomously along a track. *Angew. Chem. Int. Ed.*, 43:4906–4911, 2004.
28. B. Yurke, A.J. Turberfield, A.P. Mills, Jr., F.C. Simmel, and J.L. Neumann. A DNA-fuelled molecular machine made of DNA. *Nature*, 406:605–608, 2000.
29. M. Zuker. Mfold web server for nucleic acid folding and hybridization prediction. *Nucleic Acids Res.*, 31(13):3406–3415, 2003.

Part IV

Electronics, Nanowire and DNA

A Supramolecular Approach to Metal Array Programming Using Artificial DNA

Mitsuhiko Shionoya

Department of Chemistry, Graduate School of Science, University of Tokyo, 7-3-1 Hongo, Bunkyo-ku, 113-0033 Tokyo, Japan
mmttn-shionoya@at.wakwak.com

1 Introduction

DNA can be used as a structural basis for the arrangement of functional building blocks because of the programmability of its sequence. The genetic information of DNA is expressed by a sequence of a limited number of letters in a genetic alphabet, that is, adenine (A), thymine (T), guanine (G), and cytosine (C), and is transferred through ingeniously simple A–T and G–C base pairing to the number and sequence of 20 kinds of amino acids in proteins in a highly controlled manner. Such a bottom-up, programmable strategy has prompted the development of new tools for the array programming of molecular-scale components, leading to well-defined supermolecules with tailored chemical and physical properties through self-assembly protocols.

Metals have been used as key components in self-assembled molecular architectures. In particular, the combination of metals and chemically modified biomacromolecules is currently attracting increasing attention. This chapter describes one of our recent approaches to precise programming of metal arrays, using artificial DNAs that possess metal-ligand-type nucleobases for metal-coordination-driven base pairing.

2 Metal-Mediated Base Pairing in DNA

Among a variety of chemical approaches to DNA-based supramolecular architectures (e.g., [3]), the replacement of natural base pairs by predesigned artificial base pairs possessing a distinctive shape, size, and function shows promise for molecular arrays inside DNA in a programmable manner. For instance, if an incorporated artificial nucleobase has a metal-binding site for metal-mediated base pairing, the metal could be put on the helix axis of DNA. This strategy would allow metal alignment inside DNA and, furthermore, if more than two kinds of metal ligand-type nucleobases were incorporated site-selectively into DNA strands, metals could be arrayed heterogeneously as programmed.

Metal ions thus incorporated as metal-mediated base pairs would (1) provide a new structural motif in high-order structures of DNA such as duplex and triplex, structures that could be used to affect their thermal stability, (2) form one-dimensional metal wires with unique chemical and physical properties based on the metal sequence, and (3) serve as linkers that would assemble DNA intermolecularly to form two- or three-dimensional DNA networks. On the basis of this concept, we have recently established the first Pd^{2+}-mediated base pairing using a metal-ligand-type nucleoside bearing a phenylenediamine base [5]. We have also exploited alternative base-pairing modes such as B^{3+}-mediated base pairing using catechol [1], Pd^{2+}-mediated base pairing using 2-aminophenol [9], Ag^{+}-mediated base pairing using pyridine[8], and Cu^{2+}-mediated base pairing using hydroxypyridone [7] (Fig. 1). Among these base pairs, those with a square-planar or linear coordination geometry are expected to substitute for a flat, hydrogen-bonded natural base pair without affecting the stacking structure of base pairs. Other groups have also reported a few excellent examples of metal-ligand-type nucleobases and have stabilized double-stranded structures of DNA in the presence of suitable transition metal ions [2, 10].

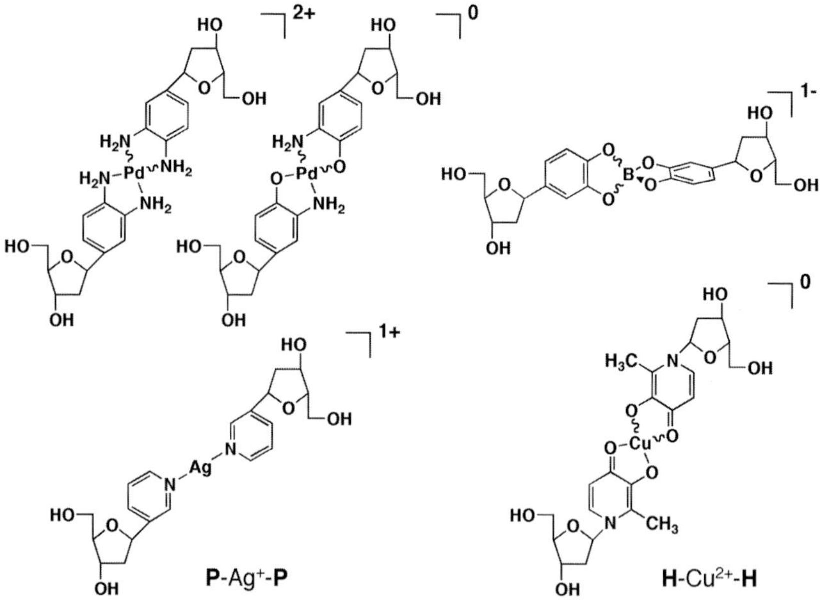

Fig. 1. Examples of metal-coordination-driven base pairs using artificial nucleosides.

Meggers et al. reported the first example of single-site incorporation of a metal-coordination-driven base pair into oligonucleotides, using a base pair with a pyridine-2,6-dicarboxylate nucleobase as a flat tridentate ligand and a pyridine nucleobase as the complementary single donor ligand [2]. For example, a 15-mer DNA duplex bearing a base pair in the middle was significantly stabilized by Cu^{2+} ions owing to the formation of a square-planar Cu^{2+} complex with the paired ligand-type nucleobases inside the DNA duplex. The stability was comparable to that of a DNA duplex containing a natural A–T base pair instead of the Cu^{2+}-mediated base pair.

Recently, we have independently reported the single-site incorporation of an Ag^+-mediated base pair and an Ag^+-mediated base triplet into a DNA duplex and a DNA triplex, respectively, using a base pair with two monodentate pyridine nucleobases (**P**) in the middle [7]. For instance, a 21-mer DNA duplex, d(5′-T_{10}**P**T_{10}-3′) · d(3′-A_{10}**P**A_{10}-5′), having a base pair in the middle, was significantly stabilized by Ag^+ ions, as shown by thermal denaturation experiments, owing to the formation of a linear, positively charged **P**–Ag^+–**P** base pair (Fig. 2). This Ag^+-dependent thermal stabilization was only slight in the case of a reference DNA duplex, d(5′-T_{21}-3′) · d(3′-A_{21}-5′). The Ag^+ complexation with the paired pyridine nucleobases inside the DNA duplex should be reinforced owing to the neighboring base pairs that are hydrogen-bonded and stacked in the hydrophobic environment within the DNA duplex. Furthermore, this pyridine base pairing occurred only with Ag^+ ions, and the addition of other metal ions, such as Cu^{2+}, Ni^{2+}, Pd^{2+}, and Hg^{2+}, showed almost no significant stabilization effect on the duplex structure.

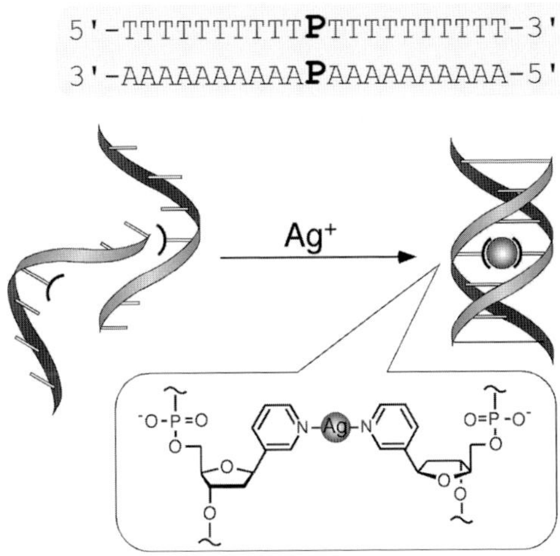

Fig. 2. Ag^+-mediated DNA duplex formation using a base pair containing pyridine bases.

Such Ag^+-dependent thermal stabilization was also significant for a DNA triplex, d(5′-T_{10}**P**T_{10}-3′) · d(3′-A_{10}**P**A_{10}-5′) · d(5′-T_{10}**P**T_{10}-3′) [5] (Fig. 3). In this case, the stabilization was due to the formation of an Ag^+-mediated base triplet in which the three N donors of the pyridyl groups bind to an Ag^+ ion within the DNA triplex.

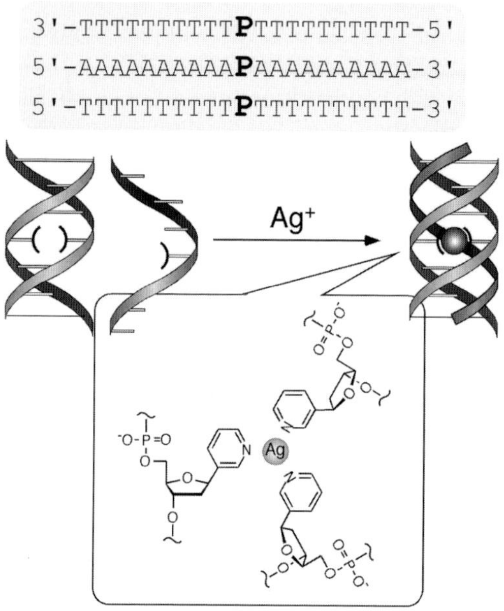

Fig. 3. Ag^+-mediated DNA triplex formation using a base triplet containing pyridine bases.

Another flat ligand base, a hydroxypyridone nucleobase (**H**), was incorporated into a 15-mer nucleotide DNA duplex, d(5′-CACATTA**H**TGTTGTA-3′) · d(3′-GTGTAAT**H**ACAACAT-5′), so that a base pair would be formed in the middle [7] (Fig. 4). In the presence of an equimolar concentration of Cu^{2+} ions, a neutral Cu^{2+}-mediated base pair of two hydroxypyridone nucleobases (**H**-Cu^{2+}-**H**) was formed quantitatively within the DNA duplex, and the duplex structure was highly stabilized compared with a reference oligoduplex, d(5′-CACATTAATGTTGTA-3′) · d(3′-GTGTAATTACAACAT-5′), containing a natural A–T base pair instead of the Cu^{2+}-mediated base pair. In addition, electron spin resonance (ESR) and circular-dichroism studies showed that a radical site was formed on the Cu^{2+} center within a right-handed duplex structure. This Cu^{2+}-mediated base pair was then used for the construction of magnetic chains within DNA duplexes.

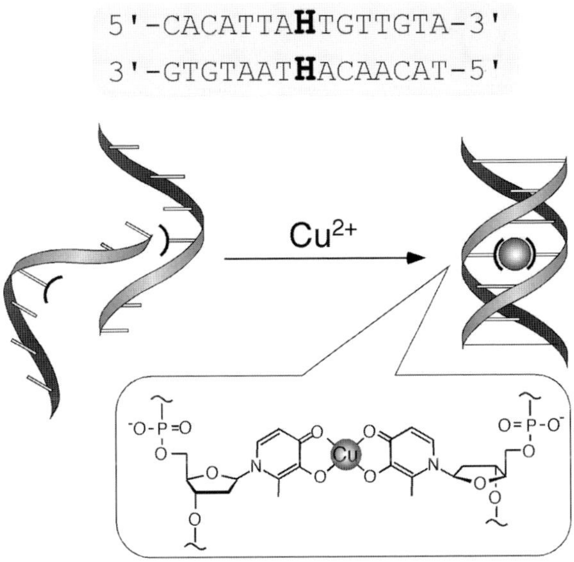

Fig. 4. Cu^{2+}-mediated DNA duplex formation using a base pair containing hydroxypyridone bases.

3 Discrete Self-assembled Metal Arrays in DNA

Multisite incorporation of metal-coordination-driven base pairs within DNA would lead to the construction of molecular-scale metal wires along with the DNA in a programmable manner. Recently, we have reported the synthesis of a series of oligonucleotides d(5′-G**H**$_n$C-3′) (n = 1–5), possessing one to five hydroxypyridone nucleobases (**H**) [6, 4]. These oligonucleotides quantitatively produced right-handed double helices of the oligonucleotides $nCu^{2+}\cdot$ d(5-G**H**$_n$C-3)$_2$ (n = 1–5), through Cu^{2+}-mediated base pairing (**H**–Cu^{2+}–**H**) (Fig. 5). In these metallo-DNA duplexes, the Cu^{2+} ions incorporated into each complex are aligned along the helix axis inside the duplex with a Cu^{2+}–Cu^{2+} distance of ca. 3.7 Å, as determined by ESR measurement. The Cu^{2+} ions are coupled to one another through unpaired d electrons to form magnetic chains. The electron spins on adjacent Cu^{2+} centers couple ferromagnetically with the accumulating Cu^{2+} ions attaining the highest spin state, as expected from the lining-up of Cu^{2+} ions. A proposed structure of a right-handed, double-stranded structure with a pentanuclear Cu^{2+} array inside the DNA duplex is drawn in Fig. 6.

4 Perspectives

The next key challenge in this study is to incorporate increasing numbers of metal ions into DNA and to provide a general tool for programming hetero-

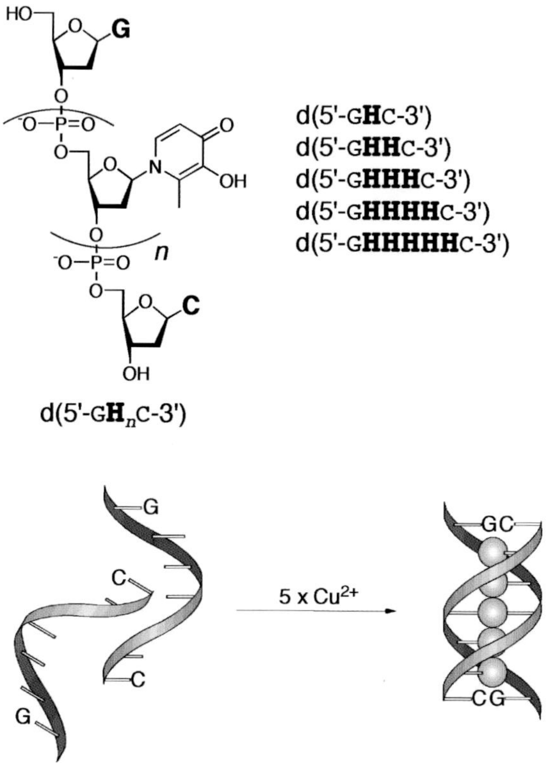

Fig. 5. Cu^{2+}-mediated DNA duplex formation from two artificial oligonucleotides bearing one to five hydroxypyridone bases.

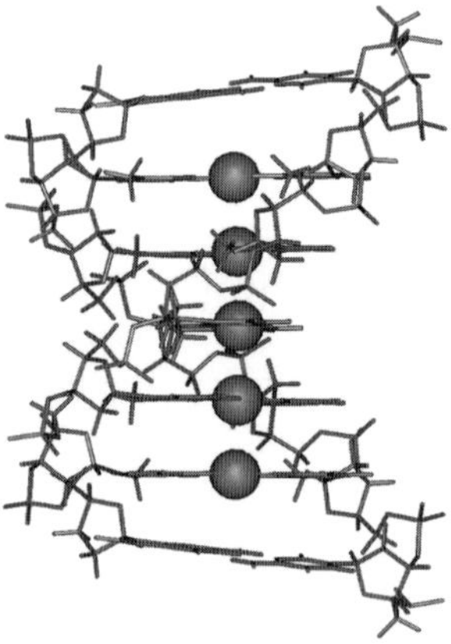

Fig. 6. A plausible structure of a pentanuclear Cu^{2+} complex aligned within a DNA duplex.

geneous metal arrays using DNA templates possessing more than two kinds of artificial nucleobases. Metal-coordination-driven base pairs in DNA would not only affect the assembly–disassembly processes of DNA double strands but also confer a variety of metal-based functions upon DNA. Such a novel structural motif in DNA will influence research in areas as diverse as nanobiotechnology, materials science, and computer science.

References

1. H. Cao, K. Tanaka, and M. Shionoya. An alternative base pairing of catechol-bearing nucleosides by borate formation. *Chem. Pharm. Bull.*, 48:1745–1748, 2000.
2. E. Meggers, P.L. Holland, W.B. Tolman, F.E. Romesberg, and P.G. Schultz. A novel copper-mediated DNA base pair. *J. Am. Chem. Soc.*, 122:10714–10715, 2000.
3. N.C. Seeman. From genes to machines: DNA nanomechanical devices. *Trends Biochem. Sci.*, 30:119–235, 2005.
4. M. Shionoya and K. Tanaka. Artificial metallo-DNA: a bio-inspired approach to metal array programming. *Curr. Opin. Chem. Biol.*, 8:592–597, 2004.
5. K. Tanaka and M. Shionoya. Synthesis of a novel nucleoside for alternative DNA base pairing through metal complexation. *J. Org. Chem.*, 64:5002–5003, 1999.
6. K. Tanaka, A. Tengeiji, T. Kato, N. Toyama, and M. Shionoya. A discrete self-assembled metal array in artificial DNA. *Science*, 299:1212–1213, 2003.
7. K. Tanaka, A. Tengeiji, T. Kato, N. Toyama, M. Shiro, and M. Shionoya. Efficient incorporation of a copper hydroxypyridone base pair in DNA. *J. Am. Chem. Soc.*, 124:12494–12498, 2002.
8. K. Tanaka, Y. Yamada, and M. Shionoya. Formation of silver(I)-mediated DNA duplex and triplex through an alternative base pair of pyridine nucleobases. *J. Am. Chem. Soc.*, 124:8802–8803, 2002.
9. M. Tasaka, K. Tanaka, M. Shiro, and M. Shionoya. A palladium-mediated DNA base pair of a β-C-nucleoside possessing a 2-aminophenol as the nucleobase. *Supramol. Chem.*, 13:671–675, 2001.
10. H. Weizman and Y. Tor. $2,2'$-Bipyridine ligandoside: a novel building block for modifying DNA with intra-duplex metal complexes. *J. Am. Chem. Soc.*, 123:3375–3376, 2001.

Multicomponent Assemblies Including Long DNA and Nanoparticles – An Answer for the Integration Problem?

Andreas Wolff, Andrea Csaki, and Wolfgang Fritzsche

Institute for Physical High Technology, Albert-Einstein-Strasse 9, 07745 Jena, Germany
andreas.wolff@ipht-jena.de
andrea.csaki@ipht-jena.de
wolfgang.fritzsche@ipht-jena.de

1 Introduction

The current focus on the combination of nanotechnology with biotechnology is an important milestone in the fabrication of novel miniaturized structures. These new devices (single-electron transistors [3, 44], nanoparticle-based molecular switches [48], and computing devices [68]) consisting of nanowires, nanotubes, nanocrystals, etc., seem to offer new electrical or magnetic features to overcome the problems presented by smaller and faster circuits. Hybrid systems of nanoparticles and biomaterials also provide new dimensions in the rapidly growing research field of bioelectronics. Here the integration of enzymes [5, 34, 101, 103], antigens antibodies [6, 87, 96] or DNA [94, 74, 73] with electronic elements is helping us to develop bioelectronic systems such as biosensors [102, 104] and artificial organs [77]. These biomolecules have dimensions in the range of a few nanometers (2 to 100 nm). In addition, their special applicable features can be used for the application of these molecules as building blocks for nanoscale architectures, as described below.

1.1 Features and Functions of Biomaterials

The most decisive point for using biomolecules is their capability for specific complementary recognition. The binding of, for example, an antigen to an antibody or a protein such as streptavidin to small molecules such as biotin, is very strong. Because of their special structure and chemical properties, biomolecules have the ability to self-assemble. Noncovalent interactions such as van der Waals, electrostatic, or hydrogen bonding lead to a structurally controlled arrangement of building blocks. The existence of several binding sites, such as the four binding domains of avidin, allows the construction of

2D or 3D architectures. Modern molecular engineering provides the ability to modify biomaterials with specific anchoring groups [71, 70].

1.2 Features of DNA

The widely established technology using DNA makes this molecule a good choice as a template or building block. The wide toolbox of enzymes (restriction endonucleases, ligases, polymerases, reverse transcriptases, etc.) and the possibility to synthesize artificial DNA sequences allow the construction of building blocks of predesigned length, shape, and base ordering. The binding of metal ions through electrostatic interactions and the intercalation of components are further unique properties can be used for the assembly of molecular functional complexes. The addressable binding of different proteins to specific sequences can act as a shield for such complex binding. This enables the fabrication of patterned structures. Besides this, DNA exhibits remarkable mechanical properties and rigidity that can be used for integration into nonbiological systems such as microstructures.

1.3 Features of Nanoparticles

Metal nanoparticles are available in sizes starting from around 1nm up to several hundred nanometers or even microns (so-called microparticles). Both silver and gold nanoparticles show plasmon absorbance in the visible spectrum. Their ability to label or stain biological material has been demonstrated, as well as, their use in the visualization of biological processes. Most properties, such as the plasmon absorbance mentioned above and also fluorescence of such particles, are size-dependent. So they represent a tunable tool for optical applications, for example in biosensors. The use of metal and semiconductor particles also offers the possibility of integration into electrical devices and circuits as active components.

2 Immobilization of DNA on Surfaces

The important step in generating nanowires or devices is the positioning of DNA on planar surfaces or microstructured devices. In this step, the DNA must be stretched and aligned along a given direction or orientation. The application of the DNA to the substrate takes place from solution, where long DNA is in a random-coiled state. A force is required that transforms the molecules to an extended form. Furthermore, there is a need for attachment sites to which the DNA can bind during the stretching process to prevent a snap-back of the molecules. For this reason, all substrates have to be functionalized. In Table 1 an overview of the important methods for stretching DNA is given. In the following we shall focus on the fluid-flow-induced methods that

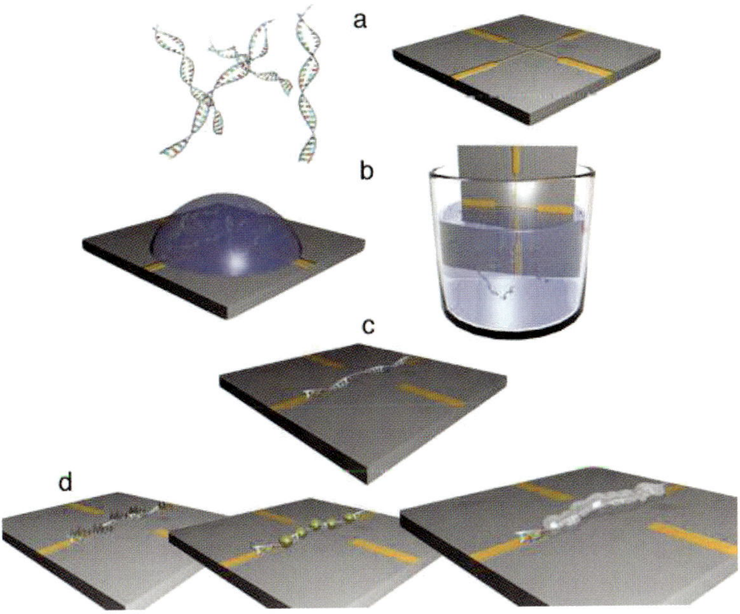

Fig. 1. Schematic illustration of the procedure for forming a chip-integrated nanowire: **(a)** DNA and microstructured chip as building blocks, **(b)** various stretching methods, **(c)** DNA positioned in an electrode gap, **(d)** binding of metal ions or particles and subsequent enhancement lead to a metallized nanowire

are used in our workgroup. The work on the optimization of the parameters, for example, DNA concentration, buffer conditions, and conditions for the surface modification steps, is very important but often tedious. Our experiments have shown that droplet sizes between 0.5 and 1 μl and DNA concentrations between 0.6 and 6 ng μl^{-1} are practicable. Also, surface modification with a simple PDMS (polydimethylsiloxane) vapor treatment in a petri dish offers a possibility to obtain suitable surfaces for the stretching and positioning of DNA strands.

2.1 Immobilization on Mica

A simple method for stretching DNA on mica has been described by Li et al. [46] and Cherny et al. [15]. The DNA solution is incubated on one side of the piece of mica for some minutes and then blown off slowly at an angle of 45°. The functionalization of the surface is provided by cleavage of the mica. Subsequently applied magnesium ions are then able to bridge the net negatively charged surface and the negative backbone of the DNA. However, mica is not suitable for further technological applications because there is no easy possibility to contact the immobilized molecules and it is not very stable

Table 1. Overview of methods for stretching DNA

Force	Method	
Magnetic	Magnetic tweezers	Smith et al. [86]
Optical	Laser tweezers	Perkins et al. [75, 76]
Moving interface	Drying droplet	Bensimon et al. [7, 8], Jing et al. [39]
	Controlled meniscus motion	Otobe and Ohtani [69]
	Moving meniscus	Michalet et al. [60]
	Sliding of a coverslip edge	Yokota et al. [107]
	Gas flow-driven droplet	Li et al. [46]
	Spin stretching	Yokota et al. [108]
Bulk fluid flow		Braun et al. [11]
Dielectric		Washizu et al. [97], Holzel et al. [36]

for handling. For this reason, there is a need for methods using substrates such as glass or silicon, which are important technological materials.

2.2 Immobilization on Glass

One of the most important impulses for stretching DNA molecules was the search for new methods for obtaining obtaining restriction maps of isolated chromatin and DNA molecules [35]. Optical methods for mapping individual DNA molecules have been described for yeast artificial chromosomes (YACs) [13, 14], restriction fragments and cosmid probes [72], and λ-DNA [98, 107], for example. The idea has been extended to human genomic DNA. One application was the mapping of microdeletions in the tuberous sclerosis 2 gene [60]. Fluorescence-stained stretched DNA can be determined optically with an accuracy of better than one micron. The literature describes many methods and variations for the immobilization of DNA in solution on planar substrates. There are many applications which use the interface between air and liquid. The DNA strands can be aligned on a surfaces either by the receding meniscus of a drying droplet or by simply pulling the substrate out of the solution in a controlled way. In the first case, the molecules are positioned radially with a large concentration of DNA in the center of the drop and few extended fibers/bundles in the peripheral region. The second approach leads to stretched strands in the direction in which the substrate is moved out of the solution. As mentioned above, all surfaces need a suitable functionalization for binding the DNA. Functionalizations usually provide a positive charge to attract and bind the DNA. The chemicals commonly used are APTES (3-aminopropyltriethoxysilane) [24], ODTS (octadecyltrichlorosilane) [92], and PMMA (polymethylmethacrylate) [30].

2.3 Immobilization on Microstructured Chips

The next step is the integration of DNA strands into technological environments such as microstructured chips with electrodes or defined electrode gaps. Bridging such gaps with molecules followed by a metallization step leads to nanowires with ohmic behavior. Such a process has to be practicable, generally applicable and reproducible. The binding of the DNA to the electrodes is typically done by interaction of complementary sequences [11] but can also be achieved by electrostatic interaction between negatively charged molecule ends and a positively charged surface on the electrode [52]. Ideally, one DNA molecule or bundle is immobilized in one gap. Thereby, bundles are made more stable for further imaging or metallization steps. In a highly parallel process, using the receding-meniscus method, precise positioning of DNA in several electrode gaps could be achieved [53]. Here the DNA follows the electrode structure, and also span the gaps.

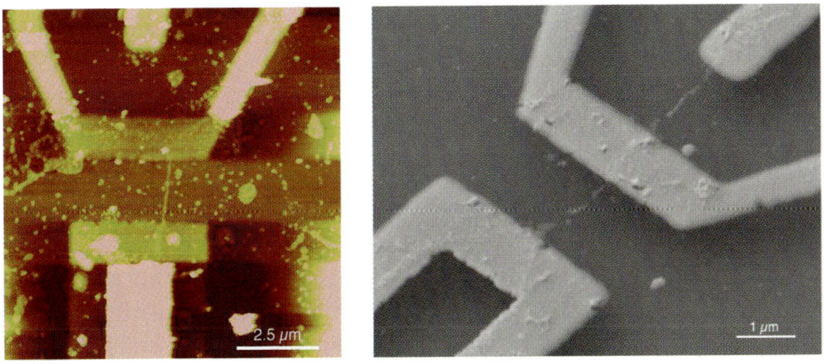

Fig. 2. Left: atomic force microscope picture of an immobilized DNA strand in an electrode gap. **Right:** Electron-scanning microscope picture of a nanoparticle-labeled DNA strand, spanning an electrode gap.

3 Nanoparticle Binding on DNA

Owing to their interesting and powerful properties, colloidal nanocrystals or nanoparticles find wide use in biology and adjacent fields, such as life sciences and nanotechnology. One main application is use as a label or a stain. Together with biological molecules [64, 61], they can work as building blocks, enabling the formation of complex patterns and assemblies. Molecular recognition leads, for example, to two-dimensional crystals and tubes [82, 83, 84]. Thereby, DNA can provide a nanoscale scaffold [65, 55], so that it is possible to build up nanowires of metallized DNA [11, 80, 79], as we shall see in Section 4. Bioconjugation between DNA and nanoparticles can be performed via simple adsorption [27] or with the well-known biotin–avidin system [85]. A very efficient and strong method is the use of thiol groups to attach DNA, as first described for thiolated oligonucleotides on planar gold surfaces and later applied to gold nanoparticles [2, 62, 23, 50, 17]. With these techniques, the nanoparticles were attached to a receptor (the oligonucleotide); now it is possible to bind these constructs to positions where a ligand (complementary sequence) is present. This leads to programmable DNA patterns and opens the way to realizing nanocircuits, possibly using nanocrystals as single-electron transistors [43] or arranging them in a desired pattern on the surface [22]. A further simple but smart method is the electrostatic binding of ligand-stabilized nanoparticles to the DNA backbone. The result is an extended linear chain-like structure or ribbon-like structure composed of parallel nanoparticles [95]. Positively charged gold nanoparticles have been used to bind to a DNA strand spanning a microstructured gap [54]. The arrangement of nanoscale building blocks on biomolecular scaffolds demonstrated in this way is a viable approach to obtaining closely spaced assemblies and a step towards biomolecular nanolithography.

4 Metallization of DNA

The last step process described above is the metallization of the aligned DNA strands. Conductivity is the main requirement for basic electronic building blocks such as wires, resistors, or p–n junctions. According to the predominant conventional wisdom, after a long period of dissent among different research groups, DNA is a poor conductor over longer distances. There are many theories that describe how such electron transport can work. Fink and Schneberger [25] reported a direct measurement of electrical current across DNA molecules that were 600 nm long. They concluded that the inner p-electrons of the base indicated that the DNA had the properties of good semiconductor. On the other hand, photoinduced electron transfer experiments showed a poor macroscopic electrical conductivity in DNA films [66, 12]. Removal of the water mantle around the double helix leads to reduced conductivity along lambda phage DNA and is strongly temperature-dependent around room temperature [91]. Electronic-structure calculations and direct measurements through λ-DNA molecules adsorbed on mica exhibit values of 10^6 Ω cm^{-1} and also show a dramatic effect on the measured conductivity which rises to high values after low-energy electron bombardment [18]. The utility of electrostatic force microscopy for probing the conductance of DNA has been demonstrated and has revealed an insulating behavior, in contrast to conducting single-wall carbon nanotubes [10]. Measurements between nanofabricated gold or platinum electrodes with different gap sizes (40500 nm) showed likewise that DNA is insulating over longer distances [88]. However, there is a consensus that charge transport takes place over the base pairs and their p-orbitals [28]. There are many advantages and disadvantages of these different methods but it seems very probable that native DNA has to be discarded for electronic circuits, and replaced by additional materials to provide the desired properties. In addition, the additional materials should have a geometry similar to that of the template DNA.

4.1 Metals and How to Obtain Wires from Them

Table 2 shows the metals frequently used in approaches to making nanowire. Interestingly, all elements applied as nanowires are members of either the first or the eighth subgroup of the periodic system of elements. All of them exhibit a very good conductivity and are noble metals so that they can be easily reduced. The idea is to attach metallic clusters or metal particles to the DNA and to form so-called "pearl chains". These can be used in the next step to form a continuous film on the biomolecule in order to achieve a conducting wire. The size of such mesoscopic clusters is in the range of the diameter of a DNA strand. This is important for homogeneous metal coverage after the enhancement step that will be discussed below. However, a chain of cationic gold colloids or Cd/S clusters [90] electrostatically bound to the anionic backbone of the DNA will not lead to a conducting wire, because the distance between the particles is too large. Therefore, growth of the clusters until they achieve spatial contact is needed. So, they are usually used as seeds in a two-step procedure. In this way, Cd/S clusters could be used to assemble an array of semiconductor nanoparticles matching the shape of the biopolymer, to form a nanowire [16]. Furthermore, selective localization of silver ions along the DNA through Ag^+/Na^+ ion exchange can be used for the seed-binding step. The recognition capabilities of DNA was used to construct a metal wire 12 mm long and 100 nm wide, connecting two electrodes [11]. DNA metallization can also be accomplished by deposition of palladium. Palladium activates the template to form a continuous palladium film after a reducing step [79, 19]. The binding of palladium complexes is very similar to the process of binding of cis-platin, which is very well understood. The use of cis-platin is very important in cancer therapy. The binding of such complexes changes the tertiary structure of the DNA. Additionally, the B-structure of the DNA becomes changed dramatically in the binding region, and base pairs are broken [47, 67].

4.2 The Final Step Towards Nanowires

The surface of the DNA is now activated with metal complexes or nanoparticles which act as seeds and subsequently as catalysts in the subsequent reducing step, which leads to a homogeneous metal-covered wire where the gaps between the centers of the metal particles are closed. In the case of palladium, Richter et al. used a mix of sodium citrate, lactid acic, and dimethylamine borane [80, 79]. They achieved wires with a diameter of about 50 nm. Some

Table 2. Commonly used metals for nanowires

Metal	Subgroup		Wire diameter (nm)	Resistance (Ω)
Gold	I	Keren et al. [41]	50–100	25
Silver	I	Braun et al. [11, 22]	100	7,000,000
Copper	VIII	Monson and Wooley [63]	3	
Platinum	VIII	Mertig et al. [57]		
Palladium	VIII	Richter et al. [80, 79]	20	

other reducing agents are hydroquinone or sodium borohydrate. Also well-known enhancement steps are the reduction of silver nitrate or silver acetate by hydroquinone [32] and the application of tetrachloroauric acid together with ammonium hydroxide [99, 100]. So, it is possible to deposit silver metal vectorially along a DNA molecule to obtain electrical functionality. The first step is the selective localization of silver ions along the DNA through silver/sodium exchange. This is selective and is restricted to the DNA template only [11]. The subsequent "development" of these aggregates is done by the standard photographic procedure, with an acidic solution of hydroquinone and silver ions under low light conditions [9, 37]. Monson and coworkers [63] have shown that copper can also be used to form nanowirelike structures with a height of 3 nm. They deposited copper metal using aqueous copper nitrate so that the copper(II) was electrostatically associated with the DNA. It was then reduced by ascorbic acid to form a metallic copper sheath around the molecule. It was demonstrated that copper nanowires were valuable as interconnects in nanoscale integrated circuitry. In ongoing experiments [29] cobalt nanoparticles have been assembled in situ on a template of double-stranded DNA to form magnetic nanowires. Palladium ions were bound to DNA and selectively reduced to zero-valence nanoclusters by deposition of Co(0) in a dimethylamine borane. The nanowires so formed were several microns long and 10 to 20 nm thick.

4.3 Sequence-Specific Molecular Lithography

A very novel tool for assembling devices into functional circuits is sequence-specific molecular lithography on DNA as a substrate, described by Keren et al. [42]. Here RecA protein binds in a sequence-specific manner and protects the DNA against metallization. This means that the lithographic information for this accurate and stable pattern is encoded in the DNA itself. The molecular lithography works with high resolution over a broad range of length scales from nanometers to many micrometers. In another approach, based on recognition between molecular building blocks, a DNA scaffold is used to localize semiconducting single-wall carbon nanotubes for the realization of a self-assembled carbon nanotube field-effect transistor operating at room temperature [41]. It has also been shown [40] that DNA can retain its biological functionality during metallization by aldehyde derivatization. So, in this case the RecA protein protects the DNA molecules in a sequence-specific manner again and allows complex patterning of molecular-scale electronic circuits.

4.4 Sequence-Specific Molecular Lithography with Nanotubes

As we have seen, carbon nanotubes also play an important role in the field of molecular nanotechnology. Single-walled carbon nanotubes (SWNTs) which have been covalently modified with DNA can hybridize selectively with complementary strands, with minimal nonspecific interactions with noncomplementary sequences [33]. These functionalized nanotubes can now act once again as molecular building blocks. Because of their interesting features and their behavior as either a metal or a semiconductor, they have emerged as important materials for nanofabrication, in both electronic devices and sensors. Controlled and selective localization of SWNTs on aligned DNA molecules on surfaces was also shown by Wooley and coworkers and could represent a route to the manipulation and positioning of SWNTs on surfaces. There are approaches to generating masks for photolithographic processes using a small number of DNA sequences to build up a structure of any size. So, it has been possible to assemble carbon nanotube transistors into circuits by using DNA [20, 21]. This provides an important tool in bottom-up biotemplated nanofabrication.

5 Concluding Remarks

Since we have been working on the integration of long DNA and nanoparticles, we have seen a great potential for these methods in new approaches to electronics. However, we have to point out that there remains a lot of work to be done. All the steps described here are well established as separate procedures. However, the combination of these steps into standard procedures has not yet been established. First of all, the problem of the parallelization of the integration of the molecules, which will be very important for commercial or forward-looking applications, has not been satisfactory solved. This is closely connected to the problem of suitable surfaces and both their modification and their functionalization. We have been working a lot on the development of simple, homogeneous surface modifications, especially on microstructured chips. But even the simple method of a drying droplet is not completely understood today. So one has in a large number of samples only a few with DNA in the desired places, leading to problems of reproducibility and throughput, and a series of established steps will not always work with the same precision and efficiency as does every separate step.

"There is plenty of room at the bottom", but there is also even more work there.

References

1. J.F. Allemand, D. Bensimon, L. Jullien, A. Bensimon, V. Croquette, pH-dependent specific binding and combing of DNA, *Biophys. J.*, **73**, 4:2064–2070, 1997.
2. A.P. Alivisatos, K.P. Johnsson, X. Peng, T.E. Wilson, C.J. Loweth, M.P. Bruchez Jr., P.G. Schultz, Organization of nanocrystal molecules using DNA, *Nature*, **382**, 6592:609–611, 1996.
3. R.P. Andres, T. Bein, M. Dorogi, S. Feng, J.I. Henderson, C.P. Kubiak, W. Mahoney, R.G. Osifchin, R. Reifenberger, Coulomb staircase at room temperature in a self-assembled molecular nanostructure, *Science*, **272**:1323–1325 1996.
4. D.P. Bancroft, L.A. Christopher L.J. Stephen, Platinum-195 NMR kinetic and mechanistic studies of cis- and trans-diamminedichloroplatinum(II) binding to DNA, *J. Am. Chem. Soc*, **112**, 19:6860–6871, 1990.
5. A. Bardea, E. Katz, A.F. Buckmann, I. Willner, NAD+-dependent enzyme electrodes: electrical contact of cofactor-dependent enzymes and electrodes, *J. Am. Chem. Soc*, **119**, 39:9114–9119, 1997.
6. A. Bardea, E. Katz, I. Willner, Probing antigen-antibody interactions on electrode supports by the biocatalyzed precipitation of an insoluble product, *Electroanalysis*, **12**, 14:1097–1106, 2000.
7. A. Bensimon, A. Simon, A. Chiffaudel, V. Croquette, F. Heslot, D. Bensimon, Alignment and sensitive detection of DNA by a moving interface, *Science*, **265**:2096–2098, 1994.
8. D. Bensimon, A.J. Simon, V. Croquette, A. Bensimon, Stretching DNA with a receding meniscus: experiments and models, *Physical Review Letters*, **74**, 23:4754–4757, 1995.
9. G.B. Birrell, D.L. Habliston, K.K. Hedberg, O.H. Griffith, Silver-enhanced colloidal gold as a cell surface marker for photoelectron microscopy, *J. Histochem. Cytochem.*, **34**, 3:339–345, 1986.
10. M. Bockrath, N. Markovic, A. Shepard, M. Tinkham, L. Gurevich, L.P. Kouwenhoven, M.W. Wu, Mingshaw, L.L. Sohn, Scanned conductance microscopy of carbon nanotubes and λ-DNA, *Nano Letters*, **2**, 3:187–190, 2002.
11. E. Braun, Y. Eichen, U. Sivan, G. Ben-Yoseph, DNA-templated assembly and electrode attachment of a conducting silver wire, *Nature*, **391**:775–778, 1998.
12. L. Cai, H. Tabata, Hitoshi, T. Kawai, Self-assembled DNA networks and their electrical conductivity, *Applied Physics Letters*, **77**, 19:3105–3106, 2000.
13. W. Cai, H. Aburatani, V.P. Jr. Stanton, D.E. Housman, Y.K. Wang, D.C. Schwartz, Ordered restriction endonuclease maps of yeast artificial chromosomes created by optical mapping on surfaces, *Proc Natl Acad Sci* USA, **92**, 11:5164–5168, 1995.
14. W. Cai, J. Jing, B. Irvin, L. Ohler, E. Rose, H. Shizuya, U.J. Kim, M. Simon, T. Anantharaman, B. Mishra, D.C. Schwartz, High-resolution restriction maps of bacterial artificial chromosomes constructed by optical mapping, *Proc. Natl. Acad. Sci.* USA, **95**, 7:3390–3395, 1998.
15. D.I. Cherny, A. Fourcade, F. Svinarchuk, P.E. Nielsen, C. Malvy, E. Delain, Analysis of various sequence-specific triplexes by electron and atomic force microscopies, *Biophys. J.*, **74**, 2:1015–1023, 1998
16. J.L. Coffer, S.R. Bigham, X. Li, R.F. Pinizzotto, Y.G. Rho, R.M. Pirtle, I.L. Pirtle, Dictation of the shape of mesoscale semiconductor nanoparticle assemblies by plasmid DNA, *Applied Physics Letters*, **96**, 25:3851–3853, 1996.

17. L.M. Demers, C.A. Mirkin, R.C. Mucic, R.A. Reynolds, R.L. Letsinger, R. Elghanian, G. Viswanadham, A fluorescence-based method for determining the surface coverage and hybridization efficiency of thiol-capped oligonucleotides bound to gold thin films and nanoparticles, *Anal. Chem.*, **72**, 22:5535–5341, 2000.
18. P.J. de Pablo, F. Moreno-Herrero, Colchero, J. Gomez Herrero, P. Herrero, A.M. Bar, P. Ordejn, J.M. Soler, E. Artacho, Absence of dc-conductivity in λ-DNA, *Physical review letters*, **85**, 23:4992–4996, 2000.
19. J. Duguid, V.A. Bloomfield, J. Benevides, G.J. Thomas Jr., Raman spectroscopy of DNA-metal complexes. I. Interactions and conformational effects of the divalent cations: Mg, Ca, Sr, Ba, Mn, Co, Ni, Cu, Pd, and Cd, *Biophys. J.*, **65**, 5:1916–1928, 1993.
20. C. Dwyer, V. Johri, M. Cheung, J. Patwardhan, A. Lebeck, D. Sorin, Design tools for a DNA-guided self-assembling carbon nanotube technology, *Nanotechnology*, **15**, 9:1240–1245(6), 2004.
21. C. Dwyer, L. Vicci, J. Poulton, D. Erie, R. Superfine, S. Washburn, R.M. Taylor II, The design of DNA self-assembled computing circuitry, *IEEE Transaction on Very Large Scale Integration (VLSI) Sytems*, **12**: 1214–1220, 2004.
22. Y. Eichen, E. Braun, U. Sivan, G. Ben-Yoseph, Self-assembly of nanoelectronic components and circuits using biological templates, *Acta Polym.*, **49**:663–670, 1998.
23. R. Elghanian, J.J. Storhoff, R.C. Mucic, R.L. Letsinger, C.A. Mirkin, Selective colorimetric detection of polynucleotides based on the distance-dependent optical properties of gold nanoparticles, *Science*, **277**, 5329:1078–1081, 1997.
24. Y. Fang, T.S. Spisz, T. Wiltshire, N.P. D'Costa, I.N. Bankman, R.H. Reewes, J.H. Hoh, Solid-state DNA sizing by atomic force microscopy, *Analytical Chemistry*, **70**, 10:2123–2129, 1998.
25. H.-W. Fink, Ch. Schneberger, Electrical conduction through DNA molecules, *Nature*, **398**:407–410, 1999.
26. W.E. Ford, O. Harnack, A. Yasuda, J.M. Wessels, Platinated DNA as precursors to templated chains of metal nanoparticles, *Advanced Materials*, **13**, 23:1793–1798, 2001.
27. L.A. Gearheart, H.J. Ploehn, C.J. Murphy, Oligonucleotide adsorption to gold nanoparticles: a surface-enhanced raman spectroscopy study of intrinsically bent DNA, *J. Phys. Chem. B*, **105**, 50:12609–12615, 2001.
28. M.W. Grinstaff, How do charges travel through DNA? An update on a current debate, *Angew. Chem. Int. Ed. Engl.*, **38**, 24:3629–3635, 1999.
29. Q. Gu, C. Cheng, D.T. Haynie, Cobalt metallization of DNA: toward magnetic nanowires, *Nanotechnology*, **16**, 8:1358–1363, 2005.
30. Z. Gueroui, C. Place, E. Freyssingeas, B. Berge, Observation by fluorescence microscopy of transcription on single combed DNA, *Proc. Nat. Acad. Sci.*, USA, **99**, 9:6005–6010, 2002
31. Z. Guo, P.J. Sadler, S.C. Tsang, Immobilization and visualization of DNA and proteins on carbon nanotubes, *Advanced Materials*, **10**, 9:701–703, 1998.
32. G.W. Hacker, G. Danscher, A.H. Graf, G. Bernatzky, A. Schiechl, L. Grimelius, The use of silver acetate autometallography in the detection of catalytic tissue metals and colloidal gold particles bound to macromolecules, *Prog. Histochem. Cytochem.*, **23**, 1-4:286–90, 1991.
33. M. Hazani, R. Naaman, F. Hennrich, M.M. Kappes, Confocal fluorescence imaging of DNA-functionalized carbon nanotubes, *Nano Letters*, **3**, 2:153–155, 2003.

34. A. Heller, Electrical connection of enzyme redox centers to electrodes, *J. Phys. Chem.*, **96**:3579–3587, 1992.
35. H.H.Q. Heng, J. Squire, L. Tsui, High-resolution mapping of mammalian genes by in situ hybridization to free chromatin, *Proc. Natl. Acad. Sci.* USA, **89**, 20:9509–9513, 1992.
36. R. Holzel, N. Gajovic-Eichelmann, F.F. Bier, Directed immobilization of nucleic acids at ultramicroelectrodes using a novel electro-deposited polymer, *Biosens Bioelectron.*, **19**, 5:417–422, 2003.
37. C.S. Holgate, P. Jackson, P.N. Cowen, C.C. Bird, Immunogold-silver staining: new method of immunostaining with enhanced sensitivity, *J. Histochem. Cytochem.*, **31**, 7:938–944, 1983.
38. R.E. Holmlin, P.J. Dandliker, J.K. Barton, Charge transfer through the DNA base stack, *Angew. Chem. Int. Ed. Engl.*, **36**, 24:2714–2730, 1997.
39. J. Jing, et al., Automated high resolution optical mapping using arrayed, fluid-fixed DNA molecules, *Proc. Natl. Acad. Sci.*, USA, **95**, 14:8046–51, 1998.
40. K. Keren, R.S. Berman, E. Braun, Patterned DNA metallization by sequence-specific localization of a reducing agent, *Nano Letters*, **4**, 2:323–326, 2004.
41. K. Keren, R.S. Berman, E. Buchstab, U. Sivan, E. Braun, DNA-templated carbon nanotube field-effect transistor, *Science*, **302**, 5649:1380–1382, 2003.
42. K. Keren, M. Krueger, R. Gilad, G. Ben-Yoseph, U. Sivan, E. Braun, Sequence-specific molecular lithography on single DNA molecules, *Science*, **297**, 5578:72–75, 2002.
43. D.L. Klein, R. Roth, A.K. Lim, A.P. Alivisatos, P.L. McEuen, A single-electron transistor made from a cadmium selenide nanocrystal, *Nature*, **389**, 6652:699–701, 1997.
44. D.L. Klein, P.L. McEuen, J.E.B. Katari, R. Roth, A.P. Alivisatos, An approach to electrical studies of single nanocrystals, *Applied Physics Letters*, **68**, 18:2574–2576, 1996.
45. A. Kumar et al., linear superclusters of colloidal gold particles by electrostatic assembly on DNA templates, *Advanced Materials*, **13**, 5:341–344, 2001.
46. J.W. Li et al., A convenient method of aligning large DNA molecules on bare mica surfaces for atomic force microscopy, *Nucleic Acid Research*, **26**, 20:4785–4786, 1998.
47. B. Lippert, Cisplatin: Chemistry and Biochemistry of a Leading Anticancer Drug, VCH, Weinheim, 1999.
48. J. Liu, M. Gomez-Kaifer, A.E. Kaifer, Switchable molecular devices: from rotaxanes to nanoparticles, *Structure and Bonding*, **99**:141–162, 2001.
49. Y. Liu, W. Meyer-Zaika, S. Franzka, G. Schmid, M. Tsoli, H. Kuhn, Gold-cluster degradation by the transition of B-DNA into A-DNA and the formation of nanowires, *Angew. Chem. Int. Ed. Engl.*, **42**, 25:2853–2857, 2003.
50. C.J. Loweth, W.B. Caldwell, X. Peng, A.P. Alivisatos, P.G. Schultz, DNA-based assembly of gold nanocrystals, *Angew. Chem. Int. Ed. Engl.*, **38**, 12:1808–1812, 1999.
51. Y. Maeda, H. Tabata, T. Kawai, Two-dimensional assembly of gold nanoparticles with a DNA network template, *Applied Physics Letters*, **79**, 8:1181–1183, 2001.
52. G. Maubach, A. Csaki, R. Seidel, M. Mertig, W. Pombe, D. Born, W. Fritzsche, Controlled positioning of one individual DNA molecule in an electrode setup based on self-assembly and microstructuring, *Nanotechnology*, **14**:546–550, 2003.

53. G. Maubach, M. Fritzsche, Precise positioning of individual DNA structures in electrode gaps by self-organization onto guiding microstructures, *Nano Letters*, **4**, 4:607–611, 2004.
54. G. Maubach, D. Born, A. Csaki, W. Fritzsche, Parallel fabrication of DNA-aligned metal nanostructures in microelectrode gaps by a self-organization process, *Small*, **6**:619–624, 2005.
55. J. Mbindyo, B.D. Reiss, B.R. Martin, C.D. Keating, M.J. Natan, T. E. Mallouk, DNA-directed assembly of gold nanowires on complementary surfaces, *Advanced Materials*, **13**, 4:249–254, 2001.
56. E. Meggers, M.E. Michel-Beyerle, B. Giese, Sequence dependent long range hole transport in DNA, *J. Am. Chem. Soc.*, **120**, 49:12950–12955, 1998.
57. M. Mertig, L.C. Ciacchi, R. Seidel, W. Pompe, A. deVita, DNA as a selective metallization template, *Nano Letters*, 2, 8:841–844, 2002.
58. M. Mertig, R. Seidel, L.C. Ciacchi, W. Pombe, Nucleation and growth of metal clusters on a DNA template, *CP633, Structural and Electronic Properties of Molecular Nanostructures*, edited by H. Kuzmany et al., 2002 American Institute of Physics 2002.
59. M. Mertig, R. Kirsch, W. Pompe, Biomolecular approach to nanotube fabrication, *Applied Physics A*, **66**:723–727, 1998.
60. X. Michalet et al., Dynamic molecular combing: stretching the whole human genome for high-resolution studies, *Science*, **277**, 5331:1518–23, 1997.
61. C.A. Mirkin, T.A. Taton, Semiconductors meet biology, *Nature*, **405**, 6787:626–627, 2000.
62. C.A. Mirkin, R.L. Letsinger, R.C. Mucic, J.J. Storhoff, A DNA-based method for rationally assembling nanoparticles into macroscopic materials, *Nature*, **382**, 6592:607–609, 1996.
63. C.F. Monson, A.T. Wooley, DNA-templated construction of copper nanowires, *Nano Letters*, **3**, 3:359–363, 2003.
64. C.M. Niemeyer, Progress in "engineering up" nanotechnology devices utilizing DNA as a construction material, *Applied Physics Letters*, **68**:119–124, 1999.
65. C.M. Niemeyer, Self-assembled nanostructures based on DNA: towards the development of nanobiotechnology, *Curr. Opin. Chem. Biol.*, **4**, 6:609–618, 2000.
66. Y. Okahata, T. Kobayashi, K. Tanaka, M. Shimomura, Anisotropic electric conductivity in an aligned DNA cast film, *Journal of the American Chemical Society*, **120**:6165–6166, 1998.
67. G.B. Onoa, G. Cervantes, V. Moreno, M.J. Prieto, Study of the interaction of DNA with cisplatin and other Pd(II) and Pt(II) complexes by atomic force microscopy, *Nucleic Acids Res.*, **26**, 6:1473–80, 1998.
68. A.O. Orlov, I. Amlani, G.H. Bernstein, C.S. Lent, G.L. Snider, Realization of a functional cell for quantum-dot cellular automata, *Science*, **277**:928–930, 1997.
69. K. Otobe, T. Ohtani, Behavior of DNA fibers stretched by precise meniscus motion control, *Nucleic Acids Res.*, **29**, 22:E109, 2001.
70. R,E. Palmer, Q. Guo, Imaging thin films of organic molecules with the scanning tunnelling microscope, *Physical Chemistry Chemical Physics*, **4**, 18:4275–4284, 2002.
71. V. Pardo-Yissar, E. Katz, I. Willner, A.B. Kotlyar, C. Sanders, H. Lill, Biomaterial engineered electrodes for bioelectronics, *Faraday Discuss*, **116**, 116:119–34; discussion 171–90, 2000.
72. I. Parra, B. Windle, High resolution visual mapping of stretched DNA by fluorescent hybridization, *Nat. Genet.*, **5**, 1:17–21, 1993.

73. F. Patolsky, E. Katz, I. Willner, Amplified DNA detection by electrogenerated biochemiluminescence and by the catalyzed precipitation of an insoluble product on electrodes in the presence of the doxorubicin intercalator, *Angew. Chem. Int. Ed. Engl.*, **41**, 18:3398–402, 2002.
74. F. Patolsky, A. Lichtenstein, I. Willner, Detection of single-base DNA mutations by enzyme-amplified electronic transduction, *Nat. Biotechnol.*, **19**, 3:253–7, 2001.
75. T.T. Perkins, S.R. Quake, D.E. Smith, S. Chu, Relaxation of a single DNA molecule observed by optical microscopy, *Science*, **264**, 5160:822–826, 1994.
76. T.T. Perkins, D.E. Smith, S. Chu, Direct observation of tube-like motion of a single polymer chain, *Science*, **264**, 5160:819–822, 1994.
77. A. Prokop, Bioartificial organs in the twenty-first century: nanobiological devices, *Annals of the New York Acad. Sci.*, **944**:472–90, 2001.
78. J. Richter, Metallization of DNA, *Physica E*, **16**, 2:157–173, 2003.
79. J. Richter, M. Mertig, W. Pompe, I. Mŏnch, H.K. Schackert, Construction of highly conductive nanowires on a DNA template, *Applied Physics Letters*, **78**, 4:536–539, 2001.
80. J. Richter et al., Nanoscale palladium metallization of DNA, *Advanced Materials*, **12**, 7:507–510, 2000.
81. G. Schmid, J.S. Bradley, Clusters and Colloids, Wiley, 1994.
82. N.C. Seeman, DNA nanotechnology: novel DNA constructions, *Annu. Rev. Biophys. Biomol. Struct.*, **27**:225–48, 1998.
83. N.C. Seeman, DNA engineering and its application to nanotechnology, *Trends Biotechnol*, **17**, 11:437–43, 1999.
84. N.C. Seeman, DNA nanotechnology, *Materials Today*, **6** Pages 7:24–30, 2003.
85. W.L. Shaiu, D.D. Larson, J. Vesenka, E. Henderson, Atomic force microscopy of oriented linear DNA molecules labeled with 5nm gold spheres, *Nucleic Acids Res.*, **21**, 1:99–103, 1993.
86. S.B. Smith, L. Finzi, C. Bustamante, Direct mechanical measurements of the elasticity of single DNA molecules by using magnetic beads, *Science*, **258**, 5085:1122–1226, 1992.
87. R.-I. Stefan, J.F. van Staden, H.Y. Aboul-Enein, Immunosensors in clinical analysis, *Fresenius' Journal of Analytical Chemistry*, **366**, 6–7, 2000.
88. A.J. Storm, J. van Noort, S. de Vries, C. Dekker, Insulating behavior for DNA molecules between nanoelectrodes at the 100 nm length scale, *Applied Physics Letters*, 79, 23:3881–3883, 2001.
89. K. Tanaka, Y. Okahata, A DNA-lipid complex in organic media and formation of an aligned cast film, *J. Am. Chem. Soc.*, **118**, 44:10679–10683, 1996.
90. T. Torimoto, M. Yamashita, S. Kuwabata, T. Sakata, H. Mori, H. Yoneyama, Hiroshi, Fabrication of CdS nanoparticle chains along DNA double strands, *J. Phys. Chem. B*, **103**, 42:8799–8803, 1999.
91. P. Tran, B. Alavi, G. Gruner, Charge transport along the λ-DNA double helix, *Physical Review Letters*, **85**, 7:15641567, 2000.
92. T. Vallant et al., Formation of self-assembled octadecylsiloxane monolayers on mica and silicon surfaces studied by atomic force microscopy and infrared spectroscopy, *Physical Chemistry B*, **102**:7190–7197, 1998.
93. G. Wang, R.W. Murray, Controlled assembly of monolayer-protected gold clusters by dissolved DNA, *Nano Letters*, **4**, 1:95–101, 2004.
94. J. Wang, Towards genoelectronics: electrochemical biosensing of DNA hybridization, *Chemistry - A European Journal*, **5**, 6:1681–1685, 1999.

95. M.G. Warner, J.E. Hutchison, Linear assemblies of nanoparticles electrostatically organized on DNA scaffolds, *Nature Materials*, **2**, 4:272–7, 2003.
96. A. Warsinke, A. Benkert, F.W. Scheller, Electrochemical immunoassays, *Fresenius' Journal of Analytical Chemistry*, **366**, 6-7:622–634, 2000.
97. M. Washizu, O. Kurosawa, Electrostatic manipulation of DNA in microfabricated structures, *IEEE Transactions on Industrial Applications*, **26**, 6:1165–1172, 1990.
98. H.U. Weier et al., Quantitative DNA fiber mapping, *Human Molecular Genetics*, **4**, 10:1903–1910, 1995.
99. Y. Weizmann, F. Patolsky, I. Willner, Amplified detection of DNA and analysis of single-base mismatches by the catalyzed deposition of gold on Au-nanoparticles, *Analyst*, **126**, 9:1502–1504, 2001.
100. Y. Weizmann, F. Patolsky, I. Popov, I. Willner, Telomerase-generated templates for the growing of metal nanowires, *Nano Letters*, **4**, 5:787–792, 2004.
101. I. Willner, V. Heleg-Shabtai, R. Blonder, E. Katz, G. Tao, A.F. Buckmann, A. Heller, Electrical wiring of glucose oxidase by reconstitution of FAD-modified monolayers assembled onto Au-electrodes, *J. Am. Chem. Soc.*, **118**, 42:10321–10322, 1996.
102. I. Willner, E. Katz, B. Willner, in Biosensors and Their Applications, V.C. Yang, T.T. Ngo, (eds.), Kluwer Academic Publishers, New York, 47–98, 2000.
103. I. Willner, A. Riklin, B. Shoham, D. Rivenzon, E. Katz, Development of novel biosensor enzyme electrodes: Glucose oxidase multilayer arrays immobilized onto self-assembled monolayers on electrodes, *Advanced Materials*, **5**, 12:912–915, 1993.
104. I. Willner, B. Willner, E. Katz, Functional biosensor systems via surface-nanoengineering of electronic elements, *Reviews in Molecular Biotechnology*, **82**, 4:325–355, 2002.
105. H. Xin, A.T. Woolley, DNA-templated nanotube localization, *J Am Chem Soc*, **125**, 29:8710–8711, 2003.
106. H. Yan, S. Park, G. Finkelstein, J.H. Reif, T. LaBean, DNA-templated self-assembly of protein arrays and highly conductive nanowires, *Science*, **301**:1882–1884, 2003.
107. H. Yokota et al., A new method for straightening DNA molecules for optical restriction mapping, *Nucleic Acids Research*, **25**, 5:1064–1070, 1997.
108. H. Yokota, J. Sunwoo, M. Sarikaya, G. van den Engh, R. Aebersold, Spin-stretching of DNA and protein molecules for detection by fluorescence and atomic force microscopy, *Anal. Chem.*, **71**, 19:4418–22, 1999.
109. D. Zanchet,C.M. Micheel, W.J. Parak, D. Gerion, A.P. Alivisatos, Electrophoretic isolation of discrete Au nanocrystal/DNA conjugates, *Nano Letters*, **1**, 1:32–35, 2001.

Molecular Electronics: from Physics to Computing

Yongqiang Xue[1] and Mark A. Ratner[2]

[1] College of Nanoscale Science and Engineering, University at Albany-State University of New York, Albany, NY 12203, USA
yxue@uamail.albany.edu
[2] Department of Chemistry and Materials Research Center, Northwestern University, Evanston, IL 60208, USA
ratner@chem.northwestern.edu

Even if Moore's Law continues to hold, it will take 250 years to fill the performance gap between present-day computers and the ultimate computer determined from the laws of physics alone. Information processing technology in the post-CMOS era will likely consist of a heterogeneous set of novel device technologies that span a broad range of materials, operational principles, data representations, logic systems and architectures. Molecular nanostructures promise to occupy a prominent role in any attempt to extend charge-based device technology beyond the projected limits of CMOS scaling. We discuss the potentials and challenges of molecular electronics and identify the fundamental knowledge gap that needs to be addressed for a successful introduction of molecule-enabled computing technology.

1 Introduction

The first functioning transistor was invented by Bardeen, Brattain, and Shockley in the late 1940s [8, 127]. It is a bipolar junction transistor (BJT) made from a small block of germanium. The first integrated circuit (IC) was invented in 1961, which combined monolithic bipolar junction transistors and passive components on a single chip. This event marked the start of the microelectronics revolution [68]. The realization of an older transistor principle – the field-effect transistor (FET) – came about with the metal-oxide-silicon (MOS) transistor in 1962. After the development of the complementary MOS (CMOS) circuits, silicon-based MOSFETs nearly completely dominated digital logic circuits due to the ease of very large-scale integration (VLSI) and low power consumption [124]. For almost 40 years, the VLSI industry has followed a steady path of constantly shrinking device geometries and increasing chip size, resulting in a history of new technology generation every two to three years, commonly refered to as "Moore's Law". The 2004 International Technology Roadmap for Semiconductors (ITRS) now extends this device scaling and increased functionality scenario to the 22-nm technology node in the year 2016, with projected minimum feature sizes below 10 nanometers and chips with more than 6 billion transistors [159].

Equally remarkable with the CMOS scaling is the fact that most of these developments have been achieved with the same basic switching element (MOSFET), the same basic circuit topology (CMOS), and with a limited number of materials (up to about 15 elements in the 1990s). In many respects, progress in these areas has been straightford following the design scaling rules in the sense that no fundamentally new inventions have been needed [29, 7]. However, there is no particular reason why Moore's Law should continue to hold: it is a law of human ingenuity, not of nature. Indeed, the current ITRS roadmap predicts that the scaling of the conventional CMOS technology will slow down or stop beyond the 22-nm node. Prior to that time, there is a large number of difficult technological challenges at the materials, device, circuit and system levels that must be met and overcome, many of which currently have no known solutions [159, 97, 162, 164]. This is because nanometer-size MOSFETs are no longer scaled short-channel devices with long-channel behavior. They are true nano-scale devices involving the creation and manufacture of objects within the regime of nanotechnology [160].

In contrast to the CMOS scaling, the technological challenges for the information processing industry in the post-CMOS era are quite different because it is far from clear what needs to be done [165, 93]. Nanostructures exhibit a variety of interesting physical, chemical and biological properties, many of which can be significantly modified by the processing and environmental conditions and are not yet fully understood. Progress in innovative device design thus often has to come hand in hand with progress in fundamental knowledge of the physics and chemistry of nanostructures. Although there is a growing consensus that the near-term extension of charge-based device technology will require a nanodevice technology that is architecturally compatible with CMOS and functionally supplementing rather than replacing CMOS, questions remain regarding the best direction to pursue for such nanoelectronics [161]. For example, what is the best functional nanostructures, carbon nanotubes, silicon/compound nanowires, or molecules, polymers? And what are the best device concepts, field-effect transistors, single-electron transistors, quantum-effect devices (resonant tunneling, quantum interference, ...), etc.? In the longer term, a new nanodevice technology may need to exploit electron and electronic charge/current in fundamentally new ways that are closely linked to the use of alternate state variables for representing information.

In general, nanostructured systems may span a broad range of materials, data representations, operational principles, and may function in different architectures and on different applications. Independent of the technology route that post-CMOS devices take, the operation and performance of nanometer-size devices will increasingly be governed by atomic-level variations in the materials/device structures and processing/environmental conditions. The prospect of device design through the bottom-up atom-engineering route of nanotechnology has far-reaching impacts on post-CMOS information processing technology either as monolithic systems or as polylithic hybrid systems interfacing to the scaled CMOS. Molecular nanostructures occupy a prominent role in any attempt to offer significant expansion in device functionality beyond the end of CMOS scaling.

In this work, we discuss opportunities for information processing based on quantum engineering of the physical states of molecules. Here we define a molecule broadly as *a unit whose physical (electrical, magnetic, mechanical, optical, ...) and/or chemical (reactivity, solubility, molecular recognition, ...) properties are sensitive to atomic-scale modification of its structure and/or environment.* Note that such a definition of molecule essentially covers all nanostructures as defined in the US National Nanotechnology Initiative (NNI) [166], including atoms, organic molecules, polymers, nanotubes, nanowires, and nanoparticles, etc. Since the only practical general-purpose information processing technology (besides the human brain) currently available is based on silicon devices which use electron and electronic charge/current to drive electronic circuits performing Boolean logic, we limit our scope here to charge-based device technology. We don't consider device technology options that use fundamentally different physical representations of computational states such as optical computing and DNA computing. We will not discuss molecular spintronics either, which forms a separate avenue of research, even though spin is intrinsically associated with electrons and nuclei.

Before we proceed to the main sections of this paper, we want to emphasize that our goal here is not to provide a complete survey of molecular electronics that have been studied or proposed so far, but rather to indentify the critial research needs in fundamental science that must be addressed in order to extend charge-based device technology through the molecular/nano-engineering route. Since any attempt in investigating the potentials of the nanotechnology route to information processing should be gauged in reference to both its ultimate physical limit and the limits of the ultimately scaled CMOS devices, we start with discussions of the ultimate physical limits to computation and the physical factors that account for the success of semiconductor technology and their limits in Sects. 2 and 3, respectively. We discuss the main topic of the paper, molecular electronics, in Sect. 4. We conclude and summarize in Sect. 5.

2 Ultimate Physical Limits to Computation

Computers are physical systems, and the laws of physics dictate what they can and cannot do [71, 79]. Much of the current activities in molecular electronics has been motivated by Feynman's pioneering work on the physical limits of miniaturization and compuation [37, 54]. Therefore we start the investigation of molecular electronics with examining the limits that the laws of physics place on the power of computers. There exists a vast literature on this topic [107, 43]. Our discussion here follows that of Lloyd [88], which explores the ultimate physical limits to the computational capacity of a computer with a mass of 1 kg occupying a volume of 1 litre (the so-called ultimate laptop computer) as determined by the speed of light c, the quantum scale $\hbar$ and the thermodynamic scale k_B.

2.1 Speed Limits

A digital computer performs computation by representing information in terms of binary digits or bits with logical states $|0\rangle$ and $|1\rangle$, and then processes that information by performing simple logical operations. Any boolean function can be constructed by repeated application of AND, NOT and FANOUT, which forms a universal set [54]. During such logical operations, the bits on which the operation is performed go from one state to another. The maximum speed per logic operation can thus be determined by how fast a quantum system can move from one distinguishable state to another, i.e., the maximum speed of dynamical evolution. Since the quantum measure of distinguishable states is the orthogonality of states involved, this is best illustrated by considering the minimum time needed for the NOT operation, which changes the $|0\rangle$ state to its orthogonal $|1\rangle$ state or vice cersa. This question is closely related to the Aharonov–Bohm interpretation of the time–energy Heisenberg uncertainty principle $\Delta E \Delta t \geq \hbar$ [2, 3]: It is not that it takes time Δt to measure the energy of a quantum system to an accuracy of ΔE, but rather that a quantum system with spread in energy ΔE takes time at least $\Delta t = h/4\Delta E$ to evolve to an orthogonal state [1]. Instead of expressing the speed of dynamical evolution in terms of the standard deviation of energy ΔE, Margolus and Levitin [95] have generalized the result to show that a quantum system with average energy E (relevant to its ground state energy) takes time at least $\Delta t = h/4E$ to evolve to an orthogonal state.

Since the simple logical operations of AND, NOT and FANOUT can all be enacted in the so-called controlled-controlled-NOT operation [41], by embedding the controlled-controlled-NOT gate in a quantum context it is easy to show that the maximum speed of logic operation is limited by the energy input to the logic gate performing the operation as $4E/h$. More complicated logic operations may involve system evolution cycling through a large number of quantum states. For evolutions that pass through an exact cycle of N mutually orthogonal states at a constant rate, it has been shown that the transition time between the orthogonal states is $\Delta t \geq \frac{N-1}{N} h/2E$, or the long-sequence asymptotic transition time is twice as long as it is for oscillation between $N = 2$ states [95]. Applying this result to a 1-kg computer with energy $E = mc^2$ shows that the ultimate laptop can perform a maximum of $4mc^2/h \approx 5.426 \times 10^{50}$ operations per second [88].

2.2 Memory Limits

A system with N accessible states can register $\log_2 N$ bits of information, so the amount of information that can be registered by a physical system is related to its thermodynamoc entroy by $I = S(E,V)/k_B \ln 2$, where $S(E,V)$ is the thermodynamic entropy of a system with expectation value for energy E confined to a volume V. When it is using all its memory space, the ultimate laptop can perform a maximum number of operations per bit per second of $\frac{4E}{h} / \frac{S}{k_B ln2} \propto k_B T/\hbar$, where $T = (\frac{\partial S}{\partial E})^{-1}$ is the operating temperature of the ultimate laptop in the maximum entropy state. A simple estimate of the maximum entropy for the 1-kg computer in a litre volume can be obtained by modeling the volume occupied by the computer as a collection of modes of elementary particles with total average energy E, and the maximum entropy $S(E,V)$ is that obtained by calculating the canonical ensemble over the modes which maximizes S for fixed energy E confined in a fixed volume V with no constraint on the spread in energy ΔE [9]. Note that this is different from the canonical ensemble normally used for open systems that interact with a thermal bath at temperature T. Consequently the temperature $T = (\frac{\partial S}{\partial E})^{-1}$ has a different role in the context of calculating the maximum entropy of a closed quantum system than it does in the case of an ordinary thermodynamic system interacting with a thermal bath.

At a particular temperature T, the entropy is dominated by the contributions from particles with mass less than $k_B T/2c^2$. The particles contribute energy $E = r\pi^2 V (k_B T)^4/30\hbar^3 c^3$ and entropy $S = 2r\pi^2 V (k_B T)^3/45\hbar^3 c^3 = 4E/3T$, where r is the number of particles/antiparticles in the species multipled by the number of polarizations multiplied by a degeneracy factor reflecting particle statistics [9]. A simple lower bound on the entropy can be obtained by assuming the energy and entropy are dominated by black-body radiation of photons, for which case $r = 2$ (a recent derivation finds the same ultimate limits for information encoded using both matter and massless fields [89]). For a 1-kg computer confined to a 1-litre volume, we find that the maximum entropy state corresponds to the operating temperature of $k_B T = 8.1 \times 10^{-15}$ J, $T = 5.87 \times 10^8$ K. The maximum entropy is $S = 2.04 \times 10^8$ J/K, which corresponds to an amount of memory space of $I = S/k_B \ln 2 = 2.13 \times 10^{31}$ bits. When the ultimate laptop is using all its memory space, it can perform $4 \ln(2) k_B E/S \approx 10^{19}$ operations per bit per second [88].

2.3 Thermodynamics of Computation

The role of thermodynamics in computation is made clear in the intimate link between information and entropy. Ordinary electronic computers are thermodynamic engines that do work and generate waste heat. Reducing the supply power and removing the heat produced have been the main technology drivers throughout the history of computing. However, contrary to intuitive thinking,

Bennett showed in his pioneering paper [10] in 1973 that it is possible to construct a general-purpose computer using only reversible, i.e., one-to-one logical operations, therefore allowing in principle dissipationless computing if we are willing to compute slowly. Energy is dissipated only when information is discarded. Landauer showed that irreversible, many-to-one operations such as AND or ERASE require heat dissipation of at least $k_B T \ln 2$ for each bit of information lost [72, 11]. A closely related but separate energy disspiation limit has been established for communicating information. Again, in the absence of noise, i.e., interaction between the physical system carrying information and another uncontrolled physical system, the energy required for transmission of a unit of information can be made arbitrarily small if we are willing to do it slowly [73]. But, as shown by Levitin, a minimum energy of kT must inevitably dissipate in order to transmit a unit of information over a noisy channel as a result of the interaction with uncontrolled degrees of freedom (environment) [81]. More recently, similar fundamental limits on the energy transfer associated with a binary switching transition have been derived in the context of semiconductor technology by Meindl and Davis [98].

Besides these fundamental energy dissipation requirements, a realistic computer will inevitably be subject to errors during its operation. Error-correcting codes can be used to detect these errors and reject them to the environment at the dissipative cost of at least $k_B T \ln 2$ per bit. Typically such error-correcting operations must be done at a high rate in order to maintain reliable operation [88, 72, 11]. The thermal load of correcting large numbers of errors alone can dictate the necessity of operating at a slower speed than the maximum allowed by the laws of physics [79, 88].

3 Limits of Semiconductor Technology

The discussion of ultimate physical limits to computation does not imply that we can construct a computer that operates at those limits. For example, it is inconceivable for present-day technology to control computers operating at $T = 5.87 \times 10^8$ K, or close to the temperature at which electrons and positrons can be produced thermally. Processing, storing and transmission of information requires that it be represented as the value of some physical quantitity, and physical laws control the materials and devices that are used to manipulate information [63, 64]. Contemporary electronic computers operate at speed, memory and energy dissipation capabilities far lower than those dictated by the consideration of physical laws alone. From the physical perspective, such computers operate in a highly redundant fashion. However, there are good technological reasons for such redundancy.

3.1 What Makes a Good Computing Device

Many ingenious proposals for better computing devices were put forward and have been the focus of well-funded development efforts as silicon microelectronics continued its relentless drive toward miniaturization in the past four decades. But the only general-purpose computers that have ever been built were built with (in chronological order) electrical relays, vacuum tubes, bipolar junction transistors and field-effect transistors. So why do so many ingenious schemes fail to realize their promise in electronic computation? The answer lies in the vast difference between the conditions in which devices are first discovered and demonstrated in the laboratories and those in a large system of many devices [65, 66].

For laboratory demonstration of a simple logic circuit, one needs only to choose a few proven devices and fine-tune their operating conditions as necessary to make them work well to perform a logic operation. But a large computer that contains tens of thousands to many millions of devices works in much less benign conditions. The output of one device is readily input to another, and so on through thousands of steps or more. A large amount of communications among the many devices is entailed. There are frequent opportunities for a signal to be altered in its passage from one device to another, suffering attenuation, diffraction, dispersion and cross-talk on the path. The multiple physical and chemical processes used in mass-production of the large numbers of components lead to small differences in device characteristics. In addition, chemical reactions and diffusion lead to additional unpredictable changes in devices over time, adding to the uncertainty inherent in manufacturing.

While the net result of the hazard factors is tolerable in a single logic operation, information must pass sequentially through a large number of stages in the computing system. Information would soon be lost if the errors introduced were allowed to propagate and accumulate from stage to stage. Digital representation of information can prevent this by resetting the output of a device to one of the standard values after each step. The output of a device may be required as input by other devices. The transmission of a signal to a multiplicity of destinations is known as fan-out and devices for computers must be able to provide fan-out. The standarization of signals and fan-out require that an electrical device controls voltage and current larger than those needed to operate it, or a device should have both current and voltage gains. Gain is essential to digital devices in order that the switching transition at the threshold occupies a small part of the signal swing and allows high noise margin. In addition, it is desired that a computation in a machine proceed in one direction, from input to final results. Each device should operate only on its inputs and not be sensitive to the actions or status of the receipients of its outputs. This property is known as input–output (I/O) isolation and is required in computer devices.

The need for I/O isolation, fan-out and high gains puts a severe limitation on the choice of devices suitable for large computing systems, which was only satisfied by electrical relays, vacuum tubes and transistors. Careful examination of all other proposed devices showed that they have difficulty in satisfying the three conditions simultaneously [65, 66]. Transistors, especially the silicon

MOSFETs, eventually win out due to their small size, fast speed, operating stability and low power consumption.

The rest of this section is thus devoted to the challenges and limits facing semiconductor technology toward the end of the ITRS roadmap as shaped by the laws of physics. Such limits can be codified at a hierarchy of levels of materials, devices, circuits and systems [97]. Many review papers have been written on various limits to silicon technology [97, 162, 164, 116, 40, 39]. We'll focus our discussion here only on those aspects of the materials and device limits of silicon technology that are likely to be relevant to the CMOS-like route to nanoelectronics through molecular/nano-engineering.

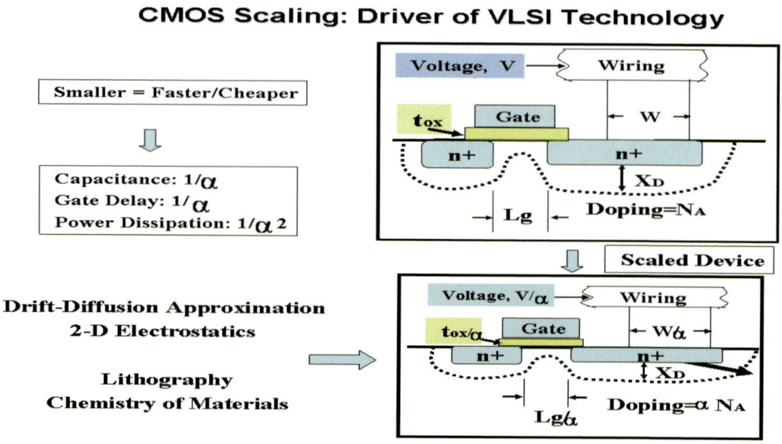

Fig. 1. Schematic illustration of MOSFET scaling

3.2 Materials Limits of Silicon Technology

Materials limits are determined by the properties of the particular semiconductor, dielectric, and metallic materials used but are essentially independent of the structural features and dimensions of particular devices. There are three key materials limits: gate stack, including both gate dielectric and gate electrode, doping in silicon, and contact formation [116].

The gate insulator in a MOSFET needs to be thin compared to the device channel length (a few percent) in order for the gate to exert dominant control over the channel potential. But quantum mechanical tunneling of carriers through the insulator increases exponentially with decreasing insulator thickness. This put the limit of silicon dioxide (SiO_2) thickness at 1.0 nm or five atomic layers thick for a sub-20 nm MOSFET operating at 1 V in order to accomodate standby power requirements in most IC applications [104]. In addition, not all the applied gate voltage is efficiently coupled to the channel due to the polysilicon depletion effects in the gate electrode and quantum confinement effects in the silicon substrate, which add aproximately 0.8 nm to the equivalent electrical thickness in the gate capacitor [116]. One solution is to introduce high-κ dielectric material which reduces the tunneling current while maintaining strong gate electrostatic control. But as the dielectric constant of the insulator increases, the band gap tends to decrease and the band lineup at the silicon–dielectric interface can be quite asymmetric. To avoid thermal emission over a Schottky barrier, a barrier height of more than 1 V is needed for both electron and hole. Another major barrier any new dielectric material will have to overcome is to achieve almost the same low-defect density as that of the native Si–SiO_2 interface. This puts a significant constraint on the choice of dielectrics and their processing steps. In addition, many of the new dielectric materials are unstable in direct contact with silicon and also in the presence of the polysilicon gate. Thus it is likely that the entire gate stack will have to be replaced with metal gates replacing polysicon, which has the advantage of lifting the gate depletion effect. But polysilicon has the advantage that it can be doped either p-type or n-type, shifting the workfunctions so that it is suitable for both NMOS and PMOS devices. In contrast, two different gate metals are needed for incorporation into a CMOS flow with workfunctions near the conduction and valence band edges, respectively, which complicates enormously the fabrication process [116].

The second issue is associated with the need of ultra-shallow source/drain juctions to reduce the parasitic resistance of the source/drain extension regions and the short-channel effect due to the drain electric field extending through the channel region. This requires increasing the doping density of the source/drain region while maintaining abrupt doping profile across the silicon body. However, the maximum dopant concentration that can be dissolved in silicon under equilibrium conditions (the solid solubility) is $\approx 2 \times 10^{21}$ atoms/cm^3) (achievable for arsenic at $\approx 1200^\circ$C) [134]. Although transient laser annealing can introduce arsenic in metastable electrically active concentrations near or above the solubility limit, there is an enormous driving force that tends to deactivate the arsenic during any subsequent thermal cycling [123]. The dominant technology used for doping silicon is ion implantation, which provides precise control of the placement and quantity of doping atoms. But the implantation process produces considerable damage in the silicon substrate as a result of the nuclear collisons involved in the stopping process. Dopants diffuse by interaction with point defects in the subsequent thermal anneal to achieve the desired doping profile. The mechanisms underlying the defect formation and dopant diffusion process are far from being fully understood [34].

The third issue is associated with the junction contact formation. Contacts in silicon technology are normally made with self-aligned silicides containing heavily doped silicon. This process provides an ohmic contact covering the area of the source/drain diffusions and minimizes the contact resistance. Further reducing the contact resistance with decrease of feature size requires increasing the silicon doping and reducing the Schottky barrier height. The doping is limited by the solid solubility as discussed earlier. Barrier height engineering in metal–silicon systems remains not fully understood despite its obvious technical importance. In addition, the silicide formation process consumes the top portion of silicon as the metal is deposited and reacted to form the silicide. This can increase the sheet resistance of the source/drain extension region and also change the dopant structure adjacent to the metal.

3.3 Device Limit of Silicon Technology

Historically MOSFET scaling has been governed by the need to preserve the good electrostatic behavior at the reduced device dimension, i.e., reducing supply voltage and gate insulator thickness and increasing doping concentration. The traditional limit of device scaling is thus determined by the effects that modify the ideal electrostatic contol. These include quantum effects due to tunneling leakage through the gate insulator, tunneling through the body-to-drain junction, direct source-to-drain tunneling, thermal effects due to thermally generated subthreshold current at room temperature and also the increasing sensitivity to minute fabrication spreads. In addition to such limits intrinsic to small device size, other limits more intimately connected to the materials and device structure of the ultimately scaled MOSFET have been proposed which, as acutely pointed out by Fischetti, suggests changing the "nature" of the nanometer-size MOSFETs moving toward the sub-10 nm regime [39].

The most fundamental one seems to be that set by the long-range Coulomb interaction between the channel electron and the "high-density" electron gas in the highly doped source, drain and gate electrodes. This is reflected in: (1) the emission and absorption of the low-frequency plasmon (on the order of magnitude of meV) in the source/drain by the channel electrons which thermalizes the hot-electron distribution in the channel and indirectly reduces the effective electron velocity; (2) the "Coulomb" drag between the channel electron and electrons in the gate (also plasmon-mediated) accross the very thin insulator results in a direct loss of momentum of the channel electrons. Both effects may contribute to the breakdown of "ballistic" transport widely assumed in a current theoretical estimate of the MOSFET scaling limit [91]: a short channel is required for "ballistic" transport, but the increased strength of the Coulomb interaction will kill it at the outset [39]. Combining with other less fundamental but equally important effects such as "remote" phonon scattering in the gate stack and scattering accompanied with substrate engineering, this may contribute further to the end-of-the-road scaling scenario that there may not be a single end point for scaling, but instead many end points each adapting optimally to its particular applications [40].

4 Molecular Electronics: from Physics to Computing

4.1 Motivation and Definition

Even if Moore's Law continues to hold, it will take about 250 years of exponential scaling to fill the gap between the ultimate laptop operating on 10^{31} bits at 10^{51} operations per second and the present-day laptop operating on 10^{11} bits at 10^{10} operations per second. Although the ultimate laptop operates at conditions that does not seem to be contollable at all from present-day technology, new physical laws may be imagined that turn today's inconceivable into tomorrow's common sense if we remember quantum physics has only 100 years' history. But we are not concerned with such exotic possibilities beyond the horizon of current understanding of physical laws. The goal of molecular electronics is instead to extend the exponential performance increase of charge-based device technology beyond that perceivable from CMOS scaling at the projected end of the ITRS roadmap as far as possible based on innovative utilization of functional nanostructures and quantum mechanical laws.

Although many technological barriers exist for which there are currently no known solutions, the past success of CMOS scaling gives us all reason to believe that the projected goal of the CMOS scaling in 2016 will be surpassed [17], at which the ultimate MOSFET will have gate oxide thickness in the 1.0-nm range, channel thickness in the 3.0-nm range, and channel length in the 9.0-nm range [3]. Since CMOS technology is the only practical general-purpose information processing technology (besides the human brain) currently available, investigation along the molecular electronics route should be gauged in close reference to the continually scaled CMOS devices in both its conventional and "nonclassical" forms [16, 76]. In addition, future devices and their target performance metrics should meet the generic criteria of: (1) that they need to be of high performance in terms of speed and density while remaining energy efficient; (2) that they should be structurally stable under room temperature operation and not be dominated by parametric variations due to processing and environmental conditions; (3) that they should be scalable through multiple generations with integer multiples of performance. In the near term, they might preferably be capable of integration on a CMOS platform, but the long-term options should be kept open (remember the 250-year span!). Consequently we shall consider materials and device issues associated with both molecular/nano-engineered devices that are structurally and/or functionally similar to CMOS devices (referred to as the CMOS route hereafter) and molecular/nano-engineered devices that are configured for information acquisition, sensing, storage and transmission in ways fundamentally different from the CMOS devices (referred to as the Non-CMOS route hereafter).

[3] Although there is no implication on the scaling limit of the channel width from the current two-dimensional scaling models, we can reasonably expect it to be comparable to the channel length.

In the preface to the first edition of his widely popular textbook on semiconductor devices published in 1969, Sze defined a semiconductor device as *a unit which consists, partially or wholly, of semiconducting materials and can perform useful functions in electronic apparatus and solid-state research* [129]. Correspondingly we define a molecular electronic device as *a system which consists, partially or wholly, of individual molecules and can perform useful functions in electronic apparatus and nanostructure research through atomic-scale control.* We only discuss molecular electronics for applications in information processing devices here and leave the discussion of molecular electronics as an "artificial" laboratory of nanoscopic physics elsewhere [149, 158].

4.2 Molecular Electronics: CMOS Routes

Molecular Transistor

Three-terminal devices, i.e., transistors, have been indispensable for building digital logic systems based on semiconductor technology due to the stringent requirement of I/O isolation, large noise margin and signal gain. Molecular field-effect transistors (MolFET), where the active part of the device is composed of quasi-one-dimensional (Q-1D) nanostructures like carbon nanotubes or nanowires, have been widely studied that are structurally and functionally similar to their CMOS analog [163, 6, 96, 22, 23]. Q-1D nanostructures offer additional advantage as alternative channel materials in the CMOS route since they can function both as active devices and interconnects and thus have the potential to provide simultaneously two of the most critical functions in any integrated nanoelectronics [25, 83, 153]. Experimental progress on single devices has been fast and useful simple circuits like inverters, mixers and decoders have been demonstrated [86, 42, 55, 100, 157, 60]. There are many points of confluence between the technologies of the scaled silicon devices and Q-1D nanostructured junctions and transistors, including the integration of high-κ gate stack, homo(pn)- and hetero-junction diodes and transistors, substrate engineering (strain) and "non-classical" transistor structures [83, 153, 57, 58]. Investigation along this route thus provides an ideal reference point both for exploring novel device design at the molecular scale and for re-examining the physical principles of semiconductor microelectronics from the bottom-up approach. Here carbon nanotubes and semiconductor nanowires offer subtle but significant differences in their prospect for post-CMOS information processing.

Carbon nanotubes (CNT) are hollow cylinders composed of one or more concentric layers of carbon atoms in a honeycomb lattice arrangement, which typically have a diameter of 1–10 nm and a length of several nanometers to several micrometers. In addition to the samll size, CNTs offer some salient features that make them attractive candidates for electronic devices [6]. (1) The quasi-1D structure implies a reduced phase space for carrier scattering by

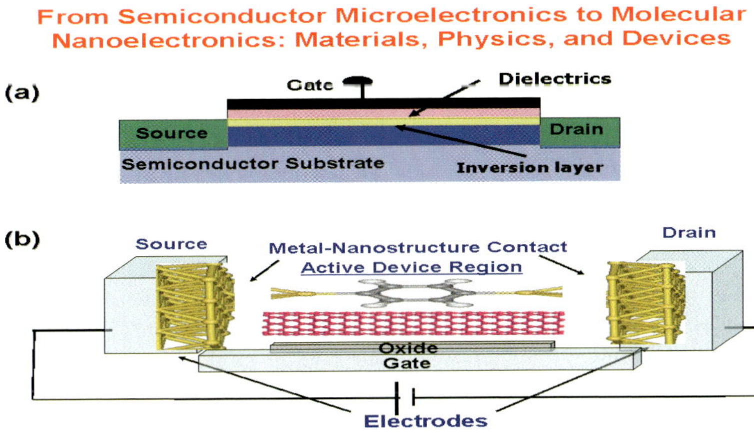

Fig. 2. Schematic illustration of the device structures of a conventional CMOS device and a typical nanodevice.

both impurity and lattice vibration. It also leads to distinctly different electrostatic behavior from the planar silicon device which affects both screening and tunneling. (2) The C-C sp^2 bonding leaves no dangling bond on the surface. In particular, for single-wall carbon nanotubes (SWNTs) all carbon atoms are surface atoms. CNT electronics are not bound to use SiO_2 as an insulator and novel transistor structures like surrounding gate transistors can be adapted. (3) The strong C-C sp^2 bonding gives CNTs high mechanical and thermal stability. Current densities $\geq 10^9$ A/cm^2 can be sustained. Several critical issues related to contact, doping and scattering remain to be sorted out for further development of CNT-based nanoelectronics.

In contrast to silicon MOSFETs, the source, drain and gate electrodes in MolFETs are made from deposited or lithographically defined metals. The Schottky barriers at the CNT–metal contacts play a significant role in determining the transport characteristics [6, 80, 150] (we can also expect that the Schottky barrier problem will play an increasingly important role as MOSFETs scale toward the sub-10nm regime, since the low-frequency plasmon in the doped source/drain region can be removed by using metal electrodes). Due to the Q-1D geometry, both the barrier height and barrier shape are important in determining the relative importance of tunneling and thermionic emission across the barrier. The recent observation of ohmic contact using Pd provides a particular challenge [57, 58] as the previous theoretical study shows a similar Schottky barrier for Pd and Au that have similar work functions. However, the model used assumes only electronic coupling across the interface with fixed atomic structure. Transition metals including both Ti and Pd are known to be chemically active attaching to the CNT surface and can form carbide immediately adjacent to the interface [163, 6]. Recent experiments have also shown that the Schottky barrier can be significantly lowered by chemical treatment of the metal–CNT interface [5]. Work will be needed that extends the theoretical model for better study of the interface chemistry including structural relaxation effects in the configuration of CNTFET with different gate structures.

Doping in semiconductors typically implies introducing a shallow impurity atom into the host lattice using ion implantation or thermal diffusion accompanied by creation of lattice defects [34]. But it may take a fundamentally different approach in CNTs. For example, doping in carbon nanotubes can be introduced chemically by exposing the CNT surface to alkali metals, by inserting C_{60} molecule inside the CNT, by surface functionalization with molecules/polymers for charge-transfer doping (which is essentially the electronic basis of sensing). In addition, the doping type can be converted between p-type and n-type by chemical treatment using, e.g. oxygen and molecular hydrogen [163, 6]. Doping in nanotubes can also be introduced physically using electrostatic gating or contact-induced charge transfer [77]. A "self-doping" mechanism for intrinsic SWNT caused by curvature induced charge redistribution has also been proposed, which shifts the Fermi-level position inside the band gap [118]. Despite its obvious importance, comprehensive experimental and theoretical study and a coherent physical picture of the various doping mechanisms, including both their electronic and structural consequences, have not yet appeared. A particularly interesting question in this regard is the optimal doping limit in carbon nanotubes through both physical and chemical doping mechanisms.

The major scattering mechanisms in CNTFET are those due to defects including dopant, gate stack and phonons. Due to the reduced phase space, the probability of back-scattering by defects and accoustic phonons is significantly reduced at low bias compared to planar silicon devices [154, 59, 109]. The absence of reactive dangling bond states at the CNT surface also make it less likely to suffer significant scattering due to the interface states and charge traps at the channel–gate interface. But it remains unclear how these favorable conditions may be modified at high bias. These include optical phonon emission by the energetic carrier, the injection of carriers into the gate dielectric and the resulting gate insulator degradation, remote phonon scattering between channel electrons and gate phonons, and the structural stability adjacent to the intrinsic or doping induced defect site. The optical phonon scattering length has been estimated at ≈ 10 nm [59], but in the absence of a realistic quantum transport model of electron–phonon coupling in CNTFET, this result should be taken with reservation [113]. Many fundamental knowledge gaps need to be addressed before we can have a convincing picture of the performance limit of CNTFETs in comparison to that of the ultimately scaled MOSFET. The recent report on suspended carbon nanotubes seems to suggest a cleaner platform for investigating many of the issues involved [14].

A different scenario applies to the nanowire FETs (NWFET), which seem to be less controversial. The vapor–liqiuid–solid phase growth process using nanoclustered catalysts pioneered by the Lieber group has led to the fabrication of single-crystal silicon nanowires [83], where the size distribition of the nanowires is determined by that of the catalyst nanoclusters. Both n-type and p-type dopants can be selectively inserted during the nanowire growing process. This has opened up the scheme of fabricating complementary logic circuits on the single silicon nanowire, where source/drain electrodes can be lithographically defined after the doped segments have been grown. Since the diameter of the nanowires is typically several tens of nanometers, well-known techniques in forming metallic contacts in planar silicon devices can be adapted leading to low barrier and low resistance contact [83, 153]. More recently, innovative techniques have been reported that solve the integrated contact and interconnect problem through selective transformation of silicon nanowires into metallic silicide nanowires [144]. The single-crystal metallic silicides have excellent high conductivity and high failure current, while being capable of forming atomically sharp metal–semiconductor heterostructures with the silicon nanowire of similar diameters. This opened up the possibility of ultra-dense integrated nanosystema that integrate both the active device area and high-performance interconnect from a single nanowire building block while inheriting all the knowledge gained in planar silicon devices (in particular the silicon-on-insulator approach) with minor modifications. In addition, different elemental, binary and ternary nanowires can be fabricated using the same vapor–liquid–solid growing process, providing significant design freedom for system designers [83, 153].

Both carbon nanotube and nanowire field-effect transistors have been demonstrated showing favorable performance compared with the state-of-the-art silicon MOSFET, while leaving substantial room for materials and device design optimization. Carbon nanotubes, even though of much smaller diamater than silicon nanowires, don't have the advantage of integrated metallic contact on the single-tube basis. This is because the reduced phase space and the correspondingly low electron density of states in the metallic SWNT doesn't allow rapid relaxation of carriers injected through the channel, which has to be connected to a larger area metal electrode to allow I/O separation and efficient heat removal. Athough this may be remedied by using bundles of metallic SWNT or metal nanowires, further materials and fabrication challenges need to be resolved in addition to the Schottky barrier problem in such interfaces. The challenge for nanowire FETs is instead to scale the nanowire to true molecular dimensions while maintaining scalable performance gain [145].

Molecular Interface to CMOS

Direct integration of molecular functionality with the scaled CMOS technology forms a starting point for hybrid top-down and bottom-up approaches. Such hybrid approaches may combine a level of advanced CMOS lithographical design patterns that represent designer-defined information and a level of molecular structures self-assembled with great precision and functional flexi-

otubes, but their larger size makes them less likely candidates for reliable room-temperature operation [117, 142]. The solution of the random background charge problem is much more difficult. Note that the electrostatic potential associated with random charged impurities in the environment is a problem for any nanoscale devices. But it poses a particularly potent problem for single-electron devices beacuse of their large charge sensitivity.

A comparison between the conventional approach and several representative single-molecule-based single-electron devices shows clearly the new physical processes introduced by the use of molecular-scale components [114, 110, 111, 82, 92, 112]. The molecular-scale dimension of the quantum dot leads to two intrinsic effects due to the ultra-small size: (1) both the wave function and the energy of the discrete electron states of the quantum dot depend on the size, shape and net charging state of the quantum dot; (2) due to the finite number of degrees of freedom and lack of an efficient relaxation mechanism on the quantum dot, the quantum dot may stay in a non-equilibrium state and self-heating may occur during the cycle of single-electron transfer. In addition, as electrons are added or removed from the molecular quantum dot, both the shape of the molecule and its position relative to the contacts may be altered. The electron states of the molecular-scale component are also sensitive to the atomic-scale change of the environment, e.g., due to presence of surface states which in turn may be modified by surface adsorption, the presence of impurities on the contact surface and/or the interaction with neighboring quantum dots. Treatment of all the above processes goes beyond the conventional theory of single-electron tunneling and is important for quantitative and realistic evaluation of their figures of merit.

So far, these devices have been formed by techniques excluding practical fabrication of integrated circuits. But there are good prospects for chemical synthesis of special molecules that would combine the structure suitable for single-electron tunneling with the ability to self-assembly from solution on prefabricated nanostructures with acceptable yield, opening a way to generically inexpensive fabrication of VLSI circuits. For logic circuits, the random background charge effects remain hard to overcome. Nevertheless, it has been suggested that the hybrid molecule-CMOS circuits, or "CMOL" circuits, that combine a CMOS stack with molecular single-elctron devices interconnected by nanowires, in defect-tolerant architectures that allow one to either tolerate or exclude bad devices, may become the basis for implementation of novel, massively parallel architectures for advanced information processing, e.g., self-evolving neuromorphic networks [85]. Such a hybrid approach can help to solve the low gain of single-electron transitors, but it remains open to demonstrate reliable high-performance digital circuits.

Molecular Quantum-Effect Devices

Intensive research on semiconductor heterostructures in the past three decades has generated many novel device concepts based on tunneling, resonant tun-

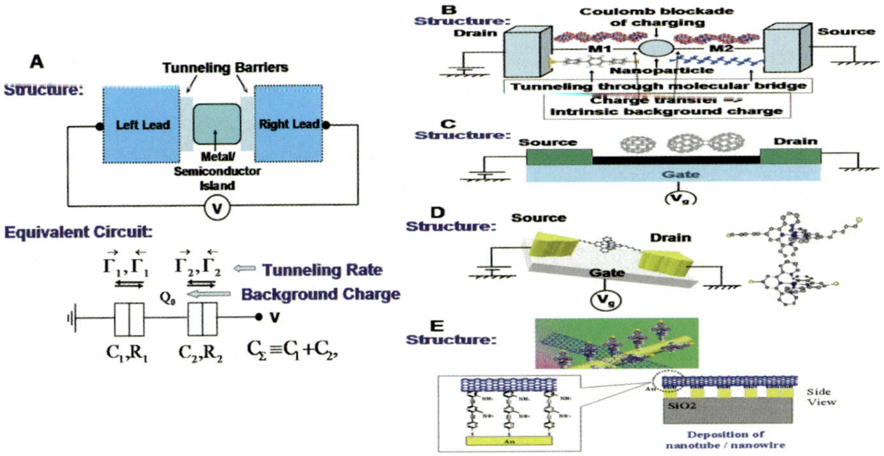

Fig. 3. (**A**) Typical structure and equivalent circuit of conventional single-electron devices. (**B**) Self-assembled or bio-directed assembly of single-electron device fabricated through synthetic routes. The nanoparticles are connected to the electrodes and/or to each other through either organic linkers or biomolecules with molecular-recognition capability [114, 151]. (**C**) A quantum dot is formed by a single C_{60} or C_{140} molecule physisorbed between two metal electrodes [110, 112]. The molecule may start oscillating as discrete charges are added to or extracted from the molecules through the contact. (**D**) The quantum dot is a single metal atom embedded within a larger molecule and connected to the metal contact pads through insulating tethers [111, 82]. (**E**) The molecule can also be adsorbed on top of a nanowire transistor which provides the source/sink of single electrons [35].

neling, real-space transfer, hot-electron transport and quantum wave interference effects, etc., in addition to creating the entire field of mesoscopic physics [15, 67, 140, 69]. Although they have not generated a real breakthrough in microelectronics, quoting a sarcastic statement from the mainstream silicon community, "heterostructure is and will be the material of the future", they provide a foundation and rich source of inspiration for going beyond the limits of conventional devices through quantum engineering of physical states in confined systems [128, 38, 143, 27, 50]. Recently they also see a rejuvenated interest as MOSFET moves toward the sub-10 nm era based on adavanced silicon-on-insulator (SOI) structures and Si–SiGe heterostructures [156].

Molecules are intrinsically heterostructures. Molecular electronics offer the ultimate testing ground for quantum-effect devices based on the atom-engineering approach to heterostructure concepts. Research in this field is intimately connected to exploiting molecular electronics as an artificial laboratory of new principles of nanoscopic physics [149, 152]. This is still a vaguely defined area and much fundamental knowledge needs to be sorted out. But molecular heterostructures already offer multiple device opportunities that are beyond the capability of or at least very difficult to achieve in scaled silicon devices. In the case of Q-1D nanostructures, this includes the possibility of fabricating metal–semiconductor and semiconductor heterojunctions with simultaneous band-gap engineering on a single nanotube and nanowire basis, and the possibility of fabricating Y-junction, T-junction, branched nanowires and superlattice devices with atomically sharp interfaces [83, 153, 144, 145, 153, 148, 139, 101, 108, 131, 102, 135, 19, 32]. Similar quantum-effect devices can also be implemented on a single-molecule basis through a synthetic chemistry approach, but can involve very different physical mechanisms and operation principles [61, 56]. Some examples are single-molecule heterostructures where a saturated molecular group can be selectively inserted between molecular groups with delocalized orbitals, complex structured molecules with three-terminal or multiple-terminal configurations and charge-transfer molecular complexes. In general, electron–vibronic coupling can be strong in such single-molecule devices, whose effects need to be sorted out. The recent surge of activity on integrating molecular functionality on a semiconductor platform also brings additional functionality through contact engineering [115, 49, 12, 141, 51] by attaching the molecule to the surface of a bulk semiconductor, semiconductor quantum well, quantum wire or quantum dots.

5 Discussion and Conclusion

Central to the vision of nanotechnology is the idea that by developing and following a common intellectual path — the bottom-up paradigm of nanoscale science and technology — it will be possible in the future to assemble virtually any kind of devices or functional systems. Much thus lies in the hands of chemists and materials scientists, where the goal is to control with atomic precision the morphology, structure, composition, and size of the nanoscale building blocks. Next, understanding the physics of nanoscale materials emerging from synthetic efforts and inserted into the device and system configurations, i.e., the effect on the operating behavior of nanostructures due to the introduction of contact, functional interface, the application of external forces and processing/environment-induced parameter variations, is a fundamental part of the bottom-up paradigm, which defines properties that may ultimately be exploited for nanotechnologies and enable us to make rational predictions and define new device concepts unique to nanoscale building blocks. Finally, to fully exploit the bottom-up paradigm, we must develop rational methods of organizing building blocks and device elements on multiple length-scales. This includes not only assembling building blocks in close-packed arrays for interconnectivity but also controlling the architecture or the spacing on multiple length-scales, i.e., hierarchical assembly, which must be done within the context of architectural design [83, 55, 100, 157, 60, 52, 13, 28].

We have focused our attention in this work on materials and memory/logic devices. But many of the materials and device structures in molecular electronics can be easily configured for applications in chemical/bio-sensors and electromechanical devices [165, 93, 158, 25, 83, 153] In addition, molecular electronics may play an important role in solving the 3-D interconnect problem in ultimately scaled nanoelectronic systems [99, 105]. Research progress in molecular electronics systems is steady and strong, which gives us cause to believe that functional molecular electronics systems may be practical in 10–15 years. Challenges to making this a reality are plentiful at every level, some naturally in the fundamental physics and chemistry of nanoelectronic materials and devices, but many in architecture and system design. These include fabricating and integrating devices, managing their power and timing, finding fault-tolerant and defect-tolerant circuits, and designing and verifying billion-gate systems. Any one of these could block practical molecular electronics if unsolved.

Acknowledgments

We are grateful for the financial support given by the US DoD-DURINT program through ARO. Y.X. has also been supported by the MARCO/DARPA Interconnect Focus Center. M.R. is supported by the NSF Network for Computational Nanotechnology and the NASA URETI program.

References

1. Y. Aharonov, J. Anandan (1990) Geometry of quantum evolution. *Phys. Rev. Lett.* 65:1697–1700.
2. Y. Aharonov, D. Bohm (1961) Time in the quantum theory and the uncertainty relation for time and energy. *Phys. Rev.* 122:1649–1658.
3. Y. Aharonov, D. Bohm (1964) Answer to Fock concerning the time energy indeterminacy relation. *Phys. Rev.* 134:1417–1418.
4. P. Allmen, K. Hess (1995) Dynamics of an atomic switch computed by first-principles molecular dynamics. *Phys. Rev.* B 52:5243.
5. S. Auvray, J. Borghetti, M.F. Goffman, A. Filoramo, V. Derycke, J.P. Bourgoin, O. Jost (2004) Carbon nanotube transistor optimization by chemical control of the nanotube-metal interface. *Appl. Phys. Lett.* 84:5106–5108.
6. Ph. Avouris, J. Appenzeller, R. Martel, S.J. Wind (2003) Carbon nanotube electronics. *Proc. IEEE* 91:1772–1784.
7. G. Baccarani, M.R. Wordeman, R.H. Dennard (1984) Submicrometer MOSFET design. *IEEE Trans. Electron Devices* ED-31:452–462.
8. J. Bardeen, W.H. Brattain (1948) The transistor, a semi-conductor triode. *Phys. Rev.* 74:230–231.
9. J.D. Beckenstein (1981) Energy cost of information transfer. *Phys. Rev. Lett.* 46:623–626.
10. C.H. Bennett (1973) Logical reversibility of computation. *IBM J. Res. Dev.* 17:525–533.
11. C.H. Bennett (1982) The thermodynamics of computation: a review. *Int. J. Theor. Phys.* 21:905–940.

12. G.A.D. Briggs, A.J. Fish (1999) STM experiment and atomistic modelling hand in hand: individual molecules on semiconductor surfaces. *Surf. Sci. Rep.* 33:1–81.
13. M. Butts, A. Dehon, S,C, Goldstein (2002) Nanoelectronics: devices, systems and tools for gigagate, gigabit chips. Proceedings of the International Conference on Computer-Aided Design, pp 433-440.
14. J. Cao, Q. Wang, D. Wang, H. Dai (2005) Suspended carbon nanotube quantum wires with two gates. *Small* 1:138–141.
15. F. Capasso, eds. (1990) *Physics of Quantum Electron Devices.* Springer, Berlin.
16. G.K. Celler, S. Cristoloveanu (2003) Frontiers of silicon-on-insulator. *J. Appl. Phys.* 93:4955–4978.
17. L. Chang, Y.K. Choi, H.A. Daewon, P. Ranade, S. Xiong, J. Bokor, C. Hu, T.J. King (2003) Extremely scaled silicon nano-CMOS devices. *Proc. IEEE* 91:1860–1873.
18. J. Chen, M.A. Reed, A.M. Rawlett, J.M. Tour (1999) Large on-off ratios and negative differential resistance in a molecular electronic device. *Science* 286:1550.
19. P.W. Chiu, M. Kaempgen, S. Roth (2004) Band-structure modulation in carbon nanotube T junctions. *Phys. Rev. Lett.* 92:246802.
20. C.P. Collier, E.W. Wong, M. Belohradsky, F.M. Raymo, J.F. Stoddart, P.J. Kuekes, R.S. Williams, J.R. Heath (1999) Electronically configurable molecular-based logic gates. *Science* 285:391.
21. C.P. Collier, G. Matterstreig, E.W. Wong, Y. Luo, K. Beverly, J. Sampaio, F.M. Raymo, J.F. Stoddart, J.R. Heath (2000) A catenane-based solid state reconfigurable switch. *Science* 289:1172.
22. Y. Cui, C.M. Lieber (2001) Functional nanoscale electronic devices assembled using silicon nanowire building blocks. *Science* 291:851–853.
23. Y. Cui, Z. Zhong, D. Wang, W.U. Wang, C.M. Lieber (2003) High performance silicon nanowire field effect transistors. *Nano Lett.* 3:149–152.
24. X.D. Cui, A. Primak, X. Zarate, J. Tomfohr, O.F. Sankey, A.L. Moore, T.A. Moore, D. Gust, G. Harris, S.M. Lindsay (2001) Reproducible measurement of single-molecule conductivity. *Science* 294:571–574.
25. H. Dai (2002) Carbon nanotubes: opportunities and challenges. *Surf. Sci.* 500:218–241.
26. S. Datta (1995) *Electronic Transport in Mesoscopic Systems.* Cambridge University Press, Cambridge.
27. S. de Hann, A. Lorke, J.P. Kotthaus, Wegscheider, M. Bichler (2004) Rectification in mesoscopic systems with broken symmetry: quasiclassical ballistic versus classical transport. *Phys. Rev. Lett.* 92:56806.
28. A. Dehon, M.J. Wilson (2004) Nanowire-based sublithographic programmable logic arrays. *Proceedings of the International Symposium on Field Programmable Gate Arrays*, pp. 123–132.
29. R.H. Dennar, F.H. Gaensslen, H.N. Yu, V.L. Ridout, E. Bassous, A.R. LeBlanc (1974) Design of ion-implanted MOSFETs with very small physical dimensions. *IEEE J. Solid-State Circuits* 9:256.
30. M.R. Diehl, D.W. Steuerman, H.R. Tseng, S.A. Vignon, A. Star, P.C. Celestre, J.F. Stoddart, J.R. Heath (2003) Single-walled carbon nanotube based molecular switch tunnel junctions. *Chem. Phys. Chem.* 4:1335–1339.

31. Z.J. Donhauser, B.A. Mantooth, K.F. Kelly, L.A. Bumm, J.D. Monnell, J.J. Stapleton, D.W. Price, A.M. Rawlett, D.L. Allara, J.M. Tour, P.S. Weiss (2001) Conductance Switching in Single Molecules Through Conformational Changes. *Science* 292:2303.
32. S.K. Doorn, M.K. O'Connell, L. Zheng, Y.T. Zhu, S. Huang, J. Liu (2005) Raman spectral imaging of a carbon nanotube intramolecular junction. *Phys. Rev. Lett.* 94:16802.
33. X. Duan, Y. Huang, C.M. Lieber (2002) Memory and Programmable logic from molecule-gated nanowires. *Nano Lett.* 2:487–490.
34. P.M. Fahey, P.B. Griffin, J.D. Plummer (1989) Point defects and dopant diffusion in silicon. *Rev. Mod. Phys.* 61:289–384.
35. C. Li, W. Fan, B. Lei, D. Zhang, S. Han, T. Tang, X. Lu, Z. Liu, S. Asano, M. Meyyappan, J. Han, C. Zhou (2004) Multilevel memory based on molecular devices. *Appl. Phys. Lett.* 84:1949.
36. C. Li, W. Fan, D.A. Straus, B. Lei, S. Asano, D. Zhang, J. Han, M. Meyyappan, C. Zhou (2004) Charge storage behavior of nanowire transistors functionalized with bis(terpyridine)-Fe(II) molecules: Dependence on Molecular Structure. *J. Am. Chem. Soc.* 126:7750.
37. R. Feynman (1960) There's plenty of room at the bottom. *Sci. Eng.* 23:22.
38. R. Fleischmann, T. Geisel (2002) Mesoscopic rectifiers based on ballistic transport. *Phys. Rev. Lett.* 89:16804.
39. M.V. Fischetti (2003) Scaling MOSFETs to the limit: A physicists's perspective. *J. Comp. Electron* 2:73–79.
40. D.J. Frank, R.H. Dennard, E. Nowak, P.M. Solomon, Y. Taur, H.S.P. Wong (2001) Device scaling limits of Si MOSFETs and their application dependencies. *Proc. IEEE* 89:259–288.
41. E. Fredkin, T. Toffoli (1982) Conservative logic. *Int. J. Theor. Phys.* 21:311–325.
42. M.S. Fuhrer, B.M. Kim, B. Duerkop, (2002) High-mobility nanotube transistor memory. *Nano. Lett.* 2:755–759.
43. A. Galindo, J.C. Martín-Delgado (2002) Information and computation: classical and quantum aspects. *Rev. Mod. Phys.* 74:347–423.
44. U. Ganguly, C. Lee, E.C. Kan (2003) Interface and oxide contamination monitoring in integration of fullerenes and carbon nanotubes with aggressively scaled CMOS gate stacks. *Mat. Res. Soc. Symp. Proc.* 789:N16.3.
45. U. Ganguly, C. Lee, E.C. Kan (2005) Experimental observation of non-volatile charge injection and molecular redox in fullerenes C60 and C70 in an EEPROM type device. *Mat. Res. Soc. Symp. Proc.* 830:D7.5.
46. U. Ganguly, E.C. Kan, Y. Zhang (2005) Carbon nanotube-based nonvolatile memory with charge storage in metal nanocrystals. *Appl. Phys. Lett.* 87:43108.
47. V. Giovannetti, S. Lloyd, L. Maccone (2003) Quantum limits to dynamical evolution. *Phys. Rev.* A 67:52109.
48. H. Grabert, M.H. Devoret, eds. (1992) *Single Charge Tunneling.* Plenum, New York.
49. N.P. Guisinger, M.E. Greene, R. Basu, A.S. Baluch, M.C. Hersam (2004) Room temperature negative differential resistance through individual organic molecules on silicon surfaces. *Nano Lett* 4:55–59.
50. B. Hackens, L. Gence, C. Gustin, X. Wallart, S. Bollaert, A. Cappy, V. Bayot (2004) Sign reversal and tunable rectification in a ballistic nanojunction. *Appl. Phys. Lett.* 4508.

95. N. Margolus, L.B. Levitin (1998) The maximum speed of dynamical evolution. *Physica D* 120:188–195.
96. P.L. McEuen, M.S. Fuhrer, H. Park (2002) Single-walled carbon nanotube electronics. *IEEE Trans. Nanotech.* 1:78–85.
97. J.D. Meindl, Q. Chen, J.A. Davis (2001) Limits on silicon nanoelectronics for terascale integration. *Science* 293:2044–2049.
98. J.D. Meindl, J.A. Davis (2000) The fundamental limit on binary switching energy for terascale integration. *IEEE J. Solid State Circuits* 35:1515–1516.
99. J.D. Meindl, J.A. Davis, P. Zarkesh-Ha, C.S. Patel, K.P. Martin, P.A. Kohl (2002) Interconnect opportunities for gigascale integration. *IBM J. Res. Dev.* 46:245.
100. N.A. Melosh, A. Boukai, F. Diana, B. Gerardot, A. Badolato, P.M. Petroff, J.R. Heath (2003) Ultrahigh-density nanowire lattices and circuits. *Science* 112–115.
101. M. Menon, D. Srivastava (1997) Carbon nanotube T junctions: nanoscale metal-semiconductor-metal contact devices. *Phys. Rev. Lett.* 79:4453–4456.
102. M. Menon, A.N. Andriotis, D. Srivastava, I. Ponomareva, L.A. Chernozatonskii (2003) Carbon nanotube T junctions: formation pathways and conductivity. *Phys. Rev. Lett.* 91:145501.
103. F. Moresco, G. Meyer, K.-H. Rieder, H. Tang, A. Gourdon, C. Joachim (2001) Conformational changes of single molecules induced by STM manipulation: a route to molecular switching. *Phys. Rev. Lett.* 86:672.
104. D.A. Muller, T. Sorsch, S. Moccio, F.H. Baumann, K. Evans-Lutterodt, G. Timp (1999) The electronic structure at the atomic scale of ultra-thin gate oxides. *Nature* 399:758–760.
105. A. Naeemi, R. Sarvari, J.D. Meindl (2005) Private communications.
106. K. Nakazato, R.J. Blaikie, H. Ahmed (1994) Single electron memory. *J. Appl. Phys.* 75:5123.
107. M.A. Nielsen, I.L. Chuang (2000) *Quantum Computation and Quantum Information.* Cambridge University Press, Cambridge.
108. C. Papadopoulos, A. Rakitin, J. Li, A.S. Vedeneev, J.M. Xu (2000) Electronic transport in Y-junction carbon nanotubes. *Phys. Rev. Lett.* 85:3476–3479.
109. J.Y. Park, S. Rosenblatt, Y. Yaish, V. Sazonova, H. Ustunel, S. Braig, T.A. Arias, P.W. Brouwer, P.L. McEuen (2004) Electron-phonon scattering in metallic single-walled carbon nanotubes. *Nano Lett.* 4: 517.
110. H. Park, J. Park, A.K.L. Lim, E.H. Anderson, A.P. Alivisatos, P.L. McEuen (2000) Nanomechanical oscillations in a single-C 60 transistor. *Nature* 407:57–60.
111. J. Park, A.N. Pasupathy, J.I. Goldsmith, C. Chang, Y. Yaish, J.R Petta, M. Rinkoski, J.P. Sethna, H.D. Abruna, P.L. McEuen, D.C. Ralph (2002) Coulomb blockade and the Kondo effect in single-atom transistors. *Nature* 417:722–725.
112. A.N. Pasupathy, J. Park, C. Chang, A.V. Soldatov, S. Lebedkin, R.C. Bialczak, J.E. Grose, L.A.K. Donev, J.P. Sethna, D.C. Ralph, P.L. McEuen (2005) Vibration-assisted electron tunneling in C140 single-molecule transistors. *Nano Lett.* 5:203–207.
113. V. Perebenios, J. Tersoff, Ph. Avouris (2005) Electron-phonon interaction and transport in semiconducting carbon nanotubes. *Phys. Rev. Lett.* 94:86802.
114. J.R. Petta, D.G. Salinas, D.C Ralph (2000) Measurements of discrete electronic states in a gold nanoparticle using tunnel junctions formed from self-assembled monolayers. *Appl. Phys. Lett.* 77:4419.

115. P.G. Piva, G.A. DiLabio, J.L. Pitters, J. Zikovsky, M. Rezeq, S. Dogel, W.A. Hofer, R.A. Wolkow (2005) Field regulation of single-molecule conductivity by a charged surface atom. *Nature* 435:658–661.
116. J.D. Plummer, P.B. Griffin (2001) Material and process limits in silicon VLSI technology. *Proc. IEEE* 89:240–258.
117. H.W. Postma, T. Teepen, Z. Yao, M. Grifoni, C. Dekker (2001) Carbon nanotube single-electron transistors at room temperature. *Science* 293:76–79.
118. A. Rakitin, C. Papadopoulos, J.M. Xu (2003) Carbon nanotube self-doping. *Phys. Rev.* B 67:33411.
119. G.K. Ramachandran, T.J. Hopson, A.M. Rawlett, L.A. Nagahara, A. Primak, S.M. Lindsay (2003) Reproducible measurement of single-molecule conductivity. *Science* 300:1413.
120. M.A. Reed, J. Chen, A.M. Rawlett, D.W. Price, J.M. Tour (2001) Room-temperature negative differential resistance in nanoscale molecular junctions. *Appl. Phys. Lett.* 78:3735.
121. S. Rosenblatt, H. Lin, V. Sazonova, S. Tiwari, P.L. McEuen (2005) Mixing at 50GHz using a single-walled carbon nanotube transistor. *Appl. Phys. Lett.* 87:153111.
122. S. Rosenblatt, Y. Yaish, J. Park, J. Gore, V. Sazonova, P.L. McEuen (2002) High performance electrolyte gated carbon nanotube transistors. *Nano Lett.* 2:869.
123. P.M. Rousseau, P.B. Griffin, J.D. Plummer (1998) Arsenic deactivation enhanced diffusion: a time, temperature and concentration study. *J. Appl. Phys.* 84:3593.
124. C.T. Sah (1988) Evolution of the MOS transistor – from conception to VLSI. *Proc. IEEE* 76:1280–1326.
125. N.C. Seeman (1998) DNA Nanotechnology: novel DNA constructions. *Annu. Rev. Bioph. Biom.* 27:225–248.
126. N.C. Seeman (2004) Nanotechnology and the double helix. *Sci. American* 290:64–75.
127. W. Shockley (1949) The theory of pn junctions in semiconductors and pn junction transistors. *Bell Syst. Tech. J.* 28:436–489.
128. A.M. Song, A. Lorke, A. Kriele, J.P. Kotthaus, W. Wegscheider, M. Bichler (1997) Nonlinear electron transport in an asymmetric microjunction: a ballistic rectifier. *Phys. Rev. Lett.* 80:3831.
129. S.M. Sze (1969) *Physics of Semiconductor Devices.* Wiley, New York.
130. S.M. Sze (1981) *Physics of Semiconductor Devices.* 2nd edition. Wiley, New York.
131. M. Terrones, F. Banhart, F. Grobert, J.C. Charlier, H. Terrones, P.M. Ajayan (2002) Molecular junctions by joining single-walled carbon nanotubes. *Phys. Rev. Lett.* 89:075505.
132. S. Tiwari, J.A. Wahl, H. Silva, F. Rana, J.J. Welser (2000) Small silicon memories: confinement, single-electron, and interface state considerations. *Appl. Phys. A* 71:403–414.
133. A. Troisi, M.A. Ratner (2004) Conformational molecular rectifiers. *Nano Lett.* 4:591–595.
134. F.A. Trumbore (1960) Solid solubilities of impurity elements in germanium and silicon. *Bell Syst. Tech. J.* 39:205.

135. M. Tzolov, B. Chang, A. Yin, D. Straus, J.M. Xu, G. Brown (2004) Electronic transport in a controllably grown carbon nanotube-silicon heterojunction array. *Phys. Rev. Lett.* 92:75505.
136. J. von Neumann (1966) Theory of Self-Reproducing Automata. University of Illinois Press, Urbana, IL.
137. Y. Wada (2001) Prospects for single molecule information processing devices. *Proc. IEEE* 89:1147–1171.
138. Y. Wada, T. Uda, M.I. Lutwyche, S. Kondo, S. Heike S (1993) A proposal of nanoscale devices based on atom/molecule switching. *J. Appl. Phys.* 74:7321.
139. D. Wang, F. Qian, C. Yang, Z. Zhong, C.M. Lieber (2004) Rational growth of branched and hyperbranched nanowire structures. *Nano Lett.* 4:871–875.
140. C. Weisbuch (1996) The future of physics of heterostructures: a glance into the crystal (quantum) ball. *Phys. Scripta* T68:102–112.
141. R.A. Wolkow (2000) Controlled molecular adsorption on Si: laying a foundation for molecular devices. *Annu. Rev. Phys. Chem.* 50:413–441.
142. M.T. Woodside, P.L. McEuen (2002) Scanned probe imaging of single-electron charge states in nanotube quantum dots. *Science* 296:1098–1101.
143. L. Worschech, S. Reitzenstein, P. Hartmann, S. Kaiser, M. Kamp, A. Forchel (2003) Self-switching of branched multiterminal junctions: a ballistic half-adder. *Appl. Phy. Lett.* 83:2462.
144. Y. Wu, J. Xiang, C. Yang, W. Lu, C.M. Lieber (2004) Single-crystal metallic nanowires and metal/semiconductor nanowire heterostructures. *Nature* 430:61–65.
145. Y. Wu, Y. Cui, L. Huynh, C.J. Barrelet, D.C. Bell, C.M. Lieber (2004) Controlled growth and structures of molecular-scale silicon nanowires. *Nano Lett.* 4:433–436.
146. F.Q. Xie, L. Nittler, Ch. Obermair, Th. Schimmel (2004) Gate-controlled atomic quantum switch. *Phys. Rev. Lett.* 93:128303.
147. B. Xu, X. Xiao, X. Yang, L. Zang, N.J. Tao (2005) Measurement and control of single molecule conductance. *J. Am. Chem. Soc.* 127:2386.
148. J. Xu (2003) Nanotube electronics: non CMOS routes. *Proc. IEEE* 91:1819-1829.
149. Y. Xue (2005) To be published.
150. Y. Xue, M.A. Ratner (2004) Scaling analysis of Schottky barrier at metal-embedded semiconducting carbon nanotube interfaces. *Phys. Rev.* B 69:161402.
151. Y. Xue, M.A. Ratner (2003) Microscopic theory of single-electron tunneling through molecular-assembled metallic nanoparticles. *Phys. Rev.* B 68:235410.
152. Y. Xue, M.A. Ratner (2005) Theoretical principles of single-molecule electronics: a chemical and mesoscopic view. *Int. J. Quantum Chem.* 102:911–924.
153. P. Yang (2005) Chemistry and physics of semiconductor nanowires. *MRS Bulletin* 30:85–91.
154. Z. Yao, C.L. Kane, C. Dekker (2000) High-field electrical transport in single-wall carbon nanotubes. *Phys. Rev. Lett.* 84:2941.
155. C. Li, D. Zhang, X. Liu, S. Han, T. Tang, C. Zhou, W. Fan, J. Koehne, J. Han, M. Meyyappan, A.M. Rawlett, D.W. Price, J.M. Tour (2003) Fabrication approach for molecule memory arrays. *Appl. Phys. Lett.* 82:645.
156. S. Luryi, A. Zaslavsky (2004) Blue sky for SOI: New opportunities for quantum and hot-electron devices. *Solid-St. Electron.* 48:877–885.

157. Z. Zhong, D. Wang, Y. Cui, M.W. Bockrath, C.M. Lieber (2003) Nanowire crossbar arrays as address decoders for integrated nanosystems. *Science* 302:1377–1379.
158. Molecular Electronics Special Feature, *Proc. Natl. Acad. Sci.* USA, Vol. 102, No. 26, 2005.
159. 2004 International Technology Roadmap for Semiconductors. http://public.itrs.net
160. Semiconductor Research Corporation (2003) Research Needs for Novel Devices. Novel Device Task Force Report, Research Triangle Park, NC 27709.
161. Semiconductor Research Corporation (2004) Silicon Nanoelectronics and Beyond: Challenges and Research Directions. Draft, Version 1.1, Research Triangle Park, NC 27709.
162. Special Issue on Limits of Semiconductor Technology, *Proc. IEEE*, Vol. 89, No. 3, 2001.
163. Special Issue on Advances in Carbon Nanotubes, MRS Bulletin, Vol. 29, No. 4, 2004.
164. Special Issue on Scaling CMOS to the Limit, *IBM J. Res. Dev.*, Vol. 46, No. 2/3, 2002.
165. Special Issue on Nanoelectronics and Nanoscale Processing, *Proc. IEEE*, Vol. 91, No. 11, 2003.
166. http://www.nano.gov/html/facts/whatIsNano.html.

Part V

Other Bio-molecules in Self-assembly

Poly- and oligo-peptides can now be genetically engineered to specifically bind to selected inorganic compounds for applications in nano- and biotechnology [1, 2, 16]. Combinatorial biological protocols, that is, bacterial cell surface and phage-display technologies, have been developed in the selection of short sequences that have an affinity to (noble) metals, semiconducting oxides and other technological compounds. Although significant advances have been made in developing protocols for the selection of surface binding polypeptides through display technologies, many questions remain before their robust genetic design and practical applications as building blocks.

These include, for example, the physical and chemical basis for genetically engineered protein recognition of inorganic surfaces and quantification of their cross-specificity for diverse materials. New avenues for extensive applications of proteins to self-assembly of molecular systems in nano- and nanobiotechnology will be opened by a better understanding of these processes.

Along these lines, there has been speculation about the possibility of constructing intricate hybrid structures composed of inorganics and proteins (Fig. 1) [9].

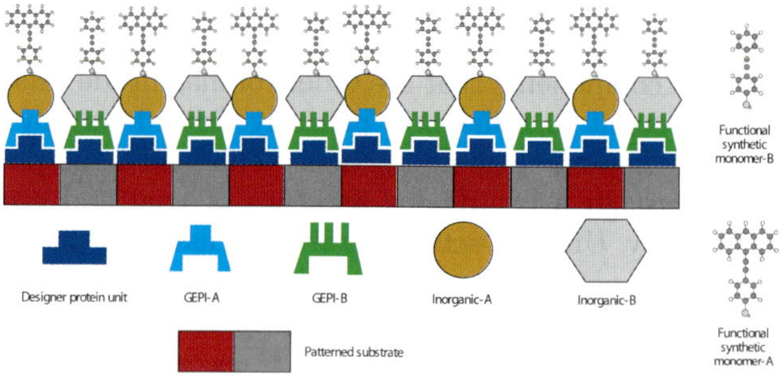

Fig. 1. The potential of using genetically engineered proteins for inorganics (GEPI). Two different GEPI proteins (GEPI-A and GEPI-B) are assembled on ordered molecular or nanoscale substrates. One could use either a designer protein, followed by genetic fusion of the respective GEPIs, or directly assemble GEPIs on the patterned substrate. The inorganic particles A and B are immobilized selectively on GEPIA and GEPI-B, respectively. Synthetic molecules (that is, conducting or photonic) are assembled using functionalized side-groups on the nanoparticles. Size, shape, separation, and distribution of nanostructural units, as well as self-assembly, are parameters unique to this approach (from [9]).

The potential of using genetically engineered proteins for inorganics (GEPI). Two different GEPI proteins (GEPI-A and GEPI-B) are assembled on ordered molecular or nanoscale substrates. One could use either a designer protein, followed by genetic fusion of the respective GEPIs, or directly assemble GEPIs on the patterned substrate. The inorganic particles A and B are immobilized selectively on GEPI-A and GEPI-B, respectively. Synthetic molecules (that is, conducting or photonic are assembled using functionalized side-groups on the nanoparticles. Size, shape, separation, and distribution of nanostructural units, as well as self-assembly, are parameters unique to this approach (from [9]).

Our understanding about recognition mechanisms between DNA and inorganic materials is still far behind with respect to the case of proteins. Nucleic acid bases are differently adsorbed on the surface of crystalline graphite. The adsorption behavior of the bases on this surface is markedly different. The equilibrium adsorption isotherms for the nucleic acid purine and pyrimidine bases dissolved in water describes an elutropic series: guanine > adenine > hypoxanthine > thymine > cytosine > uracil [15]. On the other hand, this wealth of data regarding the adsorption of the single nucleic acid bases is not significant for the issue of the DNA interaction with crystalline surfaces. In DNA the bases are buried inside the double helix, and the dominant interaction of a DNA segment with a surface is likely to be mostly sustained by the negatively charged phosphates.

2 A Crystal Surface Recognizes the T-Rich Face of Curved DNA Chains

The existence of some recognition process between a DNA chain and a crystal surface might have been inferred many years ago on the basis of the data that made it possible for [6] to determine the pitch of the DNA double helix.

They observed that a pool of nucleosomal DNA tracts of about 146 bp, obtained by chromatin differential digestion, when adsorbed on inorganic powder surfaces, show Dnase I cutting sites with 10.6 bp periodicity in foot-printing experiments.

However, in the absence of differential interactions of the sequence with the surface, a DNA chain should be adsorbed with many equally probable orientations and as a consequence the protection from the Dnase cleavage by the surface should be cylindrically averaged.

Thus, they justified the experimental evidence of the preferential phasing of the nucleosomal DNA chains under investigation by assuming the DNA phasing is due to preferential anchorage of the terminals to the surface.

An alternative explanation is possible. Nucleosomal DNAs are usually characterized by phasing of AA-rich sequences [3] that generally produces DNA curvature and, as a consequence, differential prochiral interaction with the crystal surfaces, as will be illustrated. In fact, an intrinsic DNA curvature is not only able to hinder the rotation of the chain upon its deposition on a surface, but it can also ensure a recognition mechanism of the DNA sequence by a crystal surface.

More recently, it has been recognized that the intrinsic curvature in a DNA segment defines an average plane (Figs. 2a,b) and that the two sides of faces of this plane are chemically different. Their difference is due to the different spatial distribution of the dinucleotide steps that give rise to the chain curvature. For instance, in a strongly curved fragment like that of *Crithidia* fasciculata, its almost planar structure has one A-rich face while the other is T-rich. This segregation of the complementary bases on the two faces is the result of the recurrence of the A-tract that is phased with the helical winding (see Fig. 2c).

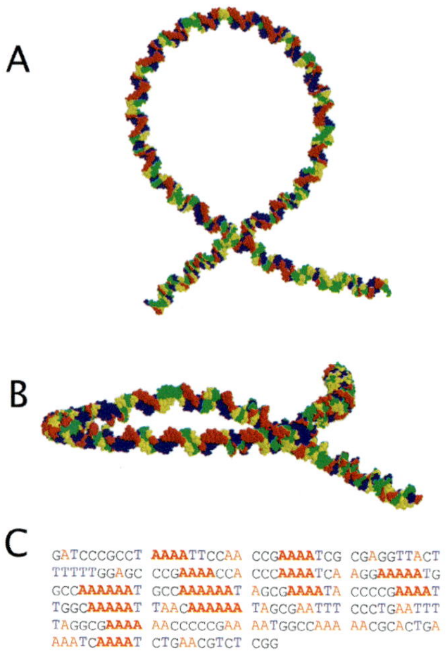

Fig. 2. Predicted 3D structure of the curved segment of *Crithidia* fasciculata (`http://archimede.chem.uniroma1.it/webdna.html`). (**a**) View along a direction nearly perpendicular to the curvature plane. (**b**) View along a direction almost parallel to the curvature plane, the segregation of adenines and thymines on the two faces of the curvature plane is clearly visible: red adenine, blue thymine, green guanine, yellow cytosine. (**c**) The base sequence of the molecule shown in (a) and (b) (from [7])

3 The Strategy of the Palindromic Dimers

A curved DNA tract can deposit on mica on either of the two faces generated by the intrinsic curvature. The direct observation (with electron microscopy, EM, or scanning force microscopy, SFM) of this segment does not allow the preferred face of adsorption, if any, to be identified, because it is not possible to read the direction of the sequence. The strategy that was used in [7] to determine the preferred adsorption face is shown schematically in Fig. 3.

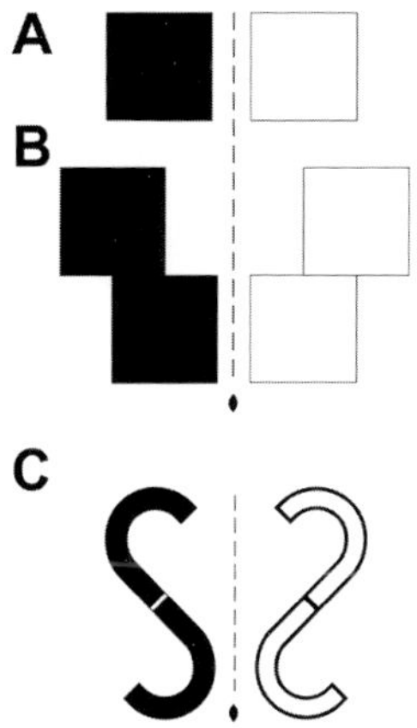

Fig. 3. (**a**) The oppositely colored faces of a thin square object can not be distinguished by form alone. (**b**) The two oppositely colored faces of a chiral object obtained by fusing two copies of an object as in (a). In this case, the face (white or black) shown can be distinguished by the morphology of the object, even if the color could not be distinguished. (**c**) A palindromic DNA dimer made with a curve DNA section is expected to have the same property as the model in (b) when flattened on a surface in an S-like shape: the face can be recognized without reading the direction of the base sequence (from [7]).

Likewise the case of the curved segment just discussed, the two faces of a square thin object (for example, a paper square), one black and one white, can only be distinguished by their colors, since they have the same shape. On the other hand, with the colored square, if a new object is built by linking two of these squares to make a chiral shape like that in Figure 3b, the two prochiral faces of this thin object can be distinguished on a purely morphological basis: just from their shape. DNA palindromic dimers of a strongly curved DNA segment are like the prochiral objects in Fig. 3b. In this case we will be able to recognize the face upon which the single palindromic molecules deposit on mica if we can read their shape. By depositing a sample of linear DNA molecules on the surface of a mica crystal, the traces of the trajectory of the double-helical axis of the individual deposited dsDNA chains can be obtained by imaging the sample by SFM. The curvatures of the single dsDNA molecules can thus be mapped along the chain. By computing the curvature from the angular chain deflections along a large number of profiles, by then averaging the values over the whole ensemble of profiles, and by plotting the resulting averaged values as a function of the (fractional) position along the chain, curvature plots are obtained for each palindromic structure under investigation [18, 11, 8].

4 Experimental Evidence of DNA Sequence Recognition by Mica Surface

Two *Crithidia* segments, like those in Fig. 3, were ligated either in the tail-to-tail (Fig. 4) or in the head-to-head (Fig. 5) orientation and two palindromic dimers were constructed.

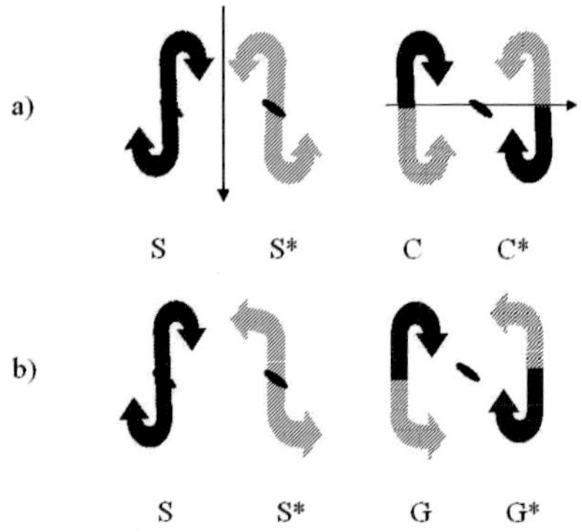

Fig. 4. Pictorial representation of the tail-to-tail DNA palindromic construct. The DNA monomers are composed in a palindromic fashion and sketched as curved ribbons with directionality defined by the sequence. The two opposite DNA faces are indicated with different gray intensities. In (**b**), contrary to (**a**), the different extent of the local curvature due to the differential interaction of the two monomeric faces with the surface is taken into account. The dyadic symmetry is thus lost and the C-like shape is, more correctly, expected to be G-like, instead.

In 3D, the dyad axis, which characterizes the averaged shape of the palindromic DNA dimers, can be oriented along any direction of space with respect to the average plane of the curved tracts. This statistical symmetry constraint also persists when the molecules are flattened on a crystal surface, such as mica in SFM images, but only two alternative directions of the dyad axis are allowed, parallel or perpendicular to the surface plane. In the former case both curved halves of the molecule have the same sign of the curvature sign; in the latter case the two curved halves have curvatures opposite in sign. We called these symmetry species C-like shape and S-like shape (or S*, the asterisk indicating the mirror image), respectively, because the curves are isomorphous with these letters (Figs. 4a, 5a). The C-like molecules will be characterized by two positive curvatures or two negative ones, depending on which end is

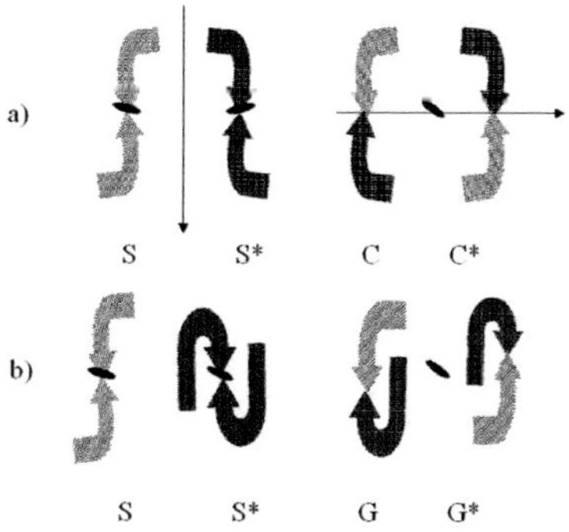

Fig. 5. Pictorial representation of the head-to-head DNA palindromic construct. See legend of Fig. 4.

chosen as the starting point of the molecule (but the sequence associated will be the same, thanks to the palindromicity). On the contrary, the two main curvatures will be oppositely signed in an S-like or an S*-like shape; either with a positive followed by a negative one (S-shape), or by a negative followed by a positive one (S* shape), independent of the direction of which end is chosen as the starting point of the molecule. These two possibilities are the result of the adhesion of the three-dimensional dimeric molecules on either of their two opposite faces. In the case of C-shaped molecules (Figs. 4a, 5a), the two faces are equivalent, instead, because within either face one half exposes a sequence complementary to that of the other.

We collected a large pool of SFM images (about 1500) of two palindromic DNA constructs containing the curved tract of *Crithidia* fasciculata, bridged head-to-head and tail-to-tail to obtain two palindromic dimers. The curvature was evaluated along all the recorded molecular profiles and averaged over all of them according to [18, 11, 8].

The average over the whole set of profiles however, did not vanish, as should be expected on the basis of equal populations of the four subclasses depicted in Figs. 4a or 5a. The two palindromic dimers exhibited the curvature profiles reported in Fig. 6: the sigmoidal shapes are similar, but oppositely signed.

The non-zero curvature profiles in Fig. 6 thus monitor the imbalance of the subpopulations of the different symmetry classes. This result proved that the surface preferentially binds one of the two different faces of the curved DNA tracts (see Fig. 3) and in principle can differently modify their curvature. This is pictorially illustrated in Figs. 4b and 5b.

We have seen that the two faces of the monomeric curved tracts expose either A-rich or T-rich sequences (Fig. 2). By analyzing the shape assumed by two large sets of the two palindromic dimers of the *Crithidia* fasciculata

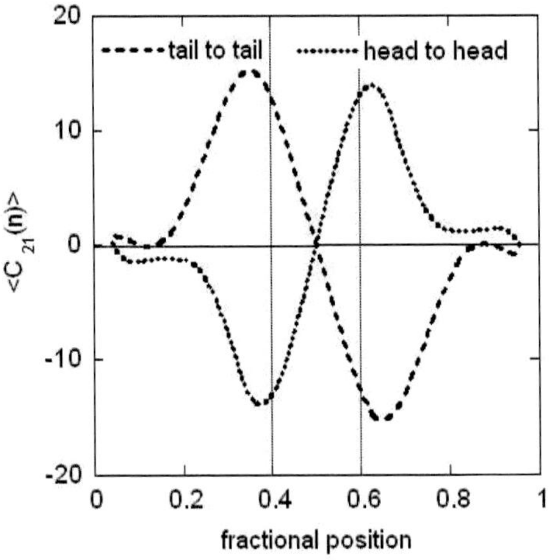

Fig. 6. SFM ensemble average curvature profiles of the head-to-head and the tail-to-tail DNA palindromic dimers. The inversion of sequence direction results in an inversion of the sign of curvature.

fragment, it turned out that the face that both dimers expose preferentially to the mica is the T-rich one [8, 17].

5 How Effective Is This Recognition Process?

In order to answer this question we must characterize the subpopulations of the different classes and compare in particular that of S with that of S*. A more precise analysis of the expected shapes for the different classes has recently been performed [10].

We have also accounted for the presence of differential interactions between the two faces on each monomer and the surface. This is expected to change the symmetry of the models depicted in Figs. 4 and 5. In fact, such differential interactions can affect the curvature and flexibility, namely the intrinsic mechanical properties of the single monomeric units within each dimer. In particular, the S and S* shapes should retain the dyad axis and the intrinsic anti-symmetry of the curvature functions but their shapes and the related curvatures will no longer be quantitatively equivalent because of the different interactions with the surface. Thus, the two faces of G-shaped molecules being physically equivalent, they must be present on the surface in the same number but their contribution to the curvature will be equivalent to those of

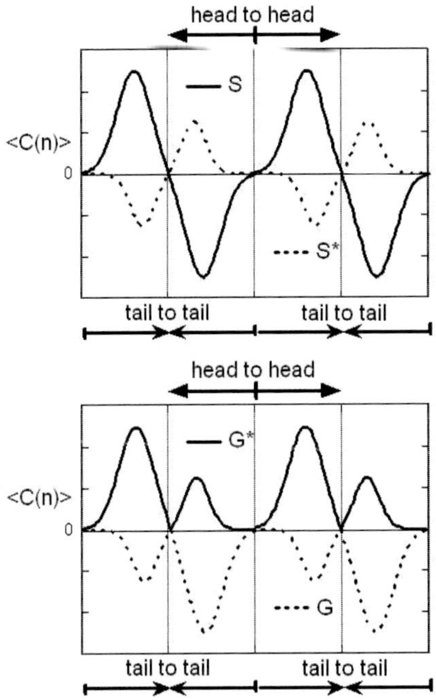

Fig. 7. Predicted curvature profiles of the S,S*, G and G* symmetry classes of tail-to-tail and head-to-head palindromic dimers as sketched in Figs. 4b and 5b. Note the inversion of curvature signs of the two halves of the molecule when the tail-to-tail dimer is formally transformed in the head-to-head dimer.

S+S* [10]. In fact, the expected curvature profiles will be as those represented in Fig. 7 for the tail-to-tail and head-to-head dimers. These predicted profiles were perfectly confirmed by those obtained experimentally (see Fig. 8) by classifying all the profiles in the four subclasses, according to their shapes, and then by averaging the curvatures plots within each subclass.

The classification of the different molecules in the subclasses indicates that the T-rich face was deposited up to 12 times more frequently than the other.

Therefore, the recognition effect is strong and the differential adhesion of DNA to mica not only privileges one face with respect to the other but also modifies its curvature.

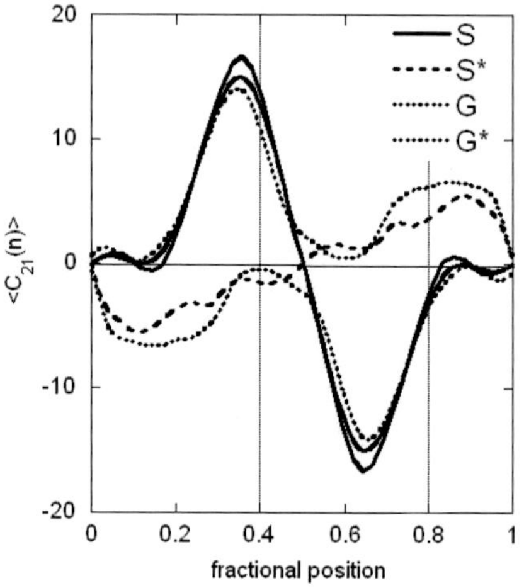

Fig. 8. SFM ensemble average curvature profiles of the tail-to-tail palindromic DNA dimer after shape classification. The curvatures of the species are coded as in the inserted legend.

6 From Statistics to Determinism

One can tailor possible applications of this recognition effect to the self-assembled integration of inorganic material in the construction of complex DNA-based nanostructures. On the other hand, this effect has been so far characterized on a statistical basis only. With the aim of building deterministically designed nano-objects, rigid DNA structures must be brought into play. These should exhibit the same segregation of complementary bases that takes place on the two faces of a curved molecule (Fig. 2). One possibility is offered by the structures composed of multiple blocked Holliday junctions developed by Seeman [5, 12, 13, 14]. If four blocked junctions are arranged in a DNA parallelogram of the kind sketched in Fig. 9, then the shape of the resulting object is determined precisely by the size of the arms and by the number of parallelograms that are assembled together in the nano-object (for instance via sticky ends).

Within each parallelogram arm, the base sequence can be made of phased A-tracts (see Fig. 8), so that all adenines will be on one side of the parallelogram plane, while all thymines will be on the other. Since the DNA parallelogram is really a 4 nm thick object (like 2 logs of wood lying at an angle on two others) then one can design the phase of the A-tracts so that one parallelogram can lie flat on either two A-rich DNA sides, or on two T-rich

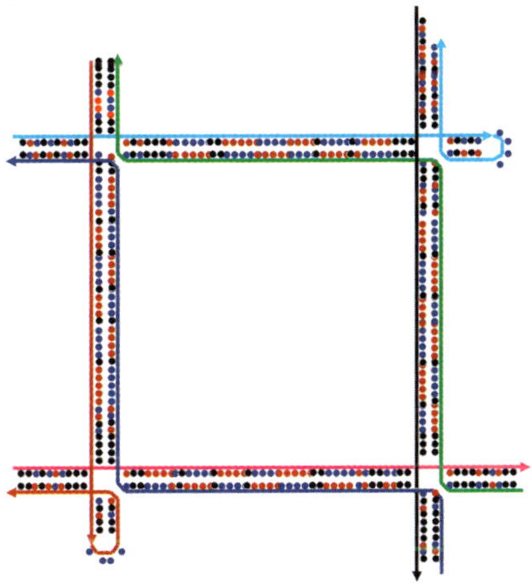

Fig. 9. Parallelogram with a segregation of A and T bases on its two faces.

DNA sides. Studies of the interaction of these structures with flat surfaces are under way in our laboratories.

References

1. S. Brown. Engineered iron oxide-adhesion mutants of the *Escherichia coli* phage lambda receptor. *Proc. Natl. Acad. Sci. USA*, 89:8651–8655, 1992.
2. S. Brown. Metal-recognition by repeating polypeptides. *Nat. Biotechnol.*, 15:269–272, 1997.
3. H.R. Drew and A.A. Travers. DNA bending and its relation to nucleosome positioning. *J. Mol. Biol.*, 186:773–790, 1985.
4. H.M. Keizer and R.P. Sijbesma. Hierarchical self-assembly of columnar aggregates. *Chem. Soc. Rev.*, 34:226–234, 2005.
5. S. Liao, C. Mao, J.J. Birktoft, S. Shuman, and N.C. Seeman. Resolution of undistorted symmetric immobile DNA junctions by vaccinia topoisomerase I. *Biochemistry*, 43:1520–1531, 2004.
6. D. Rhodes and A. Klug. Helical periodicity of DNA determined by enzyme digestion. *Nature*, 286:573–578, 1980.
7. B. Samorì and G. Zuccheri. DNA codes for nanoscience. *Angewandte Chemie Int. Ed.*, 44:1166–1181, 2005.
8. B. Sampaolese, A. Bergia, A. Scipioni, G. Zuccheri, M. Savino, B. Samorì, and P. De Santis. Recognition of the DNA sequence by an inorganic crystal surface. *Proc. Natl. Acad. Sci. (USA)*, 99:13566–13570, 2002.
9. M. Sarikaya, C. Tamerler, A.K. Jen, K. Schulten, and F. Baneyx. Molecular biomimetics: nanotechnology through biology. *Nature Mater.*, 2:577–585, 2003.

assortment of low-specificity, noncovalent contacts between discrete amino acids of the protein components, interacting with distinct nucleotide bases and the phosphate backbone of the ribosomal RNA. The structures of the ribosomal subunits have recently been resolved at atomic resolution, and the atomic structures of these subunits and their complexes with two substrate analogs have revealed that the ribosome is in fact a ribozyme [26]. Our knowledge of the atomic structure of this complex biological nanomachine not only satisfies our desire to fundamentally understand the molecular basis of life, but it also further motivates research to emulate natural systems in order to produce artificial devices of entirely novel functionality and performance.

Biological self-assembly has stimulated biomimetic "bottom-up" approaches to the development of artificial nanometer-scale elements, which are required commercially to produce microelectronic and micromechanical devices of increasingly small dimensions in the range of $\sim$ 5 to 100 nm. In this regard, Nadrian Seeman suggested early that one should fabricate synthetic nanometer-sized elements from biomolecular building blocks [60], and nowadays DNA is being extensively used as a construction material for the fabrication of nanoscale systems [61]. The simple A–T and G–C hydrogen-bonding interactions allow the convenient programming of DNA receptor moieties, which are highly specific for the complementary nucleic acid. Another very attractive feature of DNA is the great mechanical rigidity of short double helices and its comparably high physicochemical stability. Moreover, Nature provides a comprehensive toolbox of highly specific ligases, nucleases and other DNA-modifying enzymes, which can be used for processing and manipulating the DNA with atomic precision, and thus for molecular construction on the nanometer length scale.

The generation of semisynthetic DNA–protein conjugates allows one to combine the unique properties of DNA with an almost unlimited variety of protein components, which have been tailored by billions of years of evolution to perform highly specific functions, such as catalytic turnover, energy conversion, or translocation of other components. In particular, semisynthetic proteins conjugated with single-stranded DNA (ssDNA) oligomers, offer the possibility to functionalize DNA arrays with a protein content, taking advantage of the specific Watson–Crick base pairing [36, 38]. This chapter summarizes the current state of the art of the synthesis of such hybrid DNA–protein conjugates and their application in DNA nanotechnology. This approach can be used for the self-assembly of high-affinity reagents for immunoassays, nanoscale biosensor elements and the biomimetic "bottom-up" fabrication of nanostructured array devices.

2 Immobilization of Proteins by Means of DNA Hybridization

Protein microarrays are currently being explored as tools in proteome research and are highly attractive as miniaturized multianalyte immunosensors in clinical diagnostics [76, 5, 66]. The miniaturization of ligand-binding assays not only reduces costs by decreasing reagent consumption but also leads to enhanced sensitivity in comparison with macroscopic techniques. Recent applications of protein microarrays include high-throughput gene expression and antibody screening [30], analysis of antibody–antigen interactions [8], and identification of the protein targets of small molecules [31]. Whilst DNA microarrays can be easily fabricated by automated deposition techniques [44], the stepwise, robotic immobilization of multiple proteins on chemically activated surfaces is often obstructed by the instability of most biomolecules, which usually reveal a significant tendency towards denaturation. DNA-directed immobilization (DDI) provides a chemically mild process for the highly parallel binding of multiple delicate proteins to a solid support, using DNA microarrays as immobilization matrices (Fig. 1a) [51, 45, 28]. Because the lateral surface structuring is carried out at the level of stable nucleic acid oligomers, the DNA microarrays can be stored almost indefinitely, functionalized with proteins of interest via DDI immediately prior to use, and subsequent to hybridization, regenerated by alkaline denaturation of the double-helical DNA linkers. As an additional advantage of DDI in immunoassay applications, the intermolecular binding of the target antigens by antibodies can be carried out in a homogeneous solution, instead of in a less efficient heterogeneous solid-phase immunosorption process. Subsequently, the immunocomplexes formed can be site-specifically captured on the microarray by nucleic acid hybridization [54].

The reversibility and site selectivity of DDI enables a variety of applications, including the recovery and reconfiguration of biosensor surfaces, the fabrication of mixed arrays containing both nucleic acids and proteins for genome and proteome research, and the generation of miniaturized biochip elements [34]. Recent adaptations of DDI include the use of synthetic DNA analogs, namely pyranosyl–RNA oligomers, as recognition elements for the addressable immobilization of antibodies and peptides [70] and the DNA-directed immobilization of hapten groups for the immunosensing of pesticides [4]. Recently, the DDI method has been applied in functional genomics to identify the members of a small-molecule split-pool library which bind to protein targets [73, 72, 17, 11, 71]. In this approach, libraries of peptide ligands are encoded by peptide–nucleic acid (PNA) tags. After the library has been incubated with a mixture of potential binding proteins, the PNA tags are then used for the deconvolution of the library using DNA microarrays. These approaches have recently been reviewed elsewhere [27].

DDI has also been applied to inorganic gold nanoparticles thereby enabling the highly sensitive detection of nucleic acids in a DNA microarray format [65, 56]. This approach has opened the door to a large number of commercially relevant applications in the area of bioanalytics, which have recently been reviewed elsewhere [59]. With respect to further miniaturization, Demers et al. have adapted the DDI approach to surface nanostructuring by "dip-pen" nanolithography (DPN) [10]. DPN employs an SFM tip to "write" thiolated compounds with less than 30 nm linewidth resolution on gold substrates [15]. The direct writing of thiol- and acrylamide-modified oligonucleotides [9] allows the production of nanostructured DNA arrays, which can be used as an immobilization matrix for use in DDI.

Nanostructuring based on the self-assembly of DNA junctions and tile motifs might offer an even more powerful way to fabricate high-densitiy protein arrays (Fig. 1b). Initial attempts in this direction took advantage of 2D nanogrids, consisting of four-arm junctions, which contained regular arrangements of biotinyl groups. Incubation with the biotin-binding protein streptavidin (STV) led to the formation of periodic protein arrays [74, 55]. These examples demonstrate that a DNA nanoarchitecture can be used as a scaffold for arranging proteins, and it will be straightforward to extend this approach to DNA grids containing programmable sites for protein immobilization. In this respect, DDI-based techniques will be an option for fabricating nanoscale functional devices composed of DNA grids which are decorated with several different proteins.

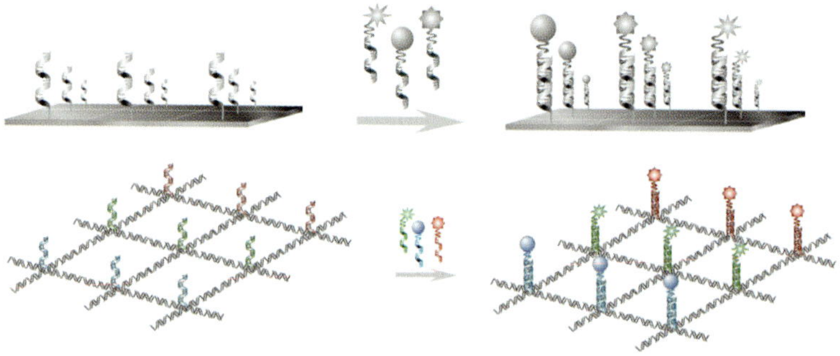

(a)

(b)

Fig. 1. (**a**) Schematic drawing of the DNA-directed immobilization (DDI) method. A microrarray of captured oligonucleotides is used as an immobilization matrix for site-specific binding of proteins tagged with complementary nucleic acid oligomers. Note that owing to the specificity of Watson–Crick base pairing, many different compounds can be site-specifically immobilized simultaneously in a single step. Reproduced from reference [27], with kind permission; (**b**) Illustration of a possible implementation of the DDI method to functionalize DNA nanoarrays with proteins.

3 Functional Multiprotein Assemblies

The concept of using DNA as a framework for the precise spatial arrangement of molecular components, initially suggested by Ned Seeman [60], was experimentally demonstrated by positioning several of the covalent DNA–STV conjugates **4** along a single-stranded nucleic acid carrier molecule containing a set of complementary sequences (Fig. 2) [51]. The covalent conjugates **4** can be used as versatile molecular adaptors because the covalent attachment of an oligonucleotide moiety to the STV provides a specific recognition domain for a complementary nucleic acid sequence in addition to the four native biotin-binding sites. For instance, supramolecular DNA–protein nanostructures (e.g., **5** in Fig. 2) have been assembled as model systems to investigate the basic principles of the DNA-directed assembly of proteins [51, 47]. These studies showed that, in particular, the formation of intramolecular secondary structures in the nucleic acid components often interferes with the effective intermolecular formation of the supramolecular DNA–protein assemblies [33].

The DNA-directed assembly of proteins can be applied to fabricate artificial, spatially well-defined multienzyme constructs, which are not accessible by conventional chemical crosslinking. In biological systems, multienzyme complexes have mechanistic advantages during the multistep catalytic transformation of a substrate because reactions limited by the rate of diffusional transport are accelerated by the immediate proximity of the catalytic centers. Furthermore, the "substrate channeling" of intermediate products avoids side reactions. As an example, STV conjugates **4** were used to assemble surface-bound bienzymic complexes, such as **8** in Fig. 2, from biotinylated luciferase and oxidoreductase [49]. The total enzymatic activities of the oxidoreductase/luciferase bienzymic complexes, which catalyze the consecutive reactions of flavinmononucleotide reduction and aldehyde oxidation, depended on the absolute and relative spatial orientation of the two enzymes. Not only are such studies useful for exploring proximity effects in biochemical pathways, but also the investigation of artificial multienzymes will allow the development of novel catalysts for enzyme process technology, capable of regenerating cofactors or of performing multistep chemical transformations of cheap precursors into drugs and fine chemicals.

With respect to synthetic nanosystems and materials science, the developments in the DNA-directed organization of semiconductor and metal nanoclusters [59, 63, 35, 22, 7, 52] have stimulated the application of DNA–STV conjugates **4** to organizing biotinylated gold nanoclusters to generate novel biometallic nanostructures, such as **6** in Fig. 2 [47]. Given that the conjugates **4** can be used like components of a molecular construction kit, functional proteins, such as immunoglobulins, can be conveniently incorporated into these biometallic nanostructures. A proof of feasibility was achieved by the assembly of the IgG-containing construct **7** (Fig. 2), capable of specifically binding to surface-immobilized complementary antigens [47].

These experiments clearly demonstrate the applicability of DNA-directed assembly to constructing functional hybrid devices from biological, bioorganic and inorganic building blocks. Similar approaches should allow the fabrication of even more complex structures, based on 2D or even 3D DNA scaffolds, such as those described by Ned Seeman and coworkers [38]. For instance, the "truncated octahedron", a DNA polyhedron containing 24 individual oligonucleotide arms at its vertices [75], could, in principle, be used as a framework for the directional positioning of 24 different proteins, inorganic nanoclusters and/or other functional molecular devices.

4 Synthesis of Semisynthetic DNA–Protein Conjugates

4.1 DNA–Streptavidin Conjugates

Motivated by the broad range of potential applications described above, our research interests have concerned the synthesis of semisynthetic DNA–protein conjugates [38]. Initial work focused on the synthesis of covalent DNA–STV conjugates **4** [51], which were accessible from thiolated oligonucleotides **2** and STV **1** using the heterobispecific crosslinker sulfosuccinimidyl 4-[p-maleimidophenyl]butyrate (sSMPB, **3**, in Fig. 2). The conjugates **4** can be used as molecular adaptors to readily modify biotinylated proteins with an ss-DNA oligomer. Hence, various applications of **4** were explored, including their use as building blocks for the fabrication of artificial protein nanostructures [51, 47, 49, 46, 39] and in protein microarray technologies [51, 45, 54, 68, 69]. The covalent conjugates **4** were also used to functionalize DNA-coated gold nanoparticles with DNA–antibody conjugates, following the same principle as implemented in the DDI technique [48]. The resulting hybrid components were used as reagents in sandwich immunoassays, and readout by gold-particle-promoted silver development allowed the spatially addressable detection of fmol amounts of chip-immobilized antigens [48, 18].

In addition, noncovalent DNA–STV conjugates obtained from biotinylated DNA and STV have been intensively investigated in our group. These DNA–STV conjugates are applicable as modular building blocks for the generation of novel immunological reagents for ultrasensitive trace analysis of proteins and other antigens by means of the immuno-PCR (IPCR) methodology [42, 53, 1]. IPCR combines the specificity of an antibody-based immunoassay with the exponential amplification power of the PCR, thereby leading to a 1000-fold-enhanced sensitivity, as compared with standard ELISA techniques. With its broad scope of application, ranging from the detection of proteins [3] to small molecules [2], IPCR is a prime example of how biomolecular nanostructures can add new performance to the well-established immunoassay methodology of biomedical diagnostics [43]. Some other applications of self-assembled noncovalent DNA–STV conjugates concern the emerging field of nanobiotechnology. For instance, these conjugates have been used as model systems for ion-

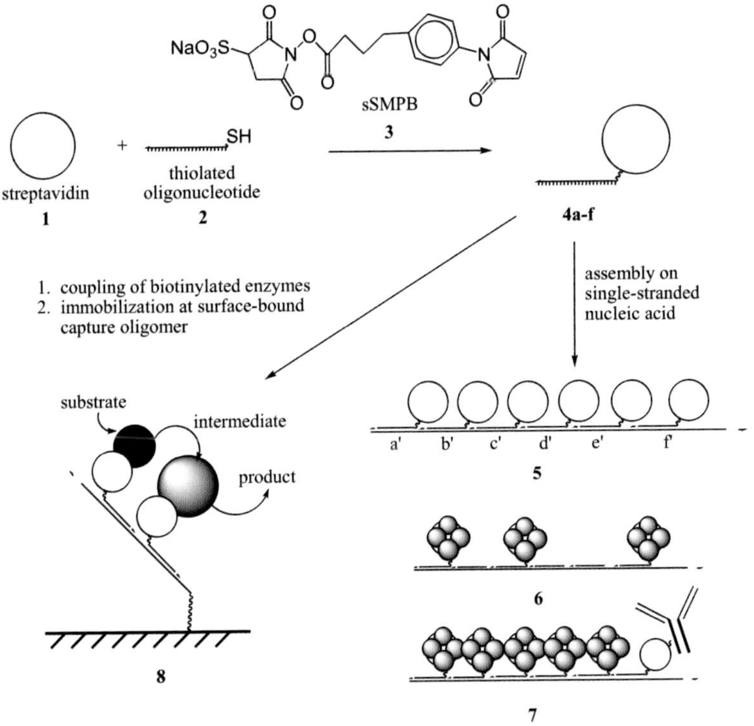

Fig. 2. Synthesis of covalent DNA–STV conjugates **6** by coupling of 5′-thiol-modified oligonucleotides **8** and streptavidin **1** using the heterobispecific crosslinker sSMPB **9**. A number of conjugates **6** with individual oligonucleotide sequences (e.g., **6**(**a**)–(**f**)) self-assemble in the presence of a single-stranded carrier nucleic acid, containing complementary sequence stretches, to form supramolecular conjugates, such as **10** [51]. Following this strategy, the biometallic aggregate **11** has been fabricated from **6** loaded with biotinylated 1.4 nm gold clusters [47]. The functional antibody-containing biometallic construct **12** was obtained from gold-labeled **6** and a conjugate obtained from **6** and a biotinylated immunoglobulin. The gold cluster conjugate and IgG-conjugate were previously coupled in separate reactions [47]. In a related approach, two STV conjugates **6** were coupled with biotinylated enzymes to allow the spatially controlled DNA-directed immobilization of the functional bienzymic complex **13** [49].

switchable nanoparticle networks [41], as nanometer-scale "soft material" calibration standards for scanning probe microscopy [14, 57] and as programmed building blocks for the rational construction of complex biomolecular architectures [40], which might be used as a template for the growth of nanometer-scale inorganic devices [37, 23].

4.2 Methodologies for Coupling Nucleic Acids to Proteins

Various methods for the chemical coupling of DNA to protein molecules have been described in the literature; these methods have been used to synthesize this type of bioconjugate, mainly for application as probes in nucleic acid analyses. These approaches have recently been reviewed elsewhere [36, 38]. Covalent coupling relies most often on the use of heterobispecific crosslinkers, such as sulfoSMPB **3** (Fig. 2), which is first reacted with primary amino groups of the protein to provide thiol-reactive maleimido groups. The maleimido-activated protein is usually purified and subsequently reacted with thiol-modified DNA oligonucleotides. This method of covalent conjugation requires extensive purification of the conjugates to remove excess protein and oligonucleotides after each chemical coupling step. For examples, see [51, 21].

In these conventional syntheses, the covalent coupling of the protein and the nucleic acid is statistical, and thus no control over the stoichiometry and the regioselectivity of the attachment site can be achieved. Directed conjugation, in particular, the regioselective and stoichiometrically controlled conjugation of nucleic acids to proteins, is much more difficult to achieve, and it is necessarily associated with the cloning and expression of recombinant proteins that contain reactive groups which can be selectively addressed by chemical means. To this end, protein engineering has been carried out to incorporate cysteine groups, which were then used for directed coupling with thiol-modified oligonucleotides by disulfide linkage or maleimide chemistry (for examples, see [6, 12, 20, 24]).

Some unconventional approaches to selective coupling of nucleic acids and proteins rely on chemically modified nucleic acid conjugates, containing, for instance, the synthetic DNA base analog 5′-Fluorocytosine [62] or the peptidyl acceptor antibiotic puromycin [32, 58]. In the first approach, several proteins can be organized along a one-dimensional dsDNA fragment, taking advantage of the specific binding of DNA (cytosine-5)-methyltransferases to distinct recognition sequences of dsDNA [62]. Covalent adducts are formed if the synthetic DNA base analog 5′-fluorocytosine (^{F}C) is present in the recognition site. The sequence-specific covalent attachment of two representative methyltransferases, M.*Hha*I and M.*Msp*I, at their target sites, G^{F}CGC and ^{F}CCGG, respectively, has been demonstrated. Since the methyltransferases can be modified with additional binding domains by recombinant techniques, this concept might be useful for generating a variety of DNA–protein conjugates, applicable as chromatin models or other functional biomolecular devices [62].

The puromycin approach is based on the in vitro translation of mRNA, covalently modified with a puromycin group at its 3′-end. The peptidyl acceptor antibiotic puromycin covalently couples the mRNA with the polypeptide chain grown on the ribosome particle, leading to a specific conjugation of the informative (mRNA) moiety with the functional (polypeptide) moiety. This approach has implications for the high-throughput screening of peptide and protein libraries, as well as for the generation of diverse protein microarrays [25].

We have recently described a novel approach for the well-controllable, site-selective linkage of nucleic acids with proteins. It is based on the expressed protein ligation (EPL) method, which utilizes the native chemical ligation of recombinant proteins containing a C-terminal thioester with cysteine conjugates of nucleic acids (Fig. 3). The EPL method had previously been used for the synthesis of a variety of proteins [19, 16]. To this end, the target protein, fused to a construct of an intein and a chitin binding domain (CBD), is expressed in *E. coli.* This latter domain allows affinity purification of the intein-fusion protein using a chitin matrix. Liberation from the column is achieved by cleaving the intein with mercaptoethanesulfonic acid (MESNA), thereby producing a C-terminal thioester of the target protein. The thioester-containing protein can be specifically ligated to cysteine–nucleic acid conjugates [28]. Owing to the convenient synthesis of PNA–Cys conjugates, as opposed to DNA–Cys conjugates, PNA was chosen as the initial model system [28], and the mild and highly efficient chemical ligation led to PNA–protein conjugates which were wel defined with respect to stoichiometric composition and regiospecific linkage. Since DNA–peptide conjugates are also available [67, 64, 29], it is now possible to produce, rapidly and automatically, both DNA–protein and PNA–protein conjugates from libraries of recombinant proteins. This method has several advantages over conventional chemical coupling techniques, and thus constitutes a major improvement for the purpose of further development of DNA-directed immobilization and the assembly of artificial multiprotein arrangements and nanostructured hybrid assemblies.

Very recently, we have explored another approach to the synthesis of well-defined, highly functional conjugates of DNA oligomers and enzymes, containing the hemin (iron protoporphyrin IX complex) prosthetic group [13]. In the first step, covalent DNA–hemin adducts ($\mathbf{hemD_1}$ and $\mathbf{hemD_2}$ in Fig. 4a) were synthesized in the solid phase. Subsequent to cleavage and purification, the **hemD** adducts were used to reconstitute apo-myoglobin (**apo-Mb**), that is, myoglobin (**Mb**) lacking its hemin prosthetic group. Reconstitution with $\mathbf{hemD_1}$ and $\mathbf{hemD_2}$ produced enzymatically active myoglobin containing one and two DNA oligomers, respectively ($\mathbf{MbD_1}$ and $\mathbf{MbD_2}$, in Fig. 4b), which we coupled to the enzyme in close proximity to its active site. To prove that the DNA oligomers could now be used as a molecular handle for selective DNA-directed assembly, the **MbD** conjugates were hybridized to complementary capture oligomers immobilized on a solid support. Subsequent analysis of the enzymatic peroxidase activity revealed that the **MbD** conjugates were far more active than native **Mb**.

Notably, the **Mb** served as a representative model system in this study for the seminal proof of the concept of generating well-defined DNA–heme enzyme conjugates by the reconstitution of apo-enzymes with covalent hemin–DNA adducts. This method should be well applicable to other heme enzymes, and thus we anticipate that this approach should open the door to a large variety

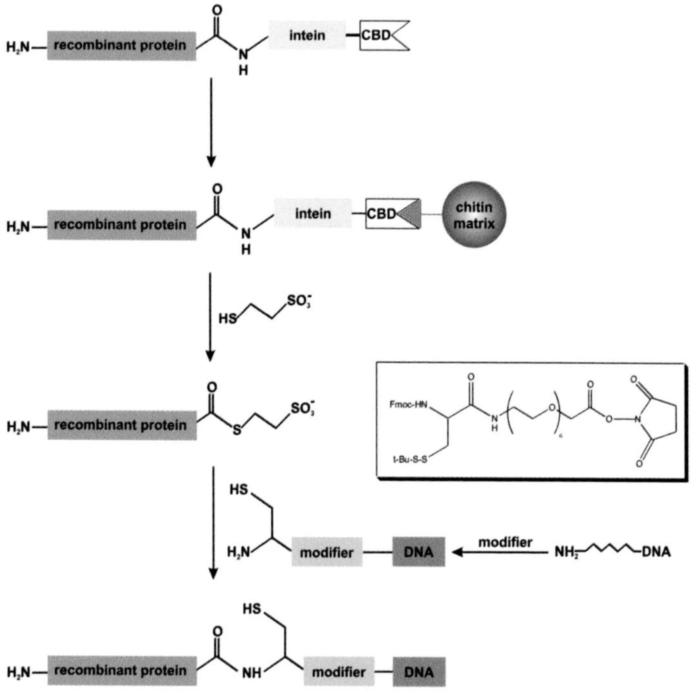

Fig. 3. Schematic drawing of the generation of DNA–protein conjugates by expressed protein ligation (EPL), using intein fusion proteins and cysteine–oligonucleotide conjugates, generated from amino-modified oligonucleotides by means of a chemical modifier (inset). Reproduced from [29], with kind permission.

of novel redox catalysts with programmable binding properties by means of their DNA moiety. This type of semisynthetic DNA–protein conjugate might therefore be useful for a broad range of applications, from biocatalysts and sensor and transducer elements to their use as building blocks for microstructured and nanostructured devices [13].

5 Conclusions

The development of semisynthetic DNA–protein conjugates has already led to a variety of commercial applications, which are currently focused on the bioanalytic sector. In addition, these chimeric components are promising for the fabrication of nanostructured molecular arrangements, and thus their development will contribute to the rapid establishment of the novel discipline often nowadays termed "nanobiotechnology" [33, 35, 50]. Future perspectives

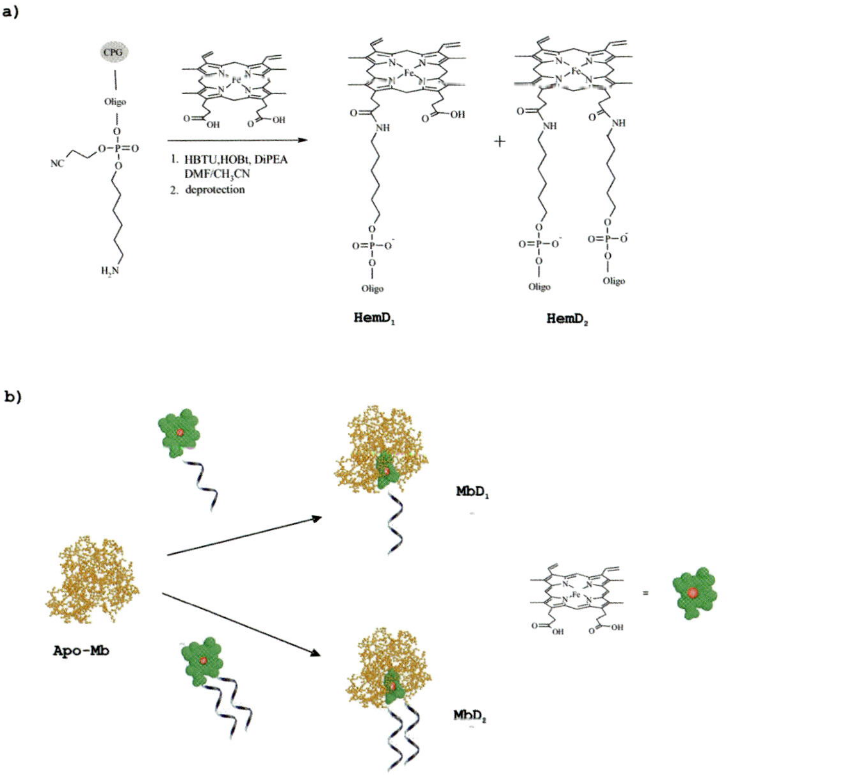

Fig. 4. Schematic representation of the synthesis and assembly of DNA–heme enzyme conjugates: (**a**) on-column synthesis of covalent hemin-DNA adducts **hemD$_1$** and **hemD$_2$**, containing one and two DNA oligomers, respectively; (**b**) reconstitution of apo-myoglobin (**apo-Mb**) with hemin–DNA adducts leads to the formation of DNA–Mb conjugates **MbD$_1$** and **MbD$_2$**. Reproduced from [13], with kind permission.

include the use of these systems as high-affinity diagnostic reagents, artificial multienzymes, light-harvesting devices or even autonomous drug delivery systems. In addition, the self-assembling nanoscale fabrication of technological elements, such as dense arrays of molecular switches, transistors and logical elements, as well as inorganic/bioorganic hybrid devices for biomedical diagnostics and interface structures between electronic and living systems might be foreseen. To realize these fascinating biotechnological perspectives, however, a variety of serious technical obstacles remain to be overcome. In particular, powerful analytical techniques and the refinement of bioconjugation and biomolecular evolution strategies are crucial for eventually attaining a comprehensive understanding of semisynthetic nucleic acid–protein conjugates and tailoring their structure and reactivity. Given that the initial steps summarized here have already clearly demonstrated the feasibility of these ideas, the future development of this new field of research promises plenty of excitement.

References

1. M. Adler, M. Langer, K. Witthohn, J. Eck, D. Blohm, and C.M. Niemeyer. Detection of rViscumin in plasma samples by immuno-PCR. *Biochem. Biophys. Res. Commun.*, 300:757–763, 2003.
2. M. Adler, R. Wacker, E. Booltink, B. Manz, and C.M. Niemeyer. Detection of femtogram amounts of biogenic amines using self-assembled DNA–protein nanostructures. *Nature Methods*, 2:147–149, 2005.
3. M. Adler, R. Wacker, and C.M. Niemeyer. A real-time immuno-PCR assay for the ultrasensitive quantification of proteins suitable for routine diagnostics. *Biochem. Biophys. Res. Commun.*, 308:240–250, 2003.
4. F.F. Bier, F. Kleinjung, E. Ehrentreich-Forster, and F.W. Scheller. Changing functionality of surfaces by directed self-assembly using oligonucleotides–the Oligo-Tag. *Biotechniques*, 27:752–756, 1999.
5. C.A. Borrebaeck. Antibodies in diagnostics – from immunoassays to protein chips. *Immunol. Today*, 21:379–382, 2000.
6. D.R. Corey and P.G. Schultz. Generation of a hybrid sequence-specific single-stranded deoxyribonuclease. *Science*, 238:1401–1403, 1987.
7. M.C. Daniel and D. Astruc. Gold nanoparticles: assembly, supramolecular chemistry, quantum-size-related properties, and applications toward biology, catalysis, and nanotechnology. *Chem. Rev.*, 104:293–346, 2004.
8. R.M. de Wildt, C.R. Mundy, B.D. Gorick, and I.M. Tomlinson. Antibody arrays for high-throughput screening of antibody–antigen interactions. *Nat. Biotechnol.*, 18:989–994, 2000.
9. L.M. Demers, D.S. Ginger, S.J. Park, Z. Li, S.W. Chung, and C.A. Mirkin. Direct patterning of modified oligonucleotides on metals and insulators by dip-pen nanolithography. *Science*, 296:1836–1838, 2002.
10. L.M. Demers, S.-J. Park, T.A. Taton, Z. Li, and C.A. Mirkin. Orthogonal assembly of nanoparticle building blocks on dip-pen nanolithographically generated templates of DNA. *Angew. Chem. Int. Ed.*, 40:3071–3073, 2001.
11. J.J. Diaz-Mochon, L. Bialy, L. Keinicke, and M. Bradley. Combinatorial libraries – from solution to 2d microarrays. *Chem. Commun.*, pages 1384–1386, 2005.
12. R.B. Fong, Z. Ding, C.J. Long, A.S. Hoffman, and P.S. Stayton. Thermoprecipitation of streptavidin via oligonucleotide-mediated self-assembly with poly(*N*-isopropylacrylamide). *Bioconjug. Chem.*, 10:720–725, 1999.
13. L. Fruk and C.M. Niemeyer. Covalent hemin–DNA adducts for generating a novel class of artifical heme enzymes. *Angew. Chem. Int. Ed.*, 44:2603–2606, 2005.
14. S. Gao, L.F. Chi, S. Lenhert, B. Anczykowsky, C.M. Niemeyer, M. Adler, and H. Fuchs. High quality mapping of DNA–protein complex by dynamic scanning force microscopy. *ChemPhysChem.*, 2:384–388, 2001.
15. D.S. Ginger, H. Zhang, and C.A. Mirkin. The evolution of dip-pen nanolithography. *Angew. Chem.-Int. Ed. Engl.*, 43:30–45, 2004.

16. R.S. Goody, K. Alexandrov, and M. Engelhard. Combining chemical and biological techniques to produce modified proteins. *ChemBioChem.*, 3:399–403, 2002.
17. J. Harris, D. E. Mason, J. Li, K. W. Burdick, B.J. Backes, T. Chen, A. Shipway, G. Van Heeke, L. Gough, A. Ghaemmaghami, F. Shakib, F. Debaene, and N. Winssinger. Activity profile of dust mite allergen extract using substrate libraries and functional proteomic microarrays. *Chem. Biol.*, 11:1361–1372, 2004.
18. P. Hazarika, B. Ceyhan, and C.M. Niemeyer. Sensitive detection of proteins using difunctional DNA–gold nanoparticles. *Small*, 1, 2005. In press.
19. R.M. Hofmann and T.W. Muir. Recent advances in the application of expressed protein ligation to protein engineering. *Curr. Opin. Biotechnol.*, 13:297–303, 2002.
20. S. Howorka, S. Cheley, and H. Bayley. Sequence-specific detection of individual DNA strands using engineered nanopores. *Nat. Biotechnol.*, 19:636–639, 2001.
21. R.D. Joerger, T.M. Truby, E.R. Hendrickson, R.M. Young, and R.C. Ebersole. Analyte detection with DNA-labeled antibodies and polymerase chain-reaction. *Clin. Chem.*, 41:1371–1377, 1995.
22. E. Katz and I. Willner. Integrated nanoparticle-biomolecule hybrid systems: synthesis, properties, and applications. *Angew. Chem. Int. Ed. Engl.*, 43:6042–6108, 2004.
23. K. Keren, M. Krueger, R. Gilad, G. Ben-Yoseph, U. Sivan, and E. Braun. Sequence-specific molecular lithography on single DNA molecules. *Science*, 297:72–75, 2002.
24. F. Kukolka and C.M. Niemeyer. Synthesis of fluorescent oligonucleotide–EYFP conjugate: towards supramolecular construction of semisynthetic biomolecular antennae. *Org. Biomol. Chem.*, 2:2203–2206, 2004.
25. M. Kurz, K. Gu, A. Al-Gawari, and P.A. Lohse. cDNA–protein fusions: covalent protein–gene conjugates for the in vitro selection of peptides and proteins. *ChemBioChem.*, 2:666–672, 2001.
26. D.M.J. Lilley. The ribosome functions as a ribozyme. *ChemBioChem.*, 2:31–35, 2001.
27. M. Lovrinovic and C.M. Niemeyer. DNA microarrays as decoding tools in combinatorial chemistry and chemical biology. *Angew. Chem. Int. Ed.*, 44:3179–3183, 2005.
28. M. Lovrinovic, R. Seidel, R. Wacker, H. Schroeder, O. Seitz, M. Engelhard, R. Goody, and C.M. Niemeyer. Synthesis of protein–nucleic acid conjugates by expressed protein ligation. *Chem. Commun.*, pages 822–823, 2003.
29. M. Lovrinovic, M. Spengler, C. Deutsch, and C.M. Niemeyer. Synthesis of covalent DNA–protein conjugates by expressed protein ligation. *Mol. BioSyst.*, 1:64–69, 2005.
30. A. Lueking, M. Horn, H. Eickhoff, K. Bussow, H. Lehrach, and G. Walter. Protein microarrays for gene expression and antibody screening. *Anal. Biochem.*, 270:103–111, 1999.
31. G. MacBeath and S.L. Schreiber. Printing proteins as microarrays for high-throughput function determination. *Science*, 289:1760–1763, 2000.
32. N. Nemoto, E. Miyamoto-Sato, Y. Husimi, and H. Yanagawa. In vitro virus: bonding of mRNA bearing puromycin at the 3′-terminal end to the C-terminal end of its encoded protein on the ribosome in vitro. *FEBS Lett.*, 414:405–408, 1997.

33. C.M. Niemeyer. Self-assembled nanostructures based on DNA: towards the development of nanobiotechnology. *Curr. Opin. Chem. Biol.*, 4:609–618, 2000.
34. C.M. Niemeyer. Bioorganic applications of semisynthetic DNA–protein conjugates. *Chem. Eur. J.*, 7:3188–3195, 2001.
35. C.M. Niemeyer. Nanoparticles, proteins, and nucleic acids: biotechnology meets materials science. *Angew. Chem. Int. Ed.*, 40:4128–4158, 2001.
36. C.M. Niemeyer. The developments of semisynthetic DNA–protein conjugates. *Trends Biotechnol.*, 20:395–401, 2002.
37. C.M. Niemeyer. Nanotechnology: tools for the biomolecular engineer. *Science*, 297:62–63, 2002.
38. C.M. Niemeyer. DNA-microarrays, In C.M. Niemeyer and C.A. Mirkin, editors, *NanoBiotechnology: Concepts, Methods and Applications*, pages 227–243. Wiley-VCH, Weinheim, 2004.
39. C.M. Niemeyer, M. Adler, S. Gao, and L.F. Chi. Nanostructured DNA–protein aggregates consisting of covalent oligonucleotide–streptavidin conjugates. *Bioconjug. Chem.*, 12:364–371, 2001.
40. C.M. Niemeyer, M. Adler, S. Gao, and L.F. Chi. Supramolecular DNA–streptavidin nanocircles with a covalently attached oligonucleotide moiety. *J. Biomol. Struct. Dyn.*, 20:223–230, 2002.
41. C.M. Niemeyer, M. Adler, S. Lenhert, S. Gao, H. Fuchs, and L.F. Chi. Nucleic-acid supercoiling as a means for construction of switchable DNA–nanoparticle networks. *ChemBioChem.*, 2:260–265, 2001.
42. C.M. Niemeyer, M. Adler, B. Pignataro, S. Lenhert, S. Gao, L.F. Chi, H. Fuchs, and D. Blohm. Self-assembly of DNA–streptavidin nanostructures and their use as reagents in immuno-PCR. *Nucleic Acids Res.*, 27:4553–4561, 1999.
43. C.M. Niemeyer, M. Adler, and R. Wacker. Immuno-PCR: high sensitivity detection of proteins by means of nucleic acid amplification. *Trends Biotechnol.*, 23:208–216, 2005.
44. C.M. Niemeyer and D. Blohm. Dna-microarrays. *Angew. Chem. Int. Ed.*, 38:2865–2869, 1999.
45. C.M. Niemeyer, L. Boldt, B. Ceyhan, and D. Blohm. DNA-directed immobilization: efficient, feversible and site-selective surface binding of proteins by means of covalent DNA–streptavidin conjugates. *Anal. Biochem.*, 268:54–63, 1999.
46. C.M. Niemeyer, L. Boldt, B. Ceyhan, and D. Blohm. Evaluation of single-stranded nucleic acids as carriers in the DNA-directed assembly of macromolecules. *J. Biomol. Struct. Dyn.*, 17:527–538, 1999.
47. C.M. Niemeyer, W. Bürger, and J. Peplies. Covalent DNA–streptavidin conjugates as building blocks for the fabrication of novel biometallic nanostructures. *Angew. Chem. Int. Ed. Engl.*, 37:2265–2268, 1998.
48. C.M. Niemeyer and B. Ceyhan. DNA-directed functionalization of colloidal gold with proteins. *Angew. Chem. Int. Ed.*, 40:3685–3688, 2001.
49. C.M. Niemeyer, J. Koehler, and C. Wuerdemann. DNA-directed assembly of bienzymic complexes from in-vivo biotinylated NAD(P)H:FMN oxidoreductase and luciferase. *ChemBioChem.*, 3:242–245, 2002.
50. C.M. Niemeyer and C.A. Mirkin, editors. *NanoBiotechnology: Concepts, Applications and Perspectives.* Wiley-VCH, Weinheim, 2004.
51. C.M. Niemeyer, T. Sano, C.L. Smith, and C.R. Cantor. Oligonucleotide-directed self-assembly of proteins: semisynthetic DNA–streptavidin hybrid molecules as connectors for the generation of macroscopic arrays and the construction of supramolecular bioconjugates. *Nucleic Acids Res.*, 22:5530–5539, 1994.

52. C.M. Niemeyer and U. Simon. DNA-based assembly of metal nanoparticles. *Eur. J. Inorg. Chem.*, 2005. In press.
53. C.M. Niemeyer, R. Wacker, and M. Adler. Hapten-functionalized DNA–streptavidin nanocircles as supramolecular reagents in a novel competitive immuno-PCR assay. *Angew. Chem. Int. Ed.*, 40:3169–3172, 2001.
54. C.M. Niemeyer, R. Wacker, and M. Adler. Combination of DNA-directed immobilization and immuno-PCR: very sensitive antigen detection by means of self-assembled DNA–protein conjugates. *Nucleic Acids Res.*, 31:e90, 2003.
55. S.H. Park, P. Yin, Y. Liu, J.H. Reif, T.H. LaBean, and H. Yan. Programmable DNA self-assemblies for nanoscale organization of ligands and proteins. *Nano Letters*, 5:729–733, 2005.
56. S.J. Park, T.A. Taton, and C.A. Mirkin. Array-based electrical detection of DNA with nanoparticle probes. *Science*, 295:1503–1506, 2002.
57. B. Pignataro, L.F. Chi, S. Gao, B. Anczykowsky, C.M. Niemeyer, M. Adler, and H. Fuchs. Dynamic scanning force microscopy study of self-assembled DNA–protein oligomers. *Appl. Phys. A*, 74:447–452, 2002.
58. R.W. Roberts and J.W. Szostak. RNA–peptide fusions for the in vitro selection of peptides and proteins. *Proc. Natl. Acad. Sci. USA*, 94:12297–12302, 1997.
59. N.L. Rosi and C.A. Mirkin. Nanostructures in biodiagnostics. *Chem. Rev.*, 105:1547–1562, 2005.
60. N.C. Seeman. Nucleic acid junctions and lattices. *J. Theor. Biol.*, 99:237–247, 1982.
61. N.C. Seeman. DNA in a material world. *Nature*, 421:427–431, 2003.
62. S.S. Smith, L.M. Niu, D.J. Baker, J.A. Wendel, S.E. Kane, and D.S. Joy. Nucleoprotein-based nanoscale assembly. *Proc. Natl. Acad. Sci. USA*, 94:2162–2167, 1997.
63. J.J. Storhoff and C.A. Mirkin. Programmed materials synthesis with DNA. *Chem. Rev.*, 99:1849–1862, 1999.
64. S. Takeda, S. Tsukiji, and T. Nagamune. Site-specific conjugation of oligonucleotides to the C-terminus of recombinant protein by expressed protein ligation. *Bioorg. Med. Chem. Lett.*, 14:2407–2410, 2004.
65. T.A. Taton, C.A. Mirkin, and R.L. Letsinger. Scanometric DNA array detection with nanoparticle probes. *Science*, 289:1757–1760, 2000.
66. M.F. Templin, D. Stoll, M. Schrenk, P.C. Traub, C.F. Vohringer, and T.O. Joos. Protein microarray technology. *Trends Biotechnol.*, 20:160–166, 2002.
67. C.H. Tung and S. Stein. Preparation and applications of peptide–oligonucleotide conjugates. *Bioconjug. Chem.*, 11:605–618, 2000.
68. R. Wacker and C.M. Niemeyer. DDI-mFIA – a readily configurable microarray-fluorescence immunoassay based on DNA-directed immobilization of proteins. *ChemBioChem.*, 5:453–459, 2004.
69. R. Wacker, H. Schroeder, and C.M. Niemeyer. Performance of antibody microarrays fabricated by either DNA-directed immobilization, direct spotting, or streptavidin–biotin attachment: a comparative study. *Anal. Biochem.*, 330:281–287, 2004.
70. N. Windhab, C. Miculka, and H.-U. Hoppe. Method for the detection and isolation of biomolecules via molecular recognition using immobilized pyranosyl nucleotide supramolecular structures (wo99/15893). *Chem. Ab.*, 130:249124x, 1999.

71. N. Winssinger, R. Damoiseaux, D.C. Tully, B.H. Geierstanger, K. Burdick, and J.L. Harris. PNA-encoded protease substrate microarrays. *Chem. Biol.*, 11:1351–1360, 2004.
72. N. Winssinger, S. Ficarro, P.G. Schultz, and J.L. Harris. Profiling protein function with small molecule microarrays. *Proc. Natl. Acad. Sci. USA*, 99:11139–11144, 2002.
73. N. Winssinger, J.L. Harris, B.J. Backes, and P.G. Schultz. From split-pool libraries to spatially addressable microarrays and its application to functional proteomic profiling. *Angew. Chem. Int. Ed.*, 40:3152–3155, 2001.
74. H. Yan, S.H. Park, G. Finkelstein, J.H. Reif, and T.H. LaBean. DNA-templated self-assembly of protein arrays and highly conductive nanowires. *Science*, 301:1882–1884, 2003.
75. Y. Zhang and N.C. Seeman. Construction of a DNA-truncated octahedron. *J. Am. Chem. Soc.*, 116:1661–1669, 1994.
76. H. Zhu and M. Snyder. Protein arrays and microarrays. *Curr. Opin. Chem. Biol.*, 5:40–45, 2001.

Bacterial Surface Layer Proteins: a Simple but Versatile Biological Self-assembly System in Nature

Dietmar Pum, Margit Sára, Bernhard Schuster, and Uwe B. Sleytr

Center for NanoBiotechnology, University of Natural Resources and Applied Life Sciences Vienna, Gregor Mendelstr. 33, 1180 Vienna, Austria
dietmar.pum@boku.ac.at

1 Introduction

"Self-assembly" is the key word for the bottom-up fabrication of molecular biomaterials. Nevertheless, it requires deep understanding of the structure and function of the individual molecular building blocks in order to build uniform, ultrasmall functional units. "Key–lock" mechanisms based on matching structural details and surface chemical groups bind the building blocks together and determine their correct assembly. Most often, weak noncovalent forces are involved. Among the well-established self-assembly systems, such as DNA lattices and junctions [48, 47] and self-assembled monolayers (SAMs) [26, 60], two-dimensional bacterial surface layer proteins (S-layer proteins) represent highly versatile assembly systems with unique features. The latter constitute the structural basis for a complete supramolecular construction kit involving all major species of biological macromolecules (proteins, lipids, glycans, and nucleic acids) (for a review, see [56, 57, 52]).

2 Occurrence of S-Layers

Although not a universal feature, many prokaryotic organisms possess crystalline arrays of proteinaceous subunits as the outermost component of their cell envelopes; these arrays are commonly referred to as surface layers or S-layers [50]; for reviews, see [51, 54, 55] (Fig. 1).

S-layers are found in members of nearly every taxonomic group in the Gram-positive and Gram-negative bacteria and archaea and in some green algae [31]. Crystalline layers similar to S-layers have also been detected in bacterial sheaths [3] and spore coats [10]. On the basis of structural and biochemical studies, S-layer-carrying cell envelope profiles can be classified into three main categories [51], namely: (i) archaeal cell envelopes, which are

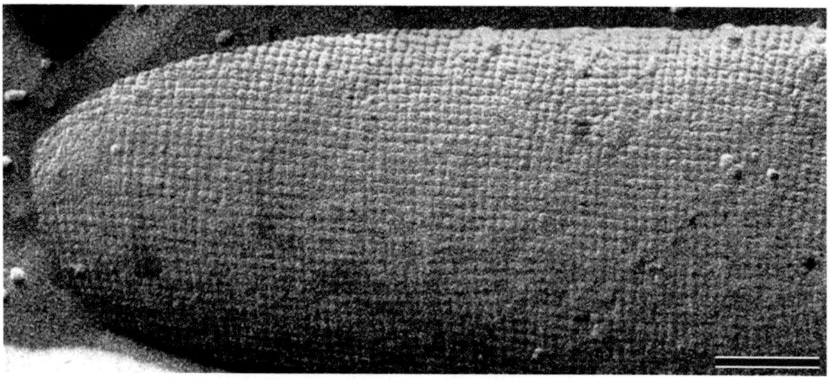

Fig. 1. Transmission electron micrograph of a freeze-etching preparation of *Desulfotomaculum nigrificans*. The S-layer exhibits a square lattice symmetry with a lattice constant of 10.2 nm. Bar, 200 nm.

composed of a cytoplasmic membrane and an associated S-layer; (ii) Gram-positive cell envelopes where the S-layer is attached either to a thick, rigid peptidoglycan-containing layer or to another polymer (e.g., pseudomurein external to the cytoplasmic membrane); and (iii) Gram-negative cell envelopes, where the S-layer is attached to the outer membrane, which is bound to a thin peptidoglycan sacculus. S-layers can be considered as the simplest biological membranes developed during evolution, as they are composed of a single molecular protein or glycoprotein species (M_r 40 to 200 kDa) and are endowed with the ability to self-assemble into two-dimensional arrays on growing and dividing cells.

3 Ultrastructure of S-Layers

S-layers exhibit either oblique ($p1$ or $p2$), square ($p4$), or hexagonal ($p3$ or $p6$) lattice symmetry with unit cell dimensions in the range of 3.5 to 35 nm (for reviews, see [51, 52, 35]) (Fig. 2).

Depending on the lattice type, one morphological unit consists of one, two, three, four, or six identical subunits. Although 17 two-dimensional plane groups are possible in theory, only $p1$, $p2$, $p3$, $p4$, and $p6$ are allowed for plane lattices of biological molecules [1]. This is because real molecules can never be related to one another by mirror planes, glide planes, or inversion centers, because they are optically asymmetric, or chiral. Thus, all those plane groups which contain one of these symmetry operators are not allowed.

While S-layer proteins reassemble into perfectly ordered crystalline arrays on flat surfaces, lattice faults (wedge disclinations and edge dislocations) are a geometrical necessity on rounded surfaces, such as those of bacterial cells, in order to maintain the integrity of the closed protein container [8, 9, 23]. Most

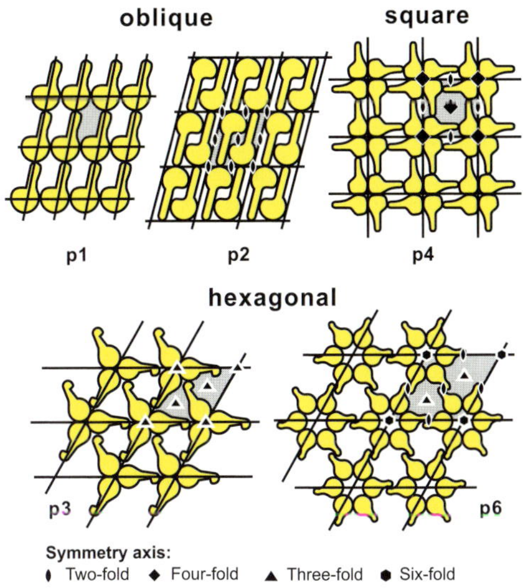

Fig. 2. Schematic drawings of possible S-layer lattice types. Owing to the chirality of proteins, space group symmetries with mirror-reflection lines or glide-reflection lines are not possible in S-layer lattices.

often, lattice faults are sites for the incorporation of new morphological units and initiation points in the cell division process [53, 58]. For example, the S-layer of the lobed archaeon *Methanocorpusculum sinense* shows hexagonal lattice symmetry with numerous lattice faults (pentagons and heptagons) [27]. Complementary pairs of pentagons and heptagons in the hexagonal S-layer are termination points of edge dislocations and function most probably as initiation points in the cell division process [20].

Bacterial S-layer lattices are generally 5 to 20 nm thick, whereas the S-layers of archaea have thicknesses up to 70 nm (for reviews, see [2, 11]). S-layers generally represent highly porous protein meshworks (30%–70% porosity), with pores of uniform size in the 2–8 nm range and of uniform morphology.

High-resolution electron and scanning force microscope studies, partially in combination with digital image processing, have revealed a smooth topography for the outer face of most S-layers and a more corrugated topography for the inner face (for reviews, see [2, 11]). Concerning the physicochemical properties of S-layers in Bacillacaea, it has been demonstrated that the outer face is usually charge-neutral, while the inner face is often net negatively charged [56, 57]. The surface charge depends on a balance of exposed carboxylic acid and amine groups or an excess of one or the other. The functional groups in the S-layer lattice are aligned in well-defined positions and orientations, which is a key condition for binding molecules and nanoparticles into ordered arrays on these protein lattices [34].

4 Secondary Cell Wall Polymers (SCWPs)

Analysis of S-layer proteins from various Bacillaceae has revealed the existence of specific lectin-type binding domains in the N-terminal parts of S-layer proteins for secondary cell wall polymers [17, 32], which are, in turn, covalently linked to the peptidoglycan matrix of the cell wall (for reviews, see [33, 35, 34, 36]). Sequence identities are extremely rare among S-layer proteins and are limited to the N-terminal region, which is responsible for anchoring the subunits to the cell surface by binding to an SCWP. In this context, three repeats of S-layer homology (SLH) motifs, consisting of 50 to 60 amino acids each, have been identified in the N-terminal parts of many S-layer proteins. Nevertheless, recent studies have shown that an additional 58-amino-acid-long SLH-like motif in the S-layer protein SbpA of *Bacillus sphaericus* CCM2177 is required for reconstituting the functional SCWP-binding domain [12]. For technological applications it is important to note that this highly specific interaction between S-layer proteins and their associated SCWPs is retained even after extraction of these heteropolysaccharides from the peptidoglycan-containing sacculi, chemical modification of the reducing end of the polymer chains, and attachment to a solid support. On SCWP-coated supports, the corresponding S-layer protein reassembles with its inner face, comprising the SLH domain, towards the support and thus exposing the outer face towards the environment. This is especially important when functional C-terminal S-layer fusion proteins are used for reassembly on solid supports [24, 25, 63]. Furthermore, the conformation of an S-layer lattice is more resistant to ethanol and acidic (pH $\sim$ 3) exposure on SCWP-coated substances compared with substrates lacking this natural surface coating [62].

5 Genetic Engineering of S-Layer Proteins

Structure–function relationships of distinct segments of various S-layer proteins have been investigated in order to gain knowledge about those amino acid positions where foreign peptide sequences can be fused without disturbing the self-assembly properties. For example, in the case of the S-layer protein SbsB from *Geobacillus stearothermophilus* PV72/p2, minimum-sized core streptavidin (118 amino acids) could be fused to the N- or C-terminal end [22]. The fusion proteins and core streptavidin were produced independently in *Escherichia coli*, isolated, purified, and refolded into heterotetramers consisting of one chain of N- or C-terminal SbsB–streptavidin fusion protein and three chains of streptavidin. The biotin-binding capacity of the heterotetramers was $\sim$ 80% in comparison with homotetramers. These findings indicated that at least three of the four streptavidin residues were accessible and active for binding biotinylated molecules. Such chimeric S-layer fusion proteins can be used as versatile templates for arranging any biotinylated compounds on the outermost surface of the protein lattice [22, 34] (Fig. 3).

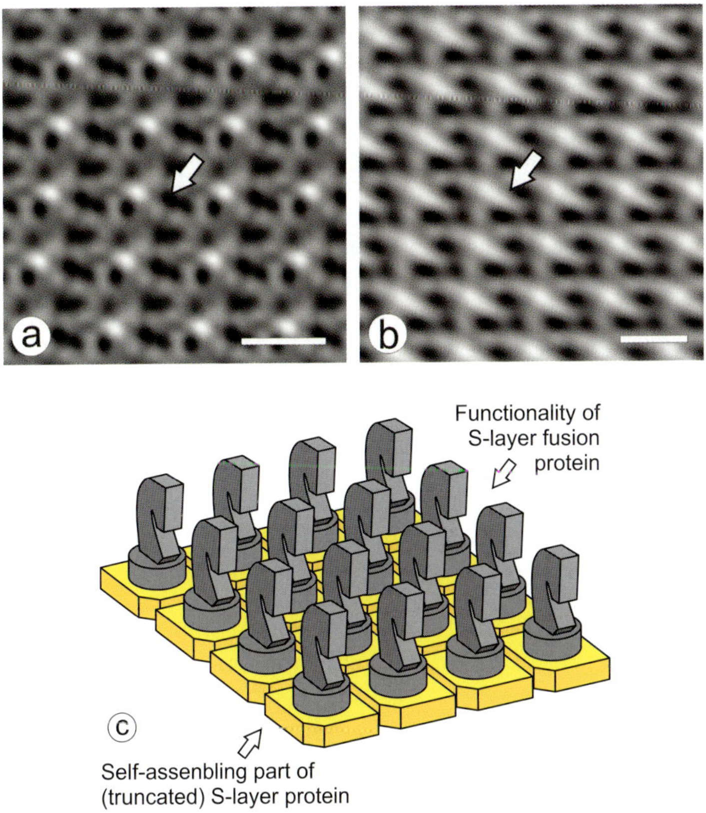

Fig. 3. Digital image reconstructions of transmission electron micrographs of negatively stained preparations of (**a**) the native S-layer protein SbsB from *Geobacillus stearothermophilus* PV72/p2 and (**b**) the streptavidin S layer fusion protein. In the lattice of the fusion protein (**b**), the streptavidin heterotetramers show up as additional mass (arrows). Bars, 10 nm. Schematic illustration of the self-assembling parts of S-layer fusion proteins and their well-oriented functional domains. (**c**) Such arrays provide, theoretically, the highest possible order (spatial control, orientation, and position) of functional domains at the nanometer scale. The knights (gray) resemble the functional domains (antigens, enzymes, antibodies, ligands, etc.) and the cut squares (yellow) represent the S-layer.

Using a similar approach, the structure–function relationship of the S-layer protein SbpA of *Bacillus sphaericus* CCM2177 has been investigated. As described above, the final aim was to construct fusion proteins with an ability to reassemble into two-dimensional arrays while presenting the introduced functional sequence or domain on the outermost surface of the protein lattice for the purpose of binding molecules, such as antibodies, antigens, ligands, or nanoparticles [14, 13, 24, 25, 63]. Up to now, the C-terminally truncated form $rSbpA_{31-1068}$, which is 1038 amino acids (aa) long, has been used as the basic molecular building block. Its C-terminal end was fused with the desired functional sequence, such as core streptavidin (118 aa), the affinity tag for streptavidin (9 aa) [14], the major birch pollen allergen Bet v1 (116 aa) [14], two copies of the IgG-binding Z domain (58 aa each) [63], green fluorescent protein (238 aa) [13], or heavy-chain camel antibody domains (117 aa) recognizing either lysoyzme [24] or prostate-specific antigen [25].

While various truncated forms of rSbpA were being screened for their ability to reassemble, it was found that a further deletion of 113 C-terminal amino acids from $rSbpA_{31-1031}$, leading to $rSpbA_{31-918}$, had a strong and unexpected impact on lattice formation [12]. In contrast to the original S-layer lattice formed by the mature and truncated forms of $rSbpA_{31-1031}$, which exhibits square symmetry with a lattice constant of 13.1 nm, a lattice with oblique lattice symmetry, base vectors of $a = 10.4$ nm and $b = 7.9$ nm, and a base angle of 81° was formed. It is interesting to note that the ultrastructure of this newly formed S-layer lattice is identical to that of SbsB [22], the S-layer protein of *G. stearothermophilus* PV72/p2. The mature SbsB comprises amino acids 32 to 920 and is only one amino acid shorter than $rSbpA_{31-918}$. Both S-layer proteins carry three SLH motifs in the N-terminal part, which showed high identity [12]. However, no sequence identities were found for the middle and C-terminal parts. Further C-terminal truncation of $rSbpA_{31-918}$ led to a complete loss of the self-assembly properties of the S-layer protein.

6 Reassembly of Native and Recombinant S-Layer Proteins

The attractiveness of isolated S-layer proteins for a broad spectrum of applications lies in their capability to form two-dimensional arrays without the bacterial cell envelope from which they have been removed (Fig. 4).

Most techniques for the isolation and purification of S-layer proteins involve mechanical disruption of the bacterial cells and subsequent differential centrifugation in order to isolate the cell wall fragments [50, 53]. Complete solubilization of S-layers into their constituent subunits and release of these subunits from supporting cell envelope layers can be achieved with high concentrations of hydrogen-bond-breaking agents (e.g., guanidine hydrochloride) or by lowering or raising the pH. Recrystallization of isolated S-layer proteins occurs upon dialysis of the disintegrating agent [58, 38]. The formation of self-assembled arrays is determined only by the amino acid sequence of the polypeptide chains and consequently the tertiary structure of the S-layer protein species [49]. Since S-layer proteins have a high proportion of nonpolar amino acids, it is most likely that hydrophobic interactions are involved in the assembly process. Some S-layers are stabilized by divalent cations interacting with acidic amino acids. Studies of the distribution of functional groups on the surface have shown that free carboxylic acid groups and amino groups

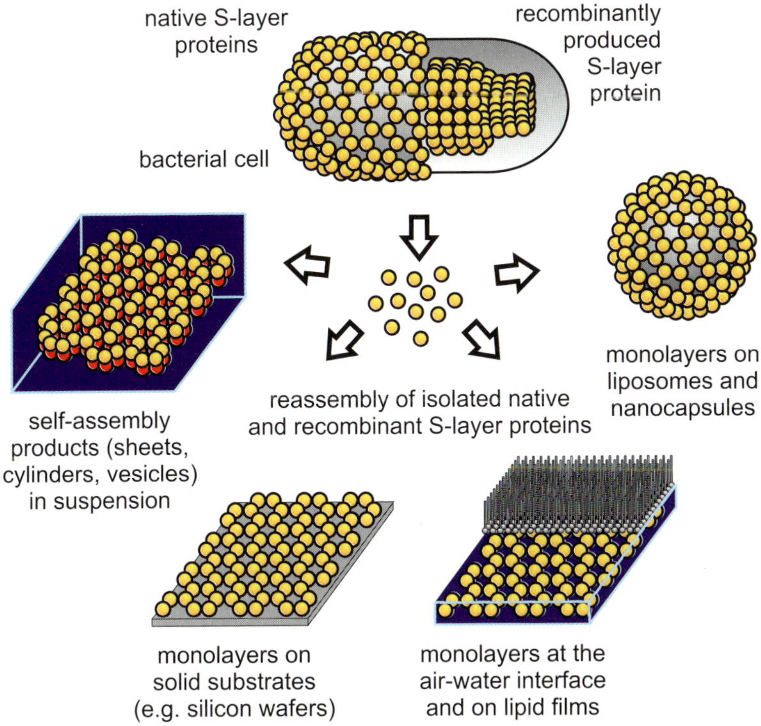

Fig. 4. Schematic drawing of the isolation of native and recombinant S-layer proteins from bacterial cells and their reassembly into crystalline arrays in suspension, on a solid support, at an air–water interface and on a planar lipid film, and on liposomes or nanocapsules. An example of S-layer proteins reassembling with a hexagonal (*p*6) lattice symmetry is shown here.

are arranged in close proximity and thus contribute to the cohesion of the proteins via electrostatic interactions [57].

6.1 Reassembly in Suspension

Depending on the specific bonding properties and the tertiary structure of the S-layer protein, either flat sheets, open-ended cylinders, or vesicles are formed [50, 53]. Both temperature and protein concentration determine the extent and rate of association. The assembly kinetics a multiphase, with a rapid initial phase and a subsequent slow rearrangement step, leading to an extended lattice [15]. Depending on the S-layer proteins used and on the environmental conditions (e.g., the ionic content and strength of the buffer solution) the self-assembly product may consist either of monolayers or of double layers. In a systematic study using the S-layer protein SgsE from *G. stearothermophilus* NRS 2004/3a [37], it was shown that two types of mono-layered and five types of double-layered assembly products with a back-to-back orientation of the constituent monolayers were formed [21]. The double layers differed in the angular displacement of their constituent S-layer sheets. As the monolayers had an inherent inclination to curve along two axes, cylindrical or flat double-layer assembly products were formed, depending on the degree of neutralization of the inherent "internal bending strain".

6.2 Reassembly on Solid Supports

Crystal growth at interfaces is initiated simultaneously at many randomly distributed nucleation points and proceeds in the plane until the crystalline domains meet, thus leading to a closed, coherent mosaic of individual S-layer domains several micrometers in size [30, 7]. A decade ago, S-layer protein monolayer formation at a liquid–air interface was studied by transmission electron microscopy (TEM) [30]. In this work, electron microscope grids were deposited on and removed from the water surface by means of a Langmuir–Schäfer transfer at regular time intervals. After staining with uranyl acetate, the samples were inspected in the microscope. In a recent study, it has been demonstrated that atomic force microscopy (AFM) is most suitable for imaging the lattice formation in real time [7]. Approximately 10 min after injection of the protein solution into the fluid cell, the first small crystalline patches became visible, and about 30 min later the silicon surface was completely covered and only small holes remained free, which were closed in due course. Extremely low loading forces ($\sim$ 100 pN) of the AFM tip were necessary in order to minimize the influence of the scanning tip on the reassembly of the proteins. The formation of coherent crystalline arrays depends strongly on the S-layer protein species, the environmental conditions of the bulk phase (e.g., temperature, pH, ion composition, and ionic strength) and, in particular, the surface properties of the substrate (hydrophobicity and surface charge) [7, 38]. Monocrystalline domains within the mosaic may be up to 15 μm in diameter.

For many technological applications of S-layers, spatial control of the reassembly is mandatory. For example, when S-layers are used as affinity matrices in the development of biochips or as templates in the fabrication of nanoelectronic devices, the S-layer must not cover the entire device area. Micromolding in capillaries allows the reassembly of the S-layer proteins to be restricted to certain areas on a solid support [6]. For this purpose, an S-layer protein solution was dropped onto a substrate in front of the channel openings of the attached mold. The solution was sucked in and the S-layer protein started to recrystallize. After removal of the mold, a patterned S-layer remained on the support. Micromolding in capillaries offers the advantage that all preparation steps may be performed under ambient conditions. In contrast, optical lithography requires drying of the protein layer prior to exposure to (deep ultraviolet) light [29]. This is a critical step, since denaturation of the protein and, consequently, loss of its structural and functional integrity cannot be excluded.

6.3 Reassembly at Lipid Interfaces

The possibility of recrystallizing isolated S layer proteins at an air/water interface or on lipid films and of handling such layers by standard Langmuir–Blodgett (LB) techniques has opened up a broad spectrum of applications in basic and applied membrane research (for reviews, see [42, 45]). It has to be stressed that S-layer-supported lipid membranes strongly resemble those archaeal envelope structures which are composed exclusively of an S-layer and a closely associated plasma membrane (for a review, see [51]). These archaea live under extreme environmental conditions, such as at $pH < 0.5$, under hydrostatic pressure, and at temperatures up to 120°C [59]. S-layer-supported LB films are able to cover holes up to 40 μm in diameter and maintain their structural and functional integrity in the course of subsequent handling procedures for a much longer period of time than for unsupported structures (e.g., black lipid membranes) (for reviews, see [42, 45]). The stabilizing effect of S-layers is explained primarily by a reduction or inhibition of disruptive horizontal vibrations of the lipid molecules. The terminology "semifluid membranes" has been coined to describe S-layer-supported membranes, since the interaction of the lipid head groups with the repetitive domains of the associated S-layer lattice significantly modulates the characteristics of the lipid film (particularly its fluidity and local order on the nanometer scale) [28]. Fluorescence-recovery-after-photobleaching (FRAP) measurements have demonstrated that the mobility of lipids in S-layer-supported bilayers was higher than in other model systems, such as hybrid bilayers or dextran-supported bilayers [5]. Neutron and X-ray reflectivity studies have clearly indicated that the S-layer protein did not penetrate or rupture the lipid monolayer [66, 65, 64]. Functional molecules such as ion channels or proton pumps may be incorporated into S-layer-stabilized lipid layers by applying well-established procedures. Voltage clamp [41, 43] and impedance spectroscopy [4] are prominent biophysical methods for characterizing the electrophysiological parameters of such composite functional biomembranes. In comparison with plain lipid bilayers, S-layer-supported lipid membranes have a decreased tendency to rupture and allow one to perform single-pore recording [43, 44, 39, 40, 46].

Furthermore, the reassembly of S-layer proteins on liposomes and nanocapsules has great technological importance [16, 18, 19, 61]. Because of their physicochemical properties, liposomes are widely used as model systems for biological membranes and as delivery systems for biologically active molecules. In general, water-soluble molecules are encapsulated within the aqueous compartment, whereas water-insoluble substances may be intercalated into the liposomal membrane. The presence of an S-layer lattice significantly enhances the stability of the liposomes against mechanical stresses such as shear forces or ultrasonication and against thermal challenges. Also, S-layer liposomes resemble the supramolecular envelope principle of a great variety of human and animal viruses [58].

7 Summary

Basic research on the structure, genetics, chemistry, morphogenesis, and function of S-layers has led to a broad spectrum of applications in molecular nanobiotechnology, which are, at least partially, now ready for exploitation in the life and nonlife sciences. A complete description would be beyond the scope of this contribution and has been published in several review articles (e.g., [55, 56, 57, 58, 52, 34]). Nevertheless, in summary, the most important applications of S-layers are found in those areas either where biologically functional molecules, such as enzymes or antibodies, have to be bound in a dense monomolecular packing or where genetically functionalized S-layer proteins themselves are used as sensing layers, as in the development of immunoassays, label-free detection systems (e.g., surface plasmon resonance spectroscopy), and affinity matrices. In addition, some emerging areas of research are in nanoelectronics, where S-layers may be used as templates for binding metallic or semiconducting nanoparticles into perfectly ordered arrays, and the field of lipid chips, where S-layers are used as stabilizing structures leading to an increased robustness and lifetime of the functional lipid membrane. Currently there is no other biological matrix known that provides the same outstanding universal self-assembly properties and patterning elements as do S-layers. The possibility to change the natural properties of S-layer proteins by genetically incorporating functional domains has opened up a new horizon for the tuning of their structural and functional features [34].

Acknowledgments

Part of this work was supported by the Austrian Science Foundation (FWF) (Projects P14419-MOB, P17170, and P16295-B10); the Volkswagen Foundation, Germany (Project I/77710); the Air Force Office of Scientific Research (grant F49620-03-1-0222); the Austrian Federal Ministry of Education, Science and Culture; the Austrian Federal Ministry of Transport, Innovation and Technology (MNA-Network); and the European Commission (Projects BIOAND IST-1999-11974 and Nanocapsules HPRN-CT-2000-00159).

References

1. L.A. Amos, R. Henderson, and P.N.T. Unwin. Three-dimensional structure determination by electron microscopy of two-dimensional crystals. *Prog. Biophys. Mol. Biol.*, 39:183–231, 1982.
2. W. Baumeister and H. Engelhardt. Three-dimensional structure of bacterial surface layers. In J.R. Harris and R.W. Horne, editors, *Electron Microscopy of Protein*, volume 6, pages 109–154. Academic Press, London, 1987.
3. T.J. Beveridge and L.L. Graham. Surface layers of bacteria. *MicroBiol. Rev.*, 55:684–705, 1991.

4. P.C. Gufler, D. Pum, U.B. Sleytr, and B. Schuster. Highly robust lipid membranes on crystalline S-layer supports investigated by electrochemical impedance spectroscopy. *Biochim. Biophys. Acta*, 1661:154–165, 2004.
5. E.S. Györvary, B. Wetzer, U.B. Sleytr, A. Sinner, A. Offenhäuser, and W. Knoll. Lateral diffusion of lipids in silane-, dextrane- and S-layer protein-supported mono- and bilayers. *Langmuir*, 15:1337–1347, 1999.
6. E.S. Györvary, A. O'Riordan, A.J. Quinn, G. Redmond, D. Pum, and U.B. Sleytr. Biomimetic nanostructure fabrication: nonlithographic lateral patterning and self-assembly of functional bacterial S-layers at silicon supports. *Nano Letters*, 3:315–319, 2003.
7. E.S. Györvary, O. Stein, D. Pum, and U.B. Sleytr. Self-assembly and recrystallization of bacterial S-layer proteins at silicon supports imaged in real time by atomic force microscopy. *J. Microsc.*, 212:300–306, 2003.
8. W.F. Harris and L.E. Scriven. Function of dislocations in cell walls and membranes. *Nature*, 228:827–829, 1970.
9. W.F. Harris and L.E. Scriven. Intrinsic disclinations as dislocation sources and sinks in surface crystals. *J. Appl. Phys.*, 42:3309–3312, 1971.
10. S.C. Holt and E.R. Leadbetter. Comparative ultrastructure of selected aerobic spore forming bacteria: a freeze-etching study. *Bacteriol Rev.*, 33:346–378, 1969.
11. S. Hovmöller, A. Sjögren, and D.N. Wang. The structure of crystalline bacterial surface layers. *Prog. Biophys. Mol. Biol.*, 51:131–163, 1988.
12. C. Huber, N. Ilk, D. Rünzler, E.M. Egelseer, S. Weigert, U.B. Sleytr, and M. Sára. The three S-layer-like homology motifs of the S-layer protein SbpA of Bacillus sphaericus CCM 2177 are not sufficient for binding to the pyruvylated secondary cell wall polymer. *Mol. Microbiol.*, 55:197–205, 2005.
13. N. Ilk, S. Küpcü, G. Moncayo, S. Klimt, R.C. Ecker, R. Hofer-Warbinek, E.M. Egelseer, U.B. Sleytr, and M. Sára. A functional chimeric S-layer/enhanced green fluorescent protein to follow the uptake of S-layer-coated liposomes into eukaryotic cells. *Biochem. J.*, 379:441–448, 2004.
14. N. Ilk, C. Völlenkle, E.M. Egelseer, A. Breitwieser, U.B. Sleytr, and M. Sára. Molecular characterization of the S-layer gene, *sbpA*, of *Bacillus sphaericus* CCM 2177 and production of a functional S-layer fusion protein with the ability to recrystallize in a defined orientation while presenting the fused allergen. *Appl. Environ. Microbiol.*, 68:3251–3260, 2002.
15. R. Jaenicke, R. Welsch, M. Sára, and U.B. Sleytr. Stability and self-assembly of the S-layer protein of the cell wall of Bacillus stearothermophilus. *Biol. Chem. Hoppe-Seyler*, 366:663–670, 1985.
16. S. Küpcü, M. Sára, and U.B. Sleytr. Liposomes coated with crystalline bacterial cell surface protein (S-layers) as immobilization structures for macromolecules. *Biochim. Biophys. Acta*, 1235:263–269, 1995.
17. C. Mader, C. Huber, D. Moll, U.B. Sleytr, and M. Sára. Interaction of the crystalline bacterial cell surface layer protein SbsB and the secondary cell wall polymer of Geobacillus stearothermophilus PV72 assessed by real-time surface plasmon resonance biosensor technology. *J. Bacteriol.*, 186:1758–1768, 2004.
18. C. Mader, S. Küpcü, M. Sára, and U.B. Sleytr. Stabilizing effect of an S-layer on liposomes towards thermal or mechanical stress. *Biochim. Biophys. Acta*, 1418:106–116, 1999.
19. C. Mader, S. Küpcü, U.B. Sleytr, and M. Sára. S-layer-coated liposomes as a versatile system for entrapping and binding target molecules. *Biochim. Biophys. Acta*, 1463:142–150, 2000.

20. P. Messner, D. Pum, M. Sára, K. Stetter, and U.B. Sleytr. Ultrastructure of the cell envelope of the archaebacteria *Thermoproteus tenax* and *Thermoproteus neutrophilus*. *J. Bacteriol.*, 166:1046–1054, 1986.
21. P. Messner, D. Pum, and U.B. Sleytr. Characterization of the ultrastructure and the self-assembly of the surface layer of *Bacillus stearothermophilus* NRS 2004/3a. *J. Ultrastruct. Mol. Struct. Res.*, 97:73–88, 1986.
22. D. Moll, C. Huber, B. Schlegel, D. Pum, U.B. Sleytr, and M. Sára. S-layer-streptavidin fusion proteins as template for nanopatterned molecular arrays. *Proc. Natl. Acad. Sci. USA*, 99:14646–14651, 2002.
23. F.R.N. Nabarro and W.F. Harris. Presence and function of disclinations in surface coats of unicellular organisms. *Nature*, 232:423, 1971.
24. M. Pleschberger, A. Neubauer, E.M. Egelseer, S. Weigert, B. Lindner, U.B. Sleytr, S. Muyldermans, and M. Sára. Generation of a functional monomolecular protein lattice consisting of an S-layer fusion protein comprising the variable domain of a camel heavy chain antibody. *Bioconj. Chem.*, 14:440–448, 2003.
25. M. Pleschberger, D. Saerens, S. Weigert, U.B. Sleytr, S. Muyldermans, M. Sára, and E.M. Egelseer. An S-layer heavy chain camel antibody fusion protein for generation of a nanopatterned sensing layer to detect the prostate-specific antigen by surface plasmon resonance technology. *Bioconj. Chem.*, 15:664–671, 2004.
26. K.L. Prime and G.M. Whitesides. Self-assembled organic monolayers: model systems for studying adsorption of proteins at surfaces. *Science*, 252:1164–1167, 1991.
27. D. Pum, P. Messner, and U.B. Sleytr. Role of the S-layer in morphogenesis and cell division of the archaebacterium *Methanocorpusculum sinense*. *J. Bacteriol.*, 173:6865–6873, 1991.
28. D. Pum and U.B. Sleytr. Large-scale reconstruction of crystalline bacterial surface layer proteins at the air–water interface and on lipids. *Thin Solid Films*, 244:882–886, 1994.
29. D. Pum, G. Stangl, C. Sponer, W. Fallmann, and U.B. Sleytr. Deep UV patterning of monolayers of crystalline S-layer protein on silicon surfaces. *Colloids Surf. B*, 8:157–162, 1997.
30. D. Pum, M. Weinhandl, C. Hödl, and U.B. Sleytr. Large-scale recrystallization of the S-layer of *Bacillus coagulans* E38-66 at the air/water interface and on lipid films. *J. Bacteriol.*, 175:2762–2766, 1993.
31. K. Roberts, G.J. Hills, and P.J. Shaw. The structure of algal cell walls. In J.R. Harris, editor, *Electron Microscopy of Proteins*, volume 3, pages 1–40. Academic Press, London, 1982.
32. D. Rünzler, C. Huber, D. Moll, G. Köhler, and M. Sára. Biophysical characterization of the entire bacterial surface layer protein SbsB and its two distinct functional domains. *J. Biol. Chem.*, 279:5207–5215, 2004.
33. M. Sára. Conserved anchoring mechanisms between crystalline cell surface S-layer proteins and secondary cell wall polymers in Gram-positive bacteria. *Trends Microbiol.*, 9:47–49, 2001.
34. M. Sára, D. Pum, B. Schuster, and U.B. Sleytr. S-layers as patterning elements for application in nanobiotechnology. *J. Nanosci. Nanotechnol.*, 2005. In press.
35. M. Sára and U.B. Sleytr. S-layer proteins. *J. Bacteriol.*, 182:859–868, 2000.
36. C. Schäffer and P. Messner. The structure of secondary cell wall polymers: how Gram-positive bacteria stick their cell walls together. *Microbiol.*, 151:643–651, 2005.

37. C. Schäffer, T. Wugeditsch, H. Kählig, A. Scheberl, S. Zayni, and P. Messner. The surface layer (S-layer) glycoprotein of *Geobacillus stearothermophilus* NRS 2004/3a. Analysis of its glycosylation. *J. Biol. Chem.*, 277:6230–6239, 2002.
38. B. Schuster, E. Györvary, D. Pum, and U.B. Sleytr. Nanotechnology with S-layer proteins. *Methods Mol. Biol.*, 300:101–124, 2005.
39. B. Schuster, D. Pum, O. Braha, H. Bayley, and U.B. Sleytr. Self-assembled alpha-hemolysin pores in an S-layer-supported lipid bilayer. *Biochim. Biophys. Acta*, 1370:280–288, 1998.
40. B. Schuster, D. Pum, M. Sára, O. Braha, H. Bayley, and U.B. Sleytr. S-layer ultrafiltration membranes: a new support for stabilizing functionalized lipid membranes. *Langmuir*, 17:499–503, 2001.
41. B. Schuster, D. Pum, and U.B. Sleytr. Voltage clamp studies on S-layer supported tetraether lipid membranes. *Biochim. Biophys. Acta*, 1369:51–60, 1998.
42. B. Schuster and U.B. Sleytr. S-layer-supported lipid membranes. *Rev. Mol. Biotechnol.*, 74:233–254, 2000.
43. B. Schuster and U.B. Sleytr. The effect of hydrostatic pressure on S-layer-supported lipid membranes. *Biochim. Biophys. Acta*, 1563:29–34, 2002.
44. B. Schuster and U.B. Sleytr. Single channel recordings of alpha-hemolysin reconstituted in S-layer-supported lipid bilayers. *Bioelectrochemistry*, 55:5–7, 2002.
45. B. Schuster and U.B. Sleytr. 2D-protein crystals (S-layers) as supports for lipid membranes. In T.H. Tien and A. Ottowa, editors, *Advances in Planar Lipid Bilayers and Liposomes*, volume 1, pages 247–293. Elsevier Science, Amsterdam, 2005.
46. B. Schuster, S. Weigert, D. Pum, M. Sára, and U.B. Sleytr. New method for generating tetraether lipid membranes on porous supports. *Langmuir*, 19:2392–2397, 2003.
47. N.C. Seeman. At the crossroads of chemistry, biology and materials: structural DNA nanotechnology. *Chem. Biol.*, 10:1151–1159, 2003.
48. N.C. Seeman. DNA in a material world. *Nature*, 421:33–37, 2003.
49. U.B. Sleytr. Heterologous reattachment of regular arrays of glycoproteins on bacterial surfaces. *Nature*, 257:400–402, 1975.
50. U.B. Sleytr. Regular arrays of macromolecules on bacterial cell walls: structure, chemistry, assembly, and function. *Int. Rev. Cytol.*, 53:1–64, 1978.
51. U.B. Sleytr and T.J. Beveridge. Bacterial S-layers. *Trends Microbiol.*, 7:253–260, 1999.
52. U.B. Sleytr, E.-M. Egelseer, D. Pum, and B. Schuster. S-layers. In C. Niemeyer and C. Mirkin, editors, *Nanobiotechnology*, pages 77–92. Wiley-VCH, Weinheim, 2004.
53. U.B. Sleytr and P. Messner. Self-assemblies of crystalline bacterial cell surface layers. In H. Plattner, editor, *Electron Microscopy of Subcellular Dynamics*, pages 13–31. CRC Press, Boca Raton, 1989.
54. U.B. Sleytr, P. Messner, D. Pum, and M. Sára. *Crystalline Bacterial Cell Surface Proteins*. R. G. Landes/Academic Press, Austin, TX, 1996.
55. U.B. Sleytr, P. Messner, D. Pum, and M. Sára. Crystalline bacterial cell surface layers (S-layers): from supramolecular cell structure to biomimetics and nanotechnology. *Angew. Chem.-Int. Ed.*, 38:1034–1054, 1999.
56. U.B. Sleytr, M. Sára, and D. Pum. Crystalline bacterial cell surface layers (S-layers): a versatile self-assembly system. In A. Ciferri, editor, *Supramolecular Polymerization*, pages 177–213. Marcel Dekker, New York, 2000.

57. U.B. Sleytr, M. Sára, D. Pum, and B. Schuster. Molecular nanotechnology and nanobiotechnology with two-dimensional protein crystals (S-layers). In M. Rosoff, editor, *Nano-Surface Chemistry*, pages 333–389. Marcel Dekker, New York, 2001.
58. U.B. Sleytr, M. Sára, D. Pum, B. Schuster, P. Messner, and C. Schäffer. Self assembly protein systems: microbial S-layers. In A. Steinbüchel, editor, *Biopolymers*, volume 3 of *Polyamides and Complex Proteinaceous Materials (Part A)*, pages 285–338. Wiley-VCH, Weinheim, 2003.
59. K.O. Stetter. Extremophiles and their adaption to hot environments. *FEBS Lett.*, 452:22–25, 1999.
60. J. Tien, Y. Xia, and G.M. Whitesides. Microcontact printing of SAMs. In A. Ulman, editor, *Thin Films*, volume 24, pages 227–253. Academic Press, 1998.
61. J.L. Toca-Herrera, R. Krastev, V. Bosio, S. Küpcü, D. Pum, A. Fery, M. Sára, and U.B. Sleytr. Recrystallization of bacterial S-layers on flat polyelectrolyte surfaces and hollow polyelectrolyte capsules. *Small*, 1:339–348, 2005.
62. J.L. Toca-Herrera, S. Moreno-Flores, J. Friedmann, D. Pum, and U.B. Sleytr. Chemical and thermal denaturation of crystalline bacterial S-layer proteins: an atomic force microscopy study. *Microsc. Res. Technol.*, 65:226–234, 2004.
63. C. Völlenkle, S. Weigert, N. Ilk, E. Egelseer, V. Weber, F. Loth, D. Falkenhagen, U.B. Sleytr, and M. Sára. Construction of a functional S-layer fusion protein comprising an immunoglobulin G-binding domain for development of specific adsorbents for extracorporeal blood purification. *Appl. Environ. Microbiol.*, 70:1514–1521, 2004.
64. M. Weygand, K. Kjaer, P.B. Howes, B. Wetzer, D. Pum, U.B. Sleytr, and M. Lösche. Structural reorganization of phospholipid headgroups upon recrystallization of an S-layer lattice. *J. Phys. Chem. B*, 106:5793–5799, 2002.
65. M. Weygand, M. Schalke, P.B. Howes, K. Kjaer, J. Friedmann, B. Wetzer, D. Pum, U.B. Sleytr, and M. Lösche. Coupling of protein sheet crystals (S-layers) to phospholipid monolayers. *J. Mater. Chem.*, 10:141–148, 2000.
66. M. Weygand, B. Wetzer, D. Pum, U.B. Sleytr, N. Cuvillier, K. Kjaer, P.B. Howes, and M. Losche. Bacterial S-layer protein coupling to lipids: X-ray reflectivity and grazing incidence diffraction studies. *Biophys. J.*, 76:458–468, 1999.

Part VI

Biomolecular Computational Models

Computing with Hairpins and Secondary Structures of DNA

Masami Hagiya[1,2], Satsuki Yaegashi[1], and Keiichiro Takahashi[3]

[1] JST CREST, Japan
hagiya@is.s.u-tokyo.ac.jp
[2] Department of Computer Science
Graduate School of Information Science and Technology, University of Tokyo, Japan
[3] Bioinformatics Division, NovusGene Inc., Japan
takahashi-k@novusgene.co.jp

1 Introduction

Since the first author became involved in research on molecular computing, we have been consistently working on molecular computations using secondary structures of DNA, whose simplest forms are hairpins. In particular, we have been interested in the computational power of formation and dissociation of such secondary structures.

In this chapter, we first review our previous research on computing with hairpins, and show that hairpin formation and dissociation have enough computational power in themselves. We then report our recent experiments on implementing logic gates by loop dissociation.

The paper is organized as follows. In Section 2, we briefly describe our research that shows the computational power of hairpin formation. We then survey the Whiplash Machine, a kind of molecular machine based on repeated formation and dissociation of hairpins, in Section 3. In Section 4, after describing our model of general-purpose molecular systems, we report the experimental results concerning the scheme of logical circuits based on loop dissociation.

2 Computing by Hairpin Formation

The most typical example of computation with hairpins was introduced in our research on solving the 3-SAT problem using hairpin formation [15]. We showed that hairpin formation can be used to enhance Adleman and Lipton's model [1, 10, 3] of molecular computation by making the step of extracting solutions from the random pool of candidates more efficient.

The SAT problem requires deciding whether a given Boolean expression is satisfiable or not, and a Boolean expression is called satisfiable if it can be made true by an appropriate assignment of truth values to its variables. For example, the Boolean expression $(x \vee y \vee \neg z) \wedge (\neg x \vee y \vee z) \wedge (\neg x \vee \neg y \vee \neg z)$ can be made true if we assign truth to x, falsity to y, and truth to z. Variables or their negations are called literals, and disjunctions of literals, such as $x \vee y \vee \neg z$ and $\neg x \vee y \vee z$, are called clauses. A Boolean expression in a clausal form is a conjunction of clauses, such as the expression mentioned above. The 3-SAT problem focuses on a clausal form in which each clause consists of three literals. Even if we restrict ourselves to such clausal forms, the SAT problem is still very difficult to solve, i.e., is said to be NP-complete [10].

Since a clausal form is a conjunction of clauses and a clause is a disjunction of its literals, satisfying a clausal form amounts to choosing at least one literal from each clause and making it true. For example, in order to satisfy the clausal form $(x \vee y \vee \neg z) \wedge (\neg x \vee y \vee z) \wedge (\neg x \vee \neg y \vee \neg z)$, we can choose x from $x \vee y \vee \neg z$, z from $\neg x \vee y \vee z$, and $\neg y$ from $\neg x \vee \neg y \vee \neg z$. This process of selection should be consistent in the sense that a variable and its negation are not chosen simultaneously. For example, choosing x from $x \vee y \vee \neg z$ and $\neg x$ from $\neg x \vee \neg y \vee \neg z$ is inconsistent.

As in the first step of Adleman and Lipton's model, i.e., that of generating a random pool of candidates, one can enumerate such selections randomly and obtain a random pool of selections. In order to solve the 3-SAT problem, it remains to extract consistent selections from the random pool. We implemented this crucial step of the algorithm by applying the computational power of hairpin formation. Remember that a variable x and its negation $\neg x$ are called complementary to each other. Representing complementary literals using complementary base sequences in the sense of Watson and Crick, we implemented the crucial step by simply letting molecules form hairpins! If a selection is inconsistent, i.e., if a selection contains both a variable and its negation, since a selection contains two subsequences that are complementary to each other, it should form a hairpin, provided that it has been made single-stranded.

During the process of hairpin formation within a single molecule, a subsequence of the molecule comprising a hairpin stem searches for its counterpart. Note that this search is performed autonomously, as in the self-assembly of ordinary double-stranded DNA molecules in the first step of Adleman and Lipton's model, or as in the self-assembly of DNA tiles in Seeman and Winfree's model of DNA computation [19, 25, 24, 18, 14]. Note that the number of experimental operations required for the second step of our algorithm is, at least theoretically, independent of the size of the problem being solved, since the computation is done by the autonomous reaction of hairpin formation.

3 Computing by Repeated Hairpin Formation and Dissociation

Prior to the above research, we devised a more sophisticated computational device, called the Whiplash Machine [5, 16, 8]. This machine is a single-stranded DNA molecule consisting of two parts. The first part is a program controlling the machine and the other part represents the current state of the machine.

The program of the machine takes the form of a state transition table and consists of a sequence of state transition rules. Each rule in the table has the form

$$5' - \mathbf{stopper} - \mathbf{state2} - \mathbf{state1} - 3',$$

where **state1** denotes the state before a transition, and **state2** the state after the transition. This rule represents the transition that allows the machine to change its state from **state1** to **state2**.

The entire machine has the form

$$5' - \mathbf{program} - \mathbf{spacer} - \mathbf{state} - 3',$$

where **program** is a sequence of state transition rules mentioned above and **state** represents the current state of the machine. This latter part is complementary to some of the states in the first part and can form a hairpin. In order to enhance hairpin formation, the **spacer** sequence of an appropriate length is inserted.

Now, assume that the current state, **state**, is complementatry to **state1** of some transition rule. We further assume that the machine is put in the polymerase buffer with polymerase, dNTP and ATP. Since **state** hybridizes with **state1** by forming a hairpin, the 3′-end of the DNA molecule is extended by the polymerase. This extension adds the sequence complementary to **state2** to the 3′-end of the molecule. We regard this reaction as a state transition from **state1** to **state2**, because the sequence at the 3′-end of the molecule changes from the sequence complementary to **state1** to that complementary to **state2**.

The sequence **stopper** plays the role of stopping the polymerase extension right after the sequence complementary to **state2** is added. There could be various ways to implement this functionality. What we have tried is to encode states using only three kinds of bases and use the remaining base for **stopper**. For example, if we use T, G and C for states, then some repetition of A, such as AAA, can be used as **stopper**. Futhermore, the polymerase buffer lacks dTTP in this case, so the polymerase extension stops right after **state2** is copied.

After the sequence complementary to **state2** is added to the 3′-end, the molecule takes the hairpin form whose stem is twice as long as the initial hairpin and therefore more stable. However, if the temperature is carefully adjusted, even this more stable hairpin can be opened with a certain probability. Once the hairpin is opened, the molecule can take another hairpin.

This means that the new current state can hybridize with the **state1** part of another transition rule, and make another state transition.

Note that even after the hairpin is opened, the molecule can take the same hairpin again. We call this hairpin formation back-hybridization. However, this reaction does no harm, because due to the **stopper** sequence, there should occur no polymerase extension and the machine does not change. It only decreases the efficiency of the next state transition [11].

We have actually implemented the Whiplash Machine by designing an appropriate thermal protocol, and succeeded in observing eight successive transitions [8].

The Whiplash Machine has various applications. For example, it can implement boolean expressions in a limited form, called μ-formulas [5]. Rose et al. propose to generate protein-encoding sequences by successive transitions of the Whiplash Machine. His ultimate goal is to apply Whiplash Machine to protein engineering by evolving protein-encoding programs *in vitro* [12, 13]. Since successive state transitions are executed autonomously in a single one-pot reaction, the Whiplash Machine can also be used to enhance Adleman and Lipton's model [8].

4 Computing by Loop Dissociation

In this chapter, we mainly report the recent experiments we have done on implementing logic gates by loop dissociation. Before describing the experiments, let us explain the model of general-purpose molecular systems we envision to implement by chain reactions of DNA and other kinds of molecules.

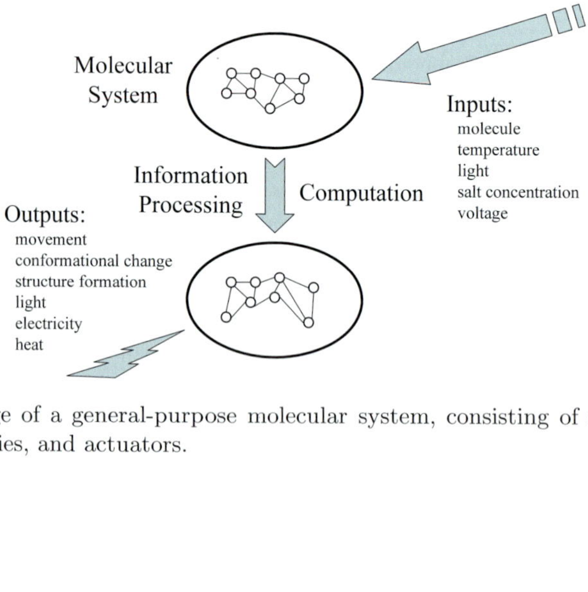

Fig. 1. Image of a general-purpose molecular system, consisting of sensors, logic gates, memories, and actuators.

4.1 General-Purpose Molecular Systems

The ultimate goal of our research is to realize general-purpose molecular systems, which accept various kinds of inputs, do some computation and finally produce outputs, also in various forms, as in Fig. 1. Furthermore, we want to prepare a set of general-purpose molecular components and establish a systematic principle for designing networks of molecular components comprising such molecular systems. The components should include various molecular sensors, logic gates, memories, and actuators, and can be combined in a systematic manner without ad hoc constraints. In other words, it should be possible to program molecular systems by simply combining those building blocks.

The most basic components we currently have at our disposal are DNA hairpins which can be opened by single-stranded DNA molecules called openers as in Fig. 2. The hairpin molecule has a single-stranded part, called the lead section, which is concatenated to one end of the hairpin stem. The opener molecule also has its lead section, which is complementary to that of the hairpin. Another part of the opener, called the invasion section, invades into the hairpin stem and opens the hairpin loop via branch migration.

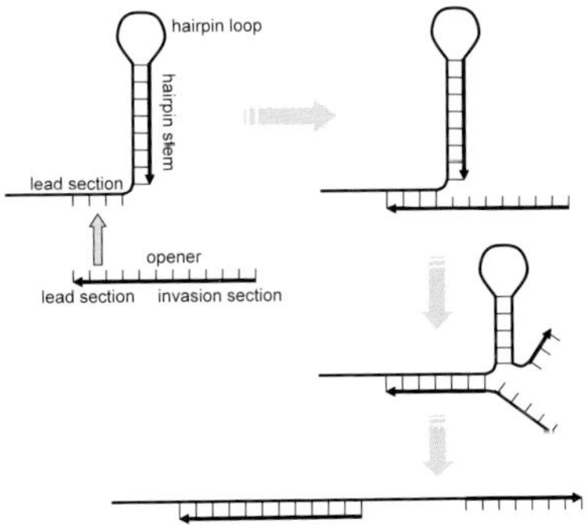

Fig. 2. Hairpin and its opener. The lead section of the opener hybridizes with that of the hairpin. The invasion section then invades into the hairpin stem via branch migration and opens the hairpin loop.

After the hairpin is opened by the opener, the hairpin loop is exposed and can interact with its complementary sequence. While the hairpin is closed, this interaction is suppressed although the hairpin loop is single-stranded. In particular, if the complementary sequence is in the middle of a longer sequence, the interaction is almost inhibited [20].

As in Fig. 3, one can imagine a chain reaction of hairpin opening. After the first hairpin is opened by its opener, it can interact with the second hairpin with its hairpin loop hybridizing with the lead section of the second hairpin. This process can be regarded as a transformation of signals represented by lead sections. While the first opener has the lead section X, the second hairpin having the lead section $\overline{Y}$ is finally opened. (The overline means complementation.) This transformation from signal X to signal Y is accomplished by the hairpin having the lead section $\overline{X}$ and the hairpin loop Y. In the next section, we further elaborate this idea into logic gates.

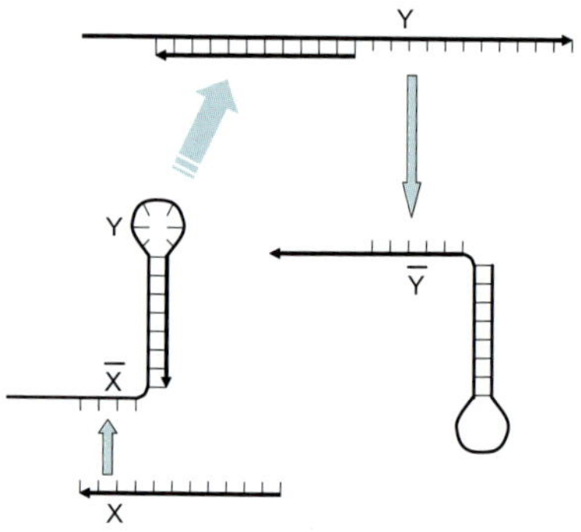

Fig. 3. Chain reaction by hairpins. The first opener, which has the lead section X, hybridizes with the hairpin and opens the harpin loop Y. The hairpin then serves as the opener of the next hairpin, whose lead section is $\overline{Y}$.

Takahashi et al. reported that hairpin opening can be controlled by openers whose activity depends on environmental conditions [21]. They introduced the opener with azobenzene molecules tethered to its side chain to photoregulate hairpin opening. The opener forms a small hairpin and cannot interact with its target hairpin while the azobenzenes are in the *trans* conformation. After the azobenzenes are irradiated with UV light, they change their conformation from *trans* to *cis*. When the azobenzenes take the *cis* transformation, the small hairpin is opened since the base stacking is destabilized. They also report the opener whose activity depends on temperature. Needless to say, these openers can serve as sensors in the model of general-purpose molecular systems mentioned above.

Kameda et al. report a molecular memory unit consisting of successive hairpins [6]. Each hairpin of a molecular unit represents a digit of the address of the unit. Only by successively opening the hairpins, can one access the memory unit. Kubota et al. extend this idea to implement a machine that has branching states, each represented by a hairpin loop [9].

Many researchers have been working in similar directions. Yurke, et al. pioneered the use of branch migration for molecular machines and established the notion of DNA fuel [26, 23]. They also pointed out the difficulty of hybridization between a loop and its complementary sequence, and showed that a single-stranded DNA molecule that opens the loop can be used as a catalyst for their hybridization [23]. Following Yurke's research, Seelig, et al. proposed logic gates based on DNA catalysts [17]. As for chain reactions of molecular machines, Dirks et al. have already proposed the scheme of hybridization chain reaction, in which hairpins successively hybridize with one another and form a huge complex [4].

4.2 Logic Gates

As mentioned in the previous section, signals represented by lead sections of openers can be transformed by chain reactions of hairpins. In Fig. 3, signal X is transformed to signal Y by the hairpin molecule. If we prepare another hairpin that can be opened by another signal Z and produces signal Y, these two hairpins comprise a system that produces signal Y with the existence of signal X or signal Z. In other words, the system acts as an OR gate whose inputs are X and Z, and whose output is Y.

Using bulge loops in addition to hairpins, we can build more complex molecular systems. It is also well known that bulge loops also prohibit hybridization with their complementary sequences [23]. In Fig. 4, two single-stranded DNA molecules form a structure containing two bulge loops. The structure accepts two inputs, which are also single-stranded DNA molecules. The first input (**input1**) hybridizes with the single-stranded section of the structure and opens the first (upper left) bulge loop via branch migration. Since the bulge loop is exposed as single-stranded, it can hybridize with the second input (**input2**), which opens the second (lower right) bulge loop via branch migration and frees the lower strand of the structure as an output. Since the output can only be produced in the presence of both inputs, this system comprises an AND gate.

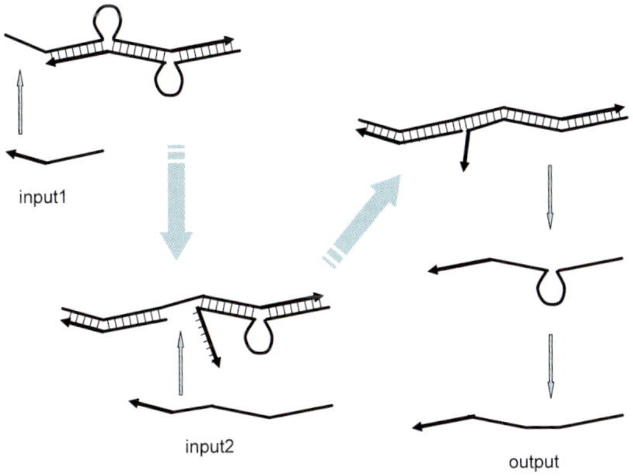

Fig. 4. AND gate, consisting of two bulge loops. When it receives the first input, the upper left bulge loop is opened and hybridizes with the second input, which then opens the lower right bulge loop and produces the output.

Note that the interactions mentioned above are the only interactions among the strands that appear in the system. The two inputs do not interact with each other without the bulge structure, and the bulge structure can only interact with the first input. The output cannot interact with any of the inputs or the bulge structure.

We reported this idea in our previous paper [22] and showed some preliminary experimental results. In this paper, we report more experimental results using redesigned sequences. In particular, we not only verified the proper behavior of the AND gate but also checked the following properties of the gate.

We prepared a hairpin molecule that can be opened by the output of the AND gate. This hairpin molecule is considered as another gate in the next stage. We therefore checked whether the output of the AND gate can trigger the next gate.

We also prepared another competing AND gate, which shares the first input with the original AND gate. In our design, since the first input of an AND gate hybridizes with the gate without the second input, the existence of the second AND gate should decrease the efficiency of the first AND gate. We checked this property by experiment.

The structures and strands used in the experiments are depicted in Fig. 5.

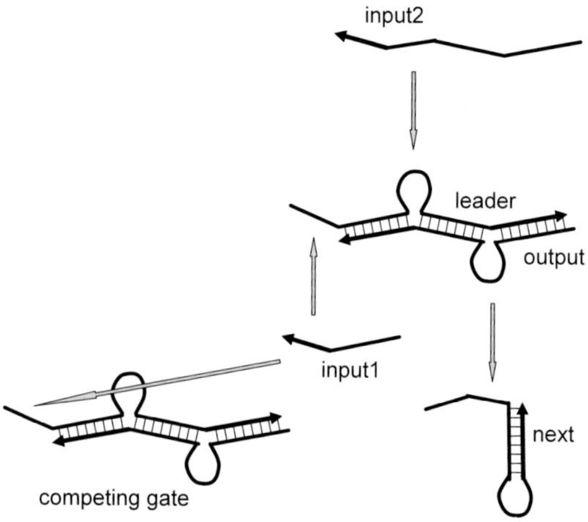

Fig. 5. Structures used in the experiments. We prepare the AND gate and its two inputs. We also prepare a hairpin molecule representing the next gate, and a competing AND gate which shares the first input with the original AND gate.

4.3 Experiments

As mentioned in the previous section, we have conducted the following three experiments on the AND gate composed of two bulge loops:

- We first checked that the AND gate works properly by giving the two inputs in various orders and observing the output by gel eletrophoresis and spectrophotoscopy. (Actually, most orderes are tried in the next experiment.)
- We prepared a hairpin molecule whose lead section is complementary to the bulge loop in the output. This hairpin is expected to be opened by the output. We checked the hybridization between the two molecules by gel eletrophoresis. We also checked that the hairpin cannot interact with the AND gate unless the two inputs are given and the output is produced.
- We prepared the second AND gate which shares the first input with the first AND gate. We observed how the output of the first gate decreased. This observation was made with various quantities of the first input.

Material and Methods

The sequences used in the experiments are given in Table 1. These sequences are designed using the extended template method [22, 2, 7]. In this extension, templates with different lengths are computed. They are then instantiated using BCH codes of those lengths, where `1` in the templates is replaced with `G` or `C`, and `0` with `T` or `A`. For the experiments, we first computed the two templates `0100111011` and `00011100100010111011`. The 10-base template guarantees at lease two-base mismatch when it is matched with any subsequence of a concatenation of the form 10*20, 20*10 or 20*20, where 10*20 means the concatenation of the 10-base template and the 20-base template in this order. Similarly, the 20-base template guarantees at lease eight-base mismatch against a concatenation of the form 10*20, 20*10 or 20*20.

In Table 1, sequences are separated into 10-base or 20-base units. Each 20-base unit is obtained from the 20-base template, while each 10-base unit is either obtained from the 10-base template or derived as a half of a sequence obtained from the 20-base template. All the bulge loops are 10 bases long and obtained from the 10-base template.

DNA molecules used in the experiments were synthesized by Sigma Aldrich Japan. For the experiments by spectrophotoscopy, the 3′-end of **leader** is modified with BHQ and the 5′-end of **output** is modified with FAM. Thus, when **output** is dissociated from **leader**, the fluorescence intensity of FAM increases.

All the experiments by spectrophotoscopy were conducted using a HITACHI F-2500 fluorescence spectrophotometer with a cell containing 400μl of 1XSSC buffer kept at 40°C by a LAUDA RC6 thermostatic bath. In each reaction, 40pmol of each species of DNA molecules was used unless otherwise stated. Thus the concentration of each species was 100nM.

Experiments by gel electrophoresis were also conducted in 1XSSC buffer incubated at 40°C and the samples were loaded on 10% polyacrylamid gel stained by SYBRGold. The concentration of each species in each reaction was also 100nM.

Table 1. Sequences used in the experiments. **leader** and **output** form the bulge structure and comprise the AND gate. **input1** and **input2** are the inputs to the AND gate. **next** represents the next gate. **leader′** and **output′** comprise another AND gate which shares the first input, **input1**.

leader:
5′-TCTAGCGTGC-ATTCGGAACTTTGTCGGTCC-GGAGGCTAGT-GGTCGGTGAAACTTCGGATA-ATACGGTTGATTGTGGGAGC-(BHQ)-3′
output:
5′-(FAM-)-GCTCCCACAATCAACCGTAT-ACATGGCACC-TATCCGAAGTTTCACCGACC-GGACCGACAAAGTTCCGAAT-3′
input1:
5′-GGACCGACAAAGTTCCGAAT-GCACGCTAGA-3′
input2:
5′-GCTCCCACAATCAACCGTAT-TATCCGAAGTTTCACCGACC-ACTAGCCTCC-AATCCGAAGA-3′
next:
5′-AACTCGCTCC-GGTGCCATGT-ATACGGTTGATTGTGGGAGC-AGTACCCTGG-GCTCCCACAATCAACCGTAT-3′
leader′:
5′-TCTAGCGTGC-ATTCGGAACTTTGTCGGTCC-CGAGGGATGT-GCTGGCAGAATCTTGGGTAA-TAAGGCATGAAAGTCCGTCC-3′
output′:
5′-GGACGGACTTTCATGCCTTA-CGACCCTAGA-TTACCCAAGATTCTGCCAGC-GGACCGACAAAGTTCCGAAT-3′

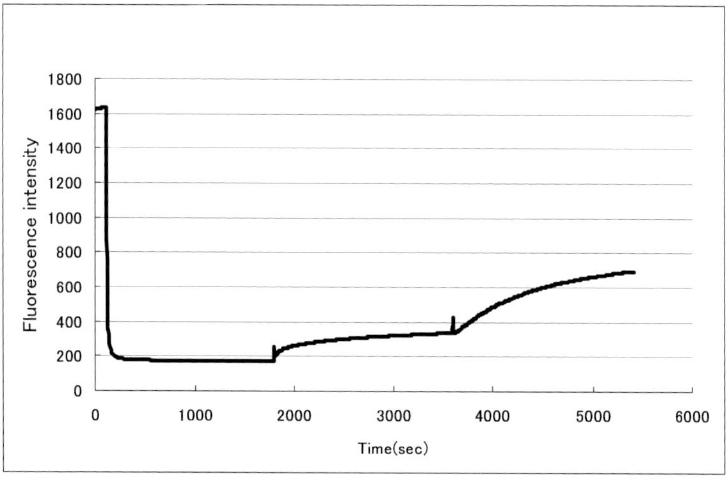

Fig. 6. Experiment 1. The molecule **output**, whose 5′-end is modified with FAM, was first applied in a cell. Then **leader**, whose 3′-end is modified with BHQ, **input2** and **input1** were applied in this order. The fluorescence intensity of FAM was observed.

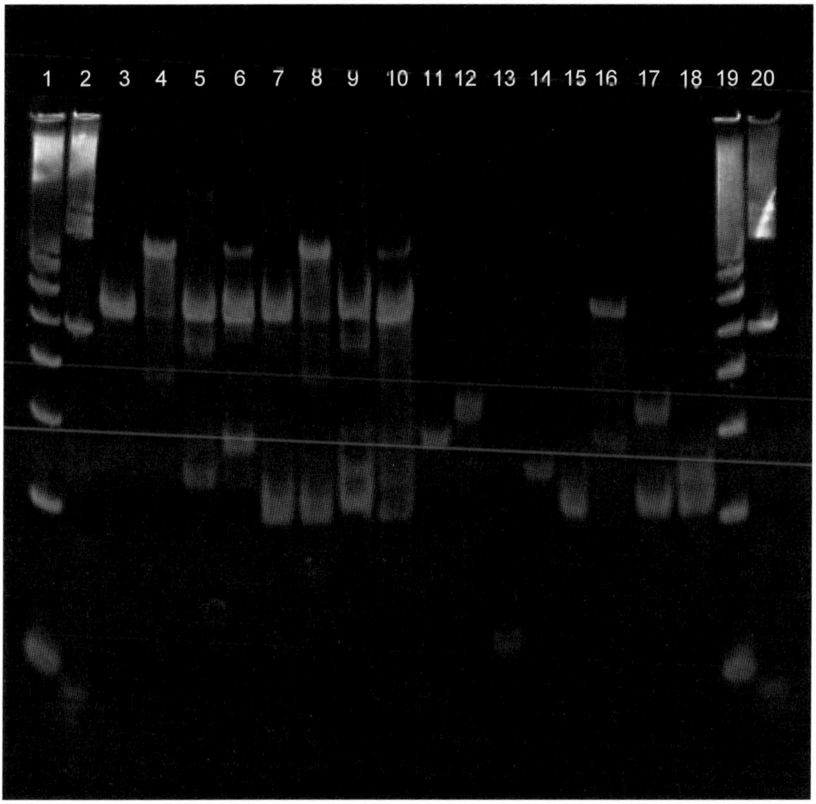

Fig. 7. Experiments 1 and 2 (gel electrophoresis). Lane 1: 20bp DNA Ladder. Lane 2: 100bp DNA Ladder. Lane 3: output+leader Lane 4: (output+leader)+input1. Lane 5: (output+leader)+input2. Lane 6: (output+leader)+input1+input2. Lane 7: (output+leader)+next. Lane 8: (output+leader)+next+input1. Lane 9: (output+leader)+next+input2. Lane 10: (output+leader)+next+input1+input2. Lane 11: output. Lane 12: leader. Lane 13: input1. Lane 14: input2. Lane 15: next. Lane 16: output+next. Lane 17: leader+next. Lane 18: input2+next. Lane 19: 20bp DNA Ladder. Lane 20: 100 bp DNA Ladder.

Experiment 1 (spectrophotoscopy): First, **output** was applied into 1XSSC buffer in a cell for the spectrophotometer, and the measurement of FAM was initiated. Then **leader**, **input2** and **input1** were applied in the cell in this order each with a 30-minute interval. The measurement of fluorescence intensity was performed at 40°C.

Experiments 1 and 2 (gel electrophoresis): Various combinations of molecular species were mixed and observed by gel electrophoresis. In each combination, **leader** and **output** were first mixed and incubated at 40°C for 30 minutes. Then other species such as **input1** and **input2** were applied.

Experiment 2 (spectrophotoscopy): The mixture of **leader** and **output** was first prepared in a cell. Then **input1**, **input2** and **next** were put in all the six possible orders each with a 30-minute interval. The measurement of fluorescence intensity was performed at 40°C.

Experiment 3 (spectrophotoscopy): The mixture of **leader** and **output** was first prepared. That of **leader′** and **output′**, which comprise the competing gate, was also prepared separately. These mixtures were put together in a cell. Then **input2** and **input1** were applied in this order each with a 30-minute interval. In this experiment, the quantity of **input1** was varied as follows: 40pmol, 80pmol and 160pmol.

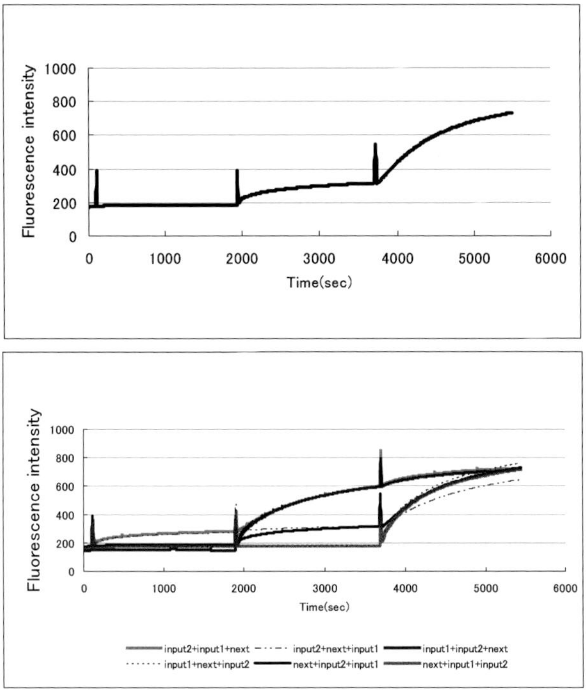

(a) (b)

Fig. 8. Experiment 2. The AND gate was first prepared in a cell. Then **input1**, **input2** and **next** were applied in all the six orders. The fluorescence intensity of FAM was observed. Panel (a) shows the specific order: **next**, **input2** and **input1**.

Results

We obtained the following results.

Experiment 1 (spectrophotoscopy): The proper behavior of the AND gate was observed as in Fig. 6. Due to the weak interaction between **input2** and **leader** without the existence of **input1**, **output** was observed slightly after **input2** was put in. However, a much stronger signal was observed after **input1** was inserted. The quantity of **output** produced by the gate is about half of the total quantify.

Experiments 1 and 2 (gel electrophoresis): The results are depicted in Fig. 7. One can clearly observe the bands for the complexes: **output** + **leader**, **output** + **leader** + **input1** and **leader** + **input1** + **input2**. The latter two are in close proximity.

In Lane 6 ((output+leader)+input1+input2), the band for output is clearly observed while in Lane 5 ((output+leader)+input2), the band is observable but much weaker.

Unfortunately, since the band for **output** + **next** observed in Lane 16 (output+next) is very close to those for **output** + **leader** + **input1** and **leader** + **input1** + **input2**, it cannot be judged whether **output** and **next** hybridize together in Lane 10 ((output+leader)+next+input1+input2). However, comparing Lanes 10 and 6, the band for **output** observed in Lane 6 disappears in Lane 10, so it can be deduced that **output** successfully hybridizes with **next**. Moreover, the smeared wide band in Lane 16 is also observed in Lane 10.

Experiment 2 (spectrophotoscopy): Figure 8 (a) shows the graph of the observed fluorescence intensity after **next**, **input2** and **input1** are put to the mixture of **output** and **leader** in this order. Since the fluorescence intensity does not change after **next** is put, we can conclude that the AND gate and the next gate **next** do not interact before the inputs to the AND gate are given. Figure 8 (b) shows the graphs for all the six orders.

Experiment 3 (spectrophotoscopy): Since the competing gate shares the first input with the original gate, it is expected that about half of the first input hybridizes with the competing gate and thus the output of the original gate decreases. Figure 9 shows the graphs for various quantifies of the first input. As the quantity of the first input increases, that of the output gets close to the level of the case without the competing gate.

4.4 Discussions

The experiments have verified the proper behavior of the AND gate. In particular, it has been verified that the output of the AND gate can serve as an input to the next gate, which does not interact with the AND gate until both the inputs to the AND gate are given.

The experiments also show the weakness of this scheme. Firstly, the output is produced only slightly with the second input. This is due to the unexpected interaction of **input2** and **leader**.

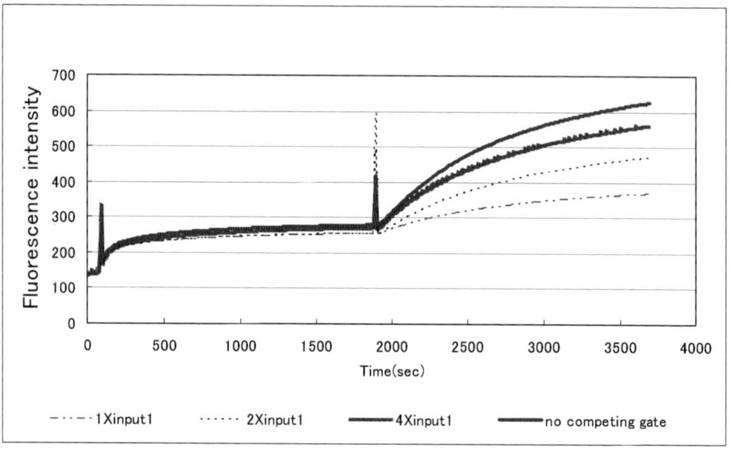

Fig. 9. Experiment 3. The AND gate and its competing AND gate were first prepared in a cell. Then **input2** and **input1** were applied in this order. The qantity of **input1** was varied as follows: **1Xinput1** for 40pmol, **2Xinput1** for 80pmol, **8Xinput1** for 160pmol. For comparison, the graph without the competing gate is also shown.

Secondly, the quantity of the output produced by the gate is only about half of the total quantity of **output**. This is an inherent limitation of this scheme. We expected that the next gate would shift the equilibrium towards the production of the output, but according to the graph in Figure 8 (b), this effect is negligible.

Thirdly, the first input was trapped by the competing gate and the level of the output was further lowered. We also predicted this behavior in our previous paper [22], and proposed a modification to the scheme. We proposed that variants of the first input which stay in a certain equilibrium with both the AND gates might migrate from the competing gate to the original gate when the second input is given to the original gate.

We actually tried the following two openers:

`AGTTCCGAAT-GCACGCTAGA    CAAAGTTCCGAAT-GCACGCT`

But unfortunately, they did not work properly as the first input even without the competing gate (data not shown).

Although the design of logic gates by Seelig et al. is much more complex than ours and it shares the problem that the first input is consumed by competing gates [17], their gates can produce more outputs than inputs because inputs serve as catalysts. We think that it is practical to combine our design and theirs, by inserting "amplifiers" based on DNA catalysts at appropriate stages of circuits composed of our simpler gates. In fact, we are currently designing such amplifiers consisting of bulge loops.

4.5 Concluding Remarks

In this chapter, we briefly described the previous contributions of the first author concerning hairpins and then reported some ongoing experiments on implementing logic gates using dissociation of loop structures of DNA. All the contributions are concerned with secondary structures of DNA, especially hairpins, and have revealed the computational power of the formation and dissociation of DNA structures. As reported in another paper [21], we are further working in this direction by introducing artificial molecules in addition to natural DNA and trying to enhance the computational power.

Acknowledgments

This work is partially supported by Grand-in-Aid for Scientific Research on Priority Area No.14085202, Ministry of Education, Culture, Sports, Science and Technology, Japan.

References

1. L.M. Adleman, Molecular computation of solutions to combinatorial problems, *Science,* Vol.266, pp. 1021–1024 (1994).
2. M. Arita, and S. Kobayashi: DNA sequence design using templates, *New Generation Computing,* Vol. 20, pp. 263–277 (2002).
3. R.S. Braich, N. Chelyapov, and C. Johnson, P. W. K. Rothemund, and L. Adleman: Solution to a 20-variable 3-SAT problem on a DNA computer, *Science,* Vol. 296, pp. 499–502 (2002).
4. R.M. Dirks and and N.A. Pierce: Triggered amplification by hybridization chain reaction, *PNAS,* Vol. 101, pp. 15275–5278 (2004).
5. M. Hagiya, M. Arita, D. Kiga, K. Sakamoto and S. Yokoyama, Towards Parallel Evaluation and Learning of Boolean μ-Formulas with Molecules, *DNA Based Computers III, DIMACS Series in Discrete Mathematics and Theoretical Computer Science,* Vol. 48, pp. 57–72 (1999).
6. A. Kameda, M. Yamamoto, H. Uejima, M. Hagiya, K. Sakamoto, and A. Ohuchi, Hairpin-Based State Machine and Conformational Addressing: Design and Experiment, *Natural Computing* Vol. 4, No. 2, pp. 103–126 (2005).
7. S. Kobayashi, T. Kondo, K. Okuda, and M. Arita: On Template Method for DNA Sequence Design, *Lecture Notes in Computer Science,* Vol. 2568, pp. 205–214 (2002).
8. K. Komiya, Ph.D. Thesis, Department of Biophysics and Biochemistry, Graduate School of Science, University of Tokyo, 2004.
9. M. Kubota, K. Ohtake, K. Komiya, K. Sakamoto, and M. Hagiya: Branching DNA Machines Based on Transitions of Hairpin Structures, *Proceedings of the 2003 Congress on EVolutionary Computation* (CEC'03), pp. 2542–2548 (2003).
10. R.J. Lipton: DNA Solution of hard computational problems, *Science,* Vol. 268, pp. 542–545 (1995).

11. J.A. Rose, R.J. Deaton, M. Hagiya, A. Suyama: An Equilibrium Analysis of the Efficiency of an Autonomous Molecular Computer, *Physical Review E,* Vol. 65, No.2-1, 021910, pp. 1–13 (2002).
12. J.A. Rose, R.J. Deaton, M. Hagiya, and A. Suyama: PNA-mediated Whiplash PCR, *Lecture Notes in Computer Science,* Vol. 2340, pp. 104–116 (2002).
13. J.A. Rose, M. Takano, M. Hagiya and A. Suyama: A DNA Computing-based Genetic Program for In Vitro Protein Evolution via Constrained Pseudomodule Shuffling, *Journal of Genetic Programming and Evolvable Machines,* Vol. 4, No.2, pp. 139–152 (2003).
14. P.W.K. Rothemund, N. Papadakis, E. Winfree: Algorithmic Self-assembly of DNA Sierpinski Triangles, *PLoS Biology,* Vol. 2, No.12, e424 (2004).
15. K. Sakamoto, H. Gouzu, K. Komiya, D. Kiga, S. Yokoyama, T. Yokomori, M. Hagiya, Molecular computation by DNA hairpin formation, *Science,* Vol. 288, pp. 1223–1226 (2000).
16. K. Sakamoto, D. Kiga, K. Komiya, H. Gouzu, S. Yokoyama, S. Ikeda, H. Sugiyama and M. Hagiya, State Transitions by Molecules, *Biosystems,* Vol. 52, pp. 81-91 (1999).
17. G. Seelig, B. Yurke, and E. Winfree: DNA Hybridization Catalysts and Catalyst Circuits, *Lecture Notes in Computer Science,* Vol. 3384, pp. 329–343 (2004).
18. N.C. Seeman: It Started with Watson and Crick, But It Sure Didn't End There: Pitfalls and Possibilities Beyond the Classic Double Helix, *Natural Computing,* Vol. 1, pp. 53–84 (2002).
19. N.C. Seeman, H. Wang, B. Liu, J. Qi, X. Li, X. Yang, F. Liu, W. Sun, Z. Shen, Y. Wang, R. Sha, C. Mao, S. Zhang, T.-J. Fu, S.M. Du, J.E. Mueller, Y. Zhang and J. Chen: The perils of polynucleotides: The experimental gap between the design and assembly of unusual DNA structures, *DNA Based Computers II, DIMACS Series in Discrete Mathematics and Theoretical Computer Science,* Vol. 44, pp. 215–233 (1999).
20. K. Takahashi and M. Hagiya: Preliminary Experiments on Hairpin Structure Dissociation for Constructing Robust DNA Machines, *Proceedings of the 2004 IEEE Conference on Cybernetics and Intelligent Systems,* Singapore, 1-3 December, 2004, pp. 285–290 (2004).
21. K. Takahashi, S. Yaegashi, H. Asanuma and M. Hagiya: Photo- and Thermoregulation of DNA Nanomachines, *DNA11, Eleventh International Meeting on DNA Based Computers, Preliminary Proceedings,* pp. 147–156 (2005).
22. K. Takahashi, S. Yaegashi, A. Kameda and M. Hagiya: Chain Reaction Systems Based on Loop Dissociation of DNA, *DNA11, Eleventh International Meeting on DNA Based Computers, Preliminary Proceedings,* pp. 343-353 (2005).
23. A.J. Turberfield, J.C. Mitchell, B. Yurke, A.P. Mills Jr, M.I. Blakey and F. C. Simmel: DNA Fuel for Free-Running Nanomachines, *Phys. Rev. Lett.,* Vol. 90, No.11, pp. 118–102 (2003).
24. E. Winfree, F. Liu, L.A Wenzler and N.C. Seeman, Design and self-assembly of two-dimensional DNA crystals, *Nature 394,* pp. 539–544 (1998).
25. E. Winfree, X. Yang, and N.C. Seeman: Universal computation via self-assembly of DNA: some theory and experiments, *DNA Based Computers II, DIMACS Series in Discrete Mathematics and Theoretical Computer Science,* Vol. 44, pp. 191–213 (1999).
26. B. Yurke, A.J. Turberfield, A.P. Mills, Jr., F.C. Simmel, and J.L. Neumann: A DNA-fuelled molecular machine made of DNA, *Nature,* Vol. 406, pp. 605–608 (2000).

Bottom-up Approach to Complex Molecular Behavior

Milan N. Stojanovic

Department of Medicine, Columbia University, New York, NY 10032, USA
mns18@columbia.edu

1 Introduction

One can quite reasonably ask why someone trained as an organic chemist would contribute to a volume dedicated to Ned Seeman's sixtieth birthday. The connection is actually quite deep, and goes back to the conversation a decade ago which I had with my Ph.D. adviser, Yoshito Kishi. In this exchange, a leading organic synthetic chemist was trying to convince me to move into a new area, nucleic acids. He was inspired in his suggestion by a conversation with Alex Rich (Ned's adviser), who had apparently lamented that a well-trained organic chemist could make a serious contribution to this field. I decided to educate myself in the area, usually neglected by traditional synthetic chemists, and, of course, stumbled immediately upon Ned's proposed synthesis of a DNA cube in a one-shot reaction [4]. This idea was so powerfully appealing to my synthetic background, that it sparked a thought process that would eventually lead to a large project with the goal of achieving complex molecular behavior through a synthetic approach using oligonucleotides.

In his classic book *Vehicles: Experiments in Synthetic Psychology*, Valentino Breitenberg [3] argued that synthetic approaches starting with simple elementary units could quickly lead to an amazing increase in complexity, leading to apparent biomimetic behavior. In a like-minded approach, helped by my childhood friend, a computer scientist Darko Stefanovic, I decided to attempt a synthetic approach with a focus on the behavior of molecules in solution. Starting from a simple and well-understood, but versatile, elementary unit, we could combine such elements in solution, building "bottom-up" complexity. What kind of properties should such an elementary unit have? First, it should have the ability to integrate the presence or absence of several inputs into a single output, that is, the unit should behave as a logic gate on the molecular scale. Second, a component should have the ability to communicate with other units through some kind of information transfer. Otherwise, we would have elements behaving completely independently in solution. Third, a unit should be commercially available and modular, that is, we did not want to start anew the construction of each component; instead we wanted to have a Lego-like approach, in which we could devise units, and obtain them within a week effortlessly. Finally, we were particularly interested in units that could communicate with sensors (that is, receive inputs from them), and produce outputs that could control some kind of drug delivery units. Our eventual applications for these systems were autonomous therapeutic and diagnostic devices; in other words, silicon-free expert systems operating at the cellular level.

2 Molecular-Scale Logic Gate as a Basic Computational Unit

After a brief consideration, we settled on a molecular unit made of allosterically modulated deoxyribozymes [7], with oligonucleotides as both inputs and outputs. Deoxyribozymes are nucleic acid catalysts, made of deoxyribonucleic acid, and in our work we use two different types: phosphodiesterases [2, 10], which cleave other oligonucleotides, with shorter products as outputs, and ligases [5], which combine two oligonucleotides into a larger product (Fig. 1). From the very beginning, it was our hypothesis that the concordance of outputs and inputs would be a key to arranging elementary units in more complex circuits. Furthermore, we assumed that we would be able to generate nearly endless numbers of units with different inputs and outputs by simply changing allosterically controlling elements to sequences complementary to new input oligonucleotide, and substrate recognition regions capable of accepting new substrates. Finally, two or even more recognition regions can be combined in the same molecular constructs to generate more complex molecular-scale logic gates.

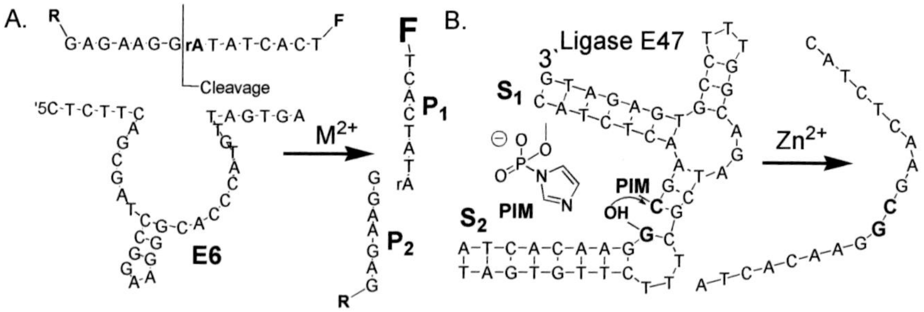

Fig. 1. Two deoxyribozymes: (**a**) Phosphodiesterases E6(5) cleaves a chimeric substrate **S** and releases two products P_1 and P_2. (**b**) Ligase E47 combines two substrates S1 and S2 into a product **P**.

Our module for allosteric regulation was based on Tyagi and Kramers molecular beacons [17]. Stoichiometric molecular beacons exist in solution

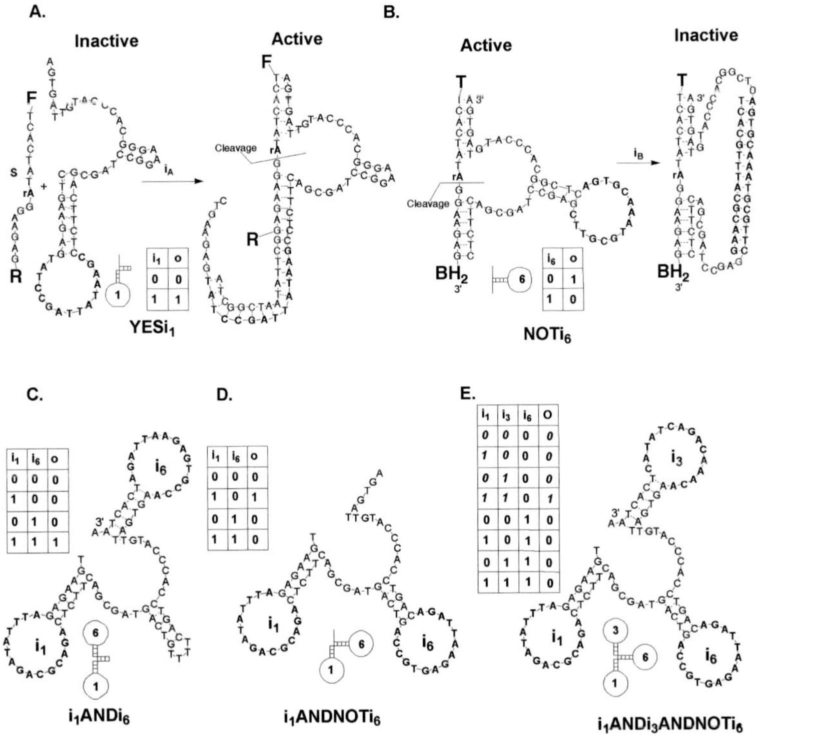

Fig. 2. Deoxyribozyme-based gates: (**a**) YES gate; (**b**) NOT gate. (**c**) AND gate; (**d**) two-input ANDNOT gate; three-input ANDANDNOT gate.

as stable stem-loop structures in which the fluorescence of a reported dye attached to the 5′ end is quenched by a proximate quencher attached to the 3′ end. In the presence of a complementary nucleic acid the stem opens, and this event is coupled to a loss of quenching and an increase in fluorescence. We combined in the YESi_1 gate a catalytic module from the core deoxyribozyme (E6 in Fig. 2), with a stem-loop module complementary to i_1 (Fig. 3). The stem-loop inhibits the catalytic module through the overlap of the stem with the 5′ (or 3′, if so desired) substrate recognition region. Hybridization of i_1 to the complementary loop opens the stem, reversing intramolecular competitive inhibition to allow substrate binding to proceed. The YESi_1 gate behaves as a two-state switch, with the active state in the presence of input [11]. In order to improve visualization of the output (cleaved oligonucleotide) in this system, we developed a fluorogenic cleavage technique method. We placed a fluorescein donor at the 5′ terminus of **S** and its fluorescence emission was partially quenched by the tetramethyl rhodamine acceptor or Black-Hole 1 quencher positioned at the 3′ terminus. Cleavage of this double end-labeled substrate to products results in at least a tenfold increase in fluorescein emission. Identical design can be used to construct ligase YES gate, however, the readout for these gates was performed initially with PAGE analysis [14].

The single-input NOTi_3 gate (Fig. 3) is inhibited by a specific oligonucleotide, and is constructed by replacing the nonconserved loop of the E6 catalytic core with a stem-loop sequence complementary to the input oligonucleotide [12]. Hybridization of the input with the loop opens the required stem structure of the core, distorts its shape and inhibits the catalytic function. Thus, the presence of the input oligonucleotide will hinder the increase in fluorescence. The NOTi_3 gate behaves as a two-state switch, with the active state in the absence of input.

To construct the AND gate, we needed a deoxyribozyme allosterically regulated by two different input oligonucleotides. We achieved this by attaching controlling elements to each end of a single catalyst. In the absence of its proper input either of the attached stem-loop structures independently inhibits output formation. Only upon hybridization of both loops to complements do both stems open, allowing recognition of S and its catalytic cleavage. The i_1ANDi_3 gate behaves as a four-state switch, with one state, in the presence of both inputs, active [12].

We also combined YES and NOT gates in a single molecule to construct an i_1ANDNOTi_3 gate [12]. We attached a stem-loop recognizing i_1 to a position at the 5′-end, in which it inhibits the catalysis, as in a YES gate, and a stem-loop recognizing i_3 to the internal position of the E6 catalytic motif, where it does not influence the catalytic reaction, until it recognizes the input oligonucleotide. The i_1ANDNOTi_3 is active only in the presence of i_1 and the absence of i_3. Finally, we combined AND gates with NOT gates in a single molecule to construct an eight-state switch with one active state, i_1ANDi_2ANDNOTi_3 [15]. Importantly, an absolutely identical method was also used to construct the complete set of ligase-based logic gates [14], because ligases have the same "hammerhead" structure as phosphodiesterases.

The YES, NOT, AND, ANDNOT, and ANDANDNOT deoxyribozyme-based logic gates we constructed represented a basic set of molecular-scale gates (AND and NOT are one possible basic set in electronics). Importantly, these gates are generic and modular, in the sense that other deoxyribozymes (or ribozymes) could be combined in similar constructs, with the expectation of similar behavior (though some limitations based on secondary structures have to be taken into account). That means that we can now, in principle at least, construct enzymatic networks that perform Boolean calculations of any complexity, that is, that we can have an arbitrary number of inputs and outputs. Monomolecular systems of more than three inputs could be envisioned, but would not be general and we were not interested in pursuing them. Instead, we opted to start developing parallel and serial arrangements of deoxyribozyme-based gates in order to achieve more complex Boolean calculations.

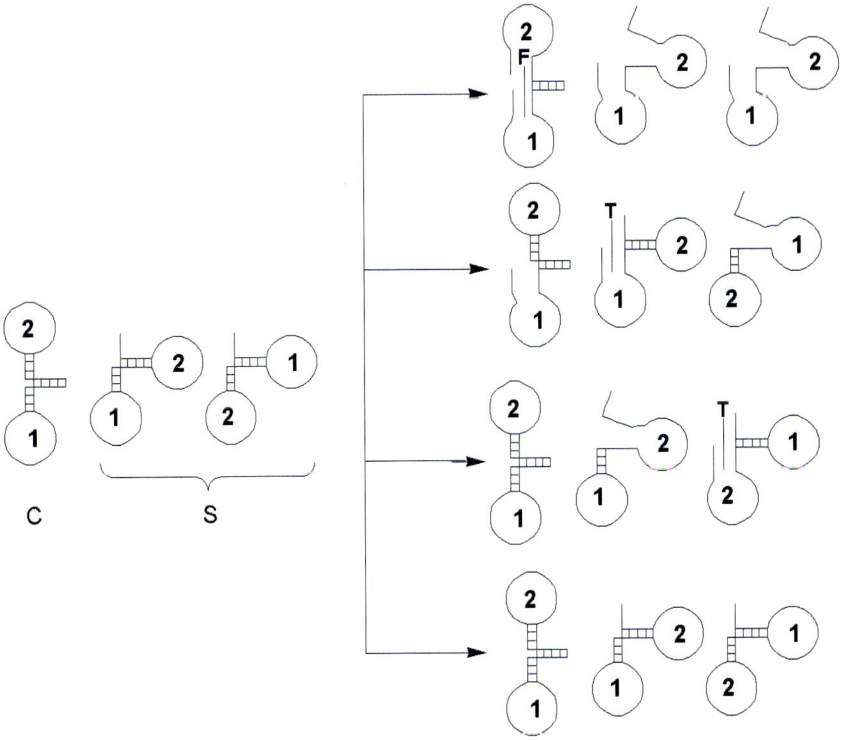

a

b

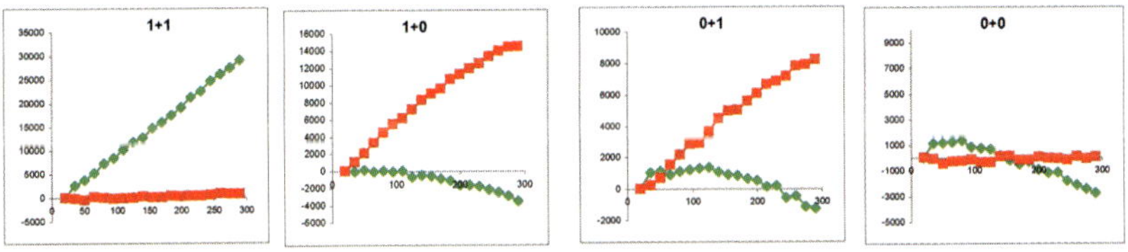

Fig. 3. (**a**) Array of deoxyribozymes behaving as a half adder. The array consists of two ANDNOT gates cleaving TAMRA-containing substrate and one AND gate cleaving fluoroescein-containing substrate. (**b**) Fluorescence changes under four different conditions, in the presence of no inputs $(0+0)$, the presence of either one of the inputs $(0+1, 1+0)$, and in the presence of both inputs $(1+1)$. Red squares represent changes in TAMRA fluorescence over times (sum digit) and green diamonds changes in fluorescein emission over time (carry digit). This behavior mimics the half-adder in engineering, if we presume that inputs represent the digits of two binary numbers being added together.

3 Initial Molecular Circuits

A parallel arrangement of molecular logic gates is called the "implicit OR function" [13]; it is accomplished by two or more logic gates sharing (and

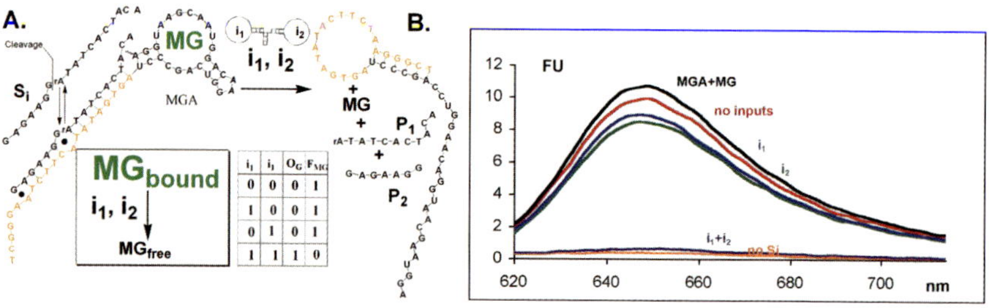

i_1	i_2	O_G	F_{MG}
0	0	0	1
1	0	0	1
0	1	0	1
1	1	1	0

Fig. 6. (**a**) A cascade consisting of an i_1**AND**i_2 gate activating the malachite green aptamer, which is allosterically regulated by its substrate Si. The malachite green aptamer exists in two conformations (binding and fluorescent, and non-binding or dark; that is, complexed with $\mathbf{S}_\mathrm{i}$ or not-complexed with $\mathbf{S}_\mathrm{i}$), which are switched based on the result of Boolean calculations from the upstream **AND** gate. The truth table contains the gate state ($\mathbf{O}_\mathrm{G}$) and the fluorescence of malachite green as outputs ($\mathbf{F}_\mathrm{MG}$). (**b**) The fluorescence intensity after 8 hours of incubation (λ ex = 600 nm) (all samples have MGA and MG): black spectra: positive control, no gate and no inputs; red spectra: both inputs absent; blue spectra: only i_1; green spectra: only i_2; maroon spectra: both inputs; orange spectra: no substrate $\mathbf{S}_\mathrm{i}$.

(**MGA**, Kd ~ 120 nM) [1]. Malachite green (**MG**) is almost non-fluorescent while free in solution, but becomes strongly fluorescent when complexed with its aptamer, and this characteristic provides us with a straightforward way of detecting the state of the aptamer, which can be binding (**1**) or non-binding (**0**). A two-state switch version of **MGA** consists of a malachite-green binding module, and an extension region; the extension can bind either $\mathbf{S}_\mathrm{i}$ or the malachite green binding module, but not both (Fig. 2a). Thus, in the presence of $\mathbf{S}_\mathrm{i}$, the aptamer would become active and bind **MG**, while upon the removal of $\mathbf{S}_\mathrm{i}$, through the catalytic reaction of a deoxyribozyme gate, the aptamer module should become inactive and release the dye.

In Fig. 6 we illustrate this approach with an upstream dual input logic gate (i_1**AND**i_2) performing a Boolean calculation based on 21-mer input oligonucleotides and releasing malachite green according to its truth table. A 90% decrease in fluorescence signal (from 8 relative units to 0.6 units) after six hours with both inputs present (Fig. 2b, maroon spectra) indicates that we effectively achieved Boolean control of the release of a small molecule through molecular scale computation elements, which is a proof-of-concept experiment for computation-triggered drug release.

6 Conclusions

It has been stated by others that our efforts fit into the general field of synthetic biology. Indeed, our desire to understand how to build complex artificial molecular networks can be viewed as part of a general push in this area. But, what are the applications of deoxyribozyme-based networks? Obviously, this type of computation cannot compete with modern computers in performing complex calculations. The enzymes are simply too slow, and even non-complementary oligonucleotides interact with each other at higher concentrations. Despite this, we see several areas of potential applications for deoxyribozyme-based and related networks in the future. First, the ability of molecules to perform, for the first time, arbitrary Boolean calculations with great precision and flexibility can be used to control molecular devices (we are reluctant to use the much abused term "nanorobots" here). For example, a series of inputs could be tied to the Boolean calculation, which would result in triggering a movement by a molecular robot made of nucleic acids (cf., our "spider molecules" or various walkers previously reported). Also along these lines, we are constructing mechanical devices, including the next generation of drug delivery elements, which could be controlled by logic gates. Second, ribozyme-based logic gates could form circuits capable of controlling cell behavior; the ease with which a deoxyribozyme-based automaton was constructed argues that such networks may have greater flexibility than corresponding protein analogs. Finally we should stress the significance of studying these networks for the theories of early life, and for mimicking early metabolic networks. Somewhat beyond the scope of this review, but definitely very intriguing, are the possibilities of interconnecting these enzymatic networks with various self-evolving systems that are used by RNA chemists. In particular, we are referring the reader to Joyce's self-evolving ribozymes [8], and in lieu of a conclusion to this chapter, we propose that such networks could be evolved to perform computational functions.

References

1. J.R. Babendure, S.R. Adams, and R.Y. Tsien. Aptamers switch on fluorescence of triphenylmethane dyes. *J. Am. Chem. Soc.*, 125:14716–14717, 2003.
2. R.R. Breaker and G.F. Joyce. A DNA enzyme with mg^{2+}-dependent RNA phosphoesterase activity. *Chemistry & Biology*, 2:655–660, 1995.
3. V. Breitenberg. *Vehicles: Experiments in Synthetic Psychology.* Bradford Books, MIT Press, Cambridge, MA, 1984.
4. J. Chen and N.C. Seeman. Synthesis from DNA of a molecule with the connectivity of a cube. *Nature*, 350:631–633, 1991.
5. B. Cuenoud and J.W. Szostak. A DNA metalloenzyme with DNA ligase activity. *Nature*, 375:611–614, 1995.
6. W.U. Dittmer, A. Reuter, and F.C. Simmel. A DNA-based machine that can cyclically bind and release thrombin. *Angew. Chem.-Int. Ed. (Engl.)*, 43:3550–3553, 2004.
7. G.M. Emilsson and R.R. Breaker. Deoxyribozymes: new activities and new applications [review]. *Cell. Mol. Life Sci.*, 59:596–607, 2002.
8. D. Ferber. Synthetic biology. Microbes made to order. *Science*, 303:158–161, 2004.

9. D. Kolpashchikov and M.N. Stojanovic. Boolean control of aptameric binding states. *J. Am. Chem. Soc.*, 127, 2005. in press.
10. S.W. Santoro and G.F. Joyce. A general purpose RNA-cleaving DNA enzyme. *Proc. Natl. Acad. Sci. USA*, 94:4262–4266, 1997.
11. M.N. Stojanovic, P. de Prada, and D.W. Landry. Catalytic molecular beacons. *J. Am. Chem. Soc.*, 2:411, 2001.
12. M.N. Stojanovic, T.E. Mitchell, and D. Stefanovic. Deoxyribozyme-based logic gates. *J. Am. Chem. Soc.*, 124:3555, 2002.
13. M.N. Stojanovic, D.B. Nikic, and D. Stefanovic. Implicit or tiling of deoxyribozymes. *J. Serb. Chem. Soc.*, 68:321, 2003.
14. M.N. Stojanovic, S. Semova, D. Kolpashchikov, J. Macdonald, C.A. Morgan, and D. Stefanovic. Deoxyribozyme ligase-based logic gates and their initial circuits. *J. Am. Chem. Soc.*, 127:6914, 2005.
15. M.N. Stojanovic and D. Stefanovic. Deoxyribozyme-based automaton. *Nature Biotech.*, 21:1069, 2003. (Widely reported in popular and scientific press)
16. M.N. Stojanovic and D. Stefanovic. Deoxyribozyme-based half-adder. *J. Am. Chem. Soc.*, 125:6637, 2003.
17. S. Tyagi and F.R. Kramer. Molecular beacons: probes that fluoresce upon hybridization. *Nature Biotechnol.*, 14:303–308, 1996.
18. B. Yurke, A.J. Turberfield, A.P.J. Mills, F.C. Simmel, and J.L. Neumann. A DNA-fuelled molecular machine made of DNA. *Nature*, 406:605–608, 2000.

Aqueous Computing: Writing on Molecules Dissolved in Water

Tom Head[1] and Susannah Gal[2]

[1] Binghamton University, Department of Mathematical Sciences, Binghamton, New York 13902-6000 USA
tom@math.binghamton.edu

[2] Binghamton University, Department of Biological Sciences, Binghamton, New York 13902-6000 USA
sgal@binghamton.edu

1 Introduction: the Aqueous Concept

The aqueous concept of computation was developed after studying Leonard Adleman's method of DNA computing in which he used the annealing, in water, of single-stranded molecules of DNA [1, 2]. In the aqueous concept it is the *water* rather than the *molecules* that is chosen as the basis for a new exploratory computing concept. Suppose that memory can somehow be contained in water and in enormous redundancy. The redundancy allows the assumption that the many representations of each item in memory are uniformly distributed throughout the water. Aqueous memory then allows two extraordinary operations: (1) *The complete memory can be replicated in unit time* (pour the aqueous memory into two containers). (2) *Two memories can be merged in unit time* (pour two into one). How can these two operations be used in computing?

A nanoscale 'writing tablet' must be chosen that can be dissolved (or suspended) in water. One must be able to 'write' on these tablets and 'read' from them. There is a seemingly endless list of potential 'tablet-writing–reading' technologies that one can imagine. Several of these choices have been prototyped in wet labs, and these are reviewed here. Whichever choice is made, each wet-lab computation begins with a test tube containing a vast number of identical nanotablets in water. Each computation consists of a *sequence of steps* in the following pattern: pour tube into two or more tubes; in each tube, write in parallel the *same* datum on *all* the tablets the tube contains (with different tubes receiving different data); and unite the tubes into a single tube. After the appropriate sequence of such steps has been carried out, the computation is completed by reading from tablets in the final tube.

Several introductions to aqueous computing have appeared and may be helpful in clarifying any points that are not made clear here [4, 8, 19].

2 At Leiden University

The first aqueous computation was initiated in the summer of 1998 at Leiden University in the laboratory of Herman Spaink with support from the Leiden Center for Natural Computing (LCNC). For use as the *tablet*, a circular double-stranded (ds) DNA molecule (a plasmid) was modified by the incorporation of a specially designed insert, INS, containing six subsegments each consisting of from 36 to 51 base pairs. Each of these subsegments (stations) was bounded at each end by a site for a restriction enzyme that had no site on the plasmid other than at these two station-bounding sites. In a computation, each of the six stations is a location at which one memory bit is represented. The initial condition of each station of the plasmid is interpreted as representing the bit *one*. *Writing* at a specific station consists of the removal of the subsegment with which the station is associated. This is done in two steps. By using the restriction enzyme, the two sites of which bound the subsegment, the plasmid is cut into two linear molecules: the short subsegment itself and the much longer linear remnant of the original circular molecule. Using a ligase, the longer linear molecule is recircularized. The deletion of the short subsegment with which the station is associated is interpreted as the writing of a *zero* at this station. Thus: *subsegment present* = 1; *subsegment absent* = 0. No provision has been made for replacing a zero by a one. This choice of writing procedure has been called 'CDL', for 'cut–delete–ligate'. The processes for *reading* in aqueous computations have been varied, even when the same writing technique has been used. Consequently reading is specified here only after the choice has beed made of the particular algorithmic problem to be treated.

The first computation done in Leiden treated an instance of the maximal-independent-set (MIS) problem of graph theory. The specific instance chosen was the one to which the maximal clique problem treated previously in [13] was reduced and then solved. The graph considered was $G = (V, E)$, having vertex set $V = \{a, b, c, d, e, f\}$ and set of *un*directed edges $E = \{(a, b), (b, c), (c, d), (d, e)\}$. Recall that the MIS problem calls for the determination of the cardinal number of a largest independent subset of V. As is usual with simple wet-lab prototype computations, the answer can be observed in advance and in this case is 4. In fact, the unique independent subset of largest cardinal number is apparently $\{a, c, e, f\}$. But how can the solution be obtained using only wet-lab operations?

The six stations of our plasmid tablet are identified with the six vertices of the graph G. The initial condition of each plasmid (six ones) is taken as a characteristic function defined on V and therefore initially represents V itself. *The initial state of our computer is a test tube T containing a vast number of these plasmid tablets.* Each of the four edges in E is treated in turn as follows: Pour the current tube T into two tubes L and R and treat the next edge (x, y) in E that has not yet been treated by writing a zero at station x in L and a zero at station y in R; unite L and R into (a new) tube T. Note that T now contains no tablet having a one at both the station x and the station y. A sequence of four such steps, one for each edge in E, gives a final tube T that contains only plasmids encoding independent subsets of V. An elementary exercise confirms that the characteristic function of any maximal independent subset of V will be encoded on plasmids in this final T. Note that the plasmids that encode the largest number of ones contain the solution to the MIS problem. We must now 'read' from T.

Recall that the six stations lie on an insert, INS, that was installed in the plasmid. This insert, was designed to have special restriction sites at each end. This allows the insert, of intermediate length, to be cut out from the much longer plasmid. Reading has been done as follows. In the final T, the inserts are cut from the plasmids leaving the short linear INS molecules and the much longer linear remnants. A gel separation is performed on the resulting contents of T. The inserts (which may have various intermediate lengths following the previous four steps of removal of selected subsegments) migrate down the gel leaving the large residues near the top of the gel. From the location of the band on the gel that contains the longest inserts, the number of ones that are encoded on these inserts can be determined. This gives the solution of this instance of the MIS problem: 4. If one wishes to determine a (the) maximum independent set itself, one may cut from the gel the band containing the molecules that encode, as a characteristic function, the maximal independent set. The six bits 101011, which specify the subset $\{a, c, e, f\}$, can be determined by DNA sequencing or by a variation of the method illustrated in the next section.

The results of the computation described here are available in [8], which gives wet-lab details not mentioned here and displays a photograph of the final gel described above.

3 At Binghamton University

In Binghamton in the late fall of 1998, with the collaboration of our guest, Masayuki Yamamura, a second aqueous computation was begun in the laboratory of Susannah Gal. The following instance of the Boolean satisfiability problem (SAT) was chosen for prototyping: Is there an assignment of the values True (1), False (0) for the logical variables p, q and r for which each of the following clauses evaluates to True (1): p OR q, p' OR q OR r', q' OR r', p' OR r? We use primes to denote negation, so that, for example, p' is True precisely when p is False. The answer can be observed in advance to be 'Yes' and it can be seen that there is a unique truth setting, namely $p = 0$, $q = 1$, $r = 0$ for which all four clauses evaluate to True. But how can the solution be obtained using only wet-lab operations?

As the tablet, a commercially available circular cloning plasmid, pBluescript, was chosen and used without modification. This plasmid contains a cluster of restriction enzyme sites called the *multiple cloning site* (MCS). Each site in the MCS is the unique location in the plasmid at which its associated restriction enzyme cuts. Six sites in the MCS were chosen to serve as the stations at which bits are represented, one bit per station. Each of the six sites chosen as a station is a site at which the corresponding enzyme cuts, producing a linear molecule having a four-base $5'$-overhanging single strand at each end. Initially each station is interpreted as representing the bit *one*. *Writing* a *zero* at a station is done in three steps: linearize the plasmid by cutting at the station with its associated restriction enzyme; complete each four-base $5'$-overhanging single strand into a double strand using a DNA polymerase; and recircularize the linear molecule into a circular plasmid using a ligase. Observe that each such alteration of a plasmid increases the circumference by four base pairs and alters the sequence at the station so that its associated restriction enzyme will no longer cut. Thus, *Can be cut at station* = 1 (True); *can't be cut* = 0 (False). No provision has been made for replacing a zero by a one. This choice of writing procedure has been called 'CEL', 'for cut–extend–ligate'.

The six chosen stations on our plasmid tablet are identified with the six literals p, p', q, q', r, r'. The computation is initiated (as always) with a tube T containing a vast number of plasmids, each of which can be cut at any one of the six stations and therefore represents six ones. This initial interpretation expresses three *logical contradictions*, the first of which is that p and p' are both true. The first three steps of the computations have the same form and each eliminates one of the contradictions: pour tube T into two tubes L and R; write a zero at p in L and a zero at p' in R; unite L and R into T. No molecule in T now contains both a one at station p and a one at station p'. Using this new T, do the same for q and q'; and with the resulting newer T treat r and r' in the same way. Observe that this results in a tube T that contains precisely the molecules that encode the eight logically consistent settings of the six literals. (Note that the number of steps of this type is the number of variables.) The next four steps of the computation have the same general form and each eliminates those molecules that encode a truth setting for which one of the clauses *fails* to evaluate to True. Pour T into L and R; add to L the enzyme that will cut (linearize) those molecules representing $p' = 1$ and add to R the enzyme that will cut (linearize) those molecules representing $q' = 1$; and unite L and R into T. Observe that this results in a tube T in which all circular molecules encode logically consistent truth settings for which p OR q evaluates to True. The next step has the same general form, but uses an extra tube: pour T into L, M and R; linearize molecules representing $p = 1$ in $L, q' = 1$ in M, and $r = 1$ in R; and unite L, M and R into T. Observe that this results in a tube T in which all circular molecules encode logically consistent truth settings for which both p OR q and p' OR q OR r' evaluate to True. After two additional similar steps a tube T results in which all circular molecules encode logically consistent truth settings for which all four clauses evaluate to True.

Assume that only the circular molecules have been saved – with the molecules linearized in the preceding steps being discarded (a standard procedure). The result can now be read as follows. The simple 'Yes' is obtained from a confirmation that there are circular molecules remaining. However, the unique truth setting that results in all four of the clauses evaluating to True is obtained as follows: *linearize the circular molecules in T by cutting with a special restriction enzyme that cuts only at a single site on the plasmid and in such a way that the segments to the left and right of the* MCS *of the plasmid differ substantially in length.* (On pBluescript the site for Sca I can be used.) Pour T into three tubes L, M and R and test to determine which of the stations p, q and r can be cut by their associated restriction enzymes. After the appropriate enzyme has been added to a tube, there will be linear molecules of two different lengths in the tube if the enzyme has cut, but only molecules of one length if cutting was not possible. Consequently when the contents of L, M and R are placed on gels to be electrically separated, a single band on a gel gives variable $= 0$ (False) and a pair of bands on the gel gives variable $= 1$ (True). The present computation was carried out and resulted in: P is False (single band), q is True (two bands) and r is False (single band), which, as previously observed, is easily verified to be the unique setting of p, q and r for which all the clauses do indeed evaluate to True. (Note that the number of steps of this type is the number of clauses.) For instances of the SAT problem that have more than one such setting there is extra lab work to do to obtain the distinct settings, but procedures are known for such cases.

The results of the computation described here are available in [6], which gives wet lab details not mentioned here and displays a photograph of the final gel described above.

Currently, at Binghamton, we are exploring the use of methylase enzymes for aqueous computing. We are attempting to carry out the same computation as described above, but using a new system for writing zeros. We wish to write zeros (i.e., prevent cutting at restriction sites) by methylating the sites, rather than by using the more complicated CEL method described above.

4 At Tokyo Institute of Technology

A third writing technique for use in aqueous computing has been developed in the laboratory of Masayuki Yamamura at Tokyo Institute of Technology. The nanotablet continues to be a dsDNA molecule and computations are initiated with a tube T containing a vast number of these tablets. For writing on the tablets, the new method uses short PNA molecules that aggressively interrupt dsDNA molecules and bind, where allowed by Watson–Crick base-pairing rules, with one of the two DNA strands. (Briefly, PNA is much like ssDNA except that the bases, A, C, G, T are held in sequence by a charge-neutral polypeptide backbone rather than the negatively charged sugar-phosphate backbone of DNA.) The number of stations at which zeros can be written by either the CDL or the CEL technique is limited by the number of distinct restriction enzyme sites that can be used. By contrast, segments of DNA of *arbitrary* sequence can be targeted for the attachment of complementary PNA molecules. Consequently, when this new method of writing with PNA is used, there is in principle no upper bound for the number of stations that can be used. However, when the option of writing with PNA at the site of a restriction enzyme is retained, reading can be done by attempting a cut at that site, since cutting at a site that has been interrupted by PNA is quite impossible. This allows elementary testing of the PNA writing technology. Alternate reading procedures have been introduced that relax the restriction of writing only at enzyme sites, and consequently an unbounded number of stations may become available if this new writing method proves to be extendable to problem instances of practical scale.

When a short PNA molecule binds to one of the two strands of a longer dsDNA molecule, the sugar–phosphate backbone of neither of the DNA strands is broken. Consequently, the short single-strand segment of DNA which has been displaced by the interrupting PNA remains secured in place by its continuations in each direction remaining bound in helical form with the companion strand. This leaves the displaced strand available for bonding with any ssDNA that one may introduce that has a complementary Watson–Crick segment. This provides a new reading technology: the bit 'zero' is represented at a station precisely if a short single strand of DNA (possibly carrying a chosen label) binds there.

Experiments that confirm successful writing into a one-bit memory using PNA and reading from that memory by using either a restriction enzyme or an ssDNA tag are presented in [20]. Wet-lab details not mentioned here are given there, along with pertinent gel photos confirming success. Whiplash PCR is also discussed as a procedure for copying PNA-based memory. The joint technique of PNA writing and restriction enzyme reading has been successfully realized in a microfluidic network [12], which has also sped up both of these processes.

5 At Hokkaido University

An interesting new approach to aqueous computing has been adopted by Azuma Ohuchi's group at Hokkaido University [18]. For the first time the option of writing both zeros and ones has been implemented. This group' choice for a nano-tablet is a specially designed ssDNA molecule having a sequence that forms hairpin structures at regular intervals. In their initial exploration, they have used a 288-base molecule consisting of the following 17 subsegments:

$$5'\text{-} - j - k - l - m - n - o - p - q - r - s - t - u - v - w - x - y - z - \text{-}3'$$

where all the subsegments are 20 bases long except for the four segments l, p, t, and x, which are only seven bases long. The 20-base segments flanking the four seven-base segments l, p, t and x on the left and right, k and m, o and q, s and u and w and y, respectively, have sequences chosen so that each pair anneals to form a helix of 20 base pairs (bp) when placed in water. This then forces each of l, p, t and x to become the seven-base loop of a hairpin structure. The remaining subsegments j, n, r, v and z serve as spacing segments separating the four hairpin structures. The four hairpins of this molecule provide four stations on the tablet at which writing can be done. Computations begin with an aqueous solution containing a vast number of these molecules in a tube T. The original four-hairpin configuration is considered to represent four ones.

The procedure for writing a zero at the hairpin nearest the $5'$-end is to add a 40-base ssDNA, call it f, that will base pair with the 40-base segment pair $j - k$ by attaching initially to the 20 bases of segment j and then displacing the 20 bases of the segment m from its attachment to k. Note that this opens the hairpin allowing this first station to now be regarded as a representation of a zero. Writing a zero at each of the other three stations is done in the same way – by adding the appropriate 40 base ssDNA, which attaches first to the spacer to the *left* of the hairpin and then invades the double-stranded base of the hairpin. *A very special feature of this tablet choice is that zeros can be rewritten into ones.* Suppose a zero has been written at the first station as described above. There is then a 40 bp helix, consisting of the strand f bonded to $j - k$, that is preventing the bonding of segments k and m which would create a hairpin. To restore the hairpin add the 40 bp complement, call it g, of f with a biotin attachment that will allow the helices formed by g and f to be withdrawn from tube T. Hairpins then reform at the first station as the $g - f$ helices are removed from the tube.

Experiments that confirm success in writing zeros and then rewriting them as ones at each of the four stations of these tablets are presented in [18]. See this reference also for the reading techniques developed for this new nano-tablet. Wet-lab details not mentioned here are given there along with pertinent gel photos confirming success.

6 At Leiden Again: a Sample from Henkel's Dissertation

In his recent Leiden University dissertation [9], Christiaan Henkel has developed several new wet lab techniques that constitute a major contribution to aqueous computing and to DNA computing in general. One of these is the generation of proteins for use in reading the results of computations. Chapter 4 of [9] is devoted to an exposition of this process which is illustrated with a wet-lab solution of an instance of the minimal-dominating-set (MDS) problem of graph theory.

The graph considered is $G = (V, E)$ having vertex set $V = \{a, b, c, d, e, f\}$ and set of *un*directed edges $E = \{(a, c), (b, c), (c, d), (d, e), (d, f)\}$. Recall that the MDS problem calls for the determination of the cardinal number of a smallest subset D of V which dominates G in the sense that the neighborhood of each vertex v of G contains an element of D. The *neighborhood* of a vertex v of G is defined to be the set $N(v) = \{x \text{ in } V \mid x = v \text{ or } (x, v) \text{ is in } E\}$. Consequently, in the present case there are six neighborhoods: $N(a) = \{a, c\}$, $N(b) = \{b, c\}$, $N(e) = \{e, d\}$, $N(f) = \{f, d\}$, $N(c) = \{a, b, c, d\}$ and $N(d) = \{c, d, e, f\}$. A subset S of V is a *dominating* set for G if it contains at least one element from each of these six neighborhoods. Moreover, if S contains at least one element from each neighborhood that is not a subset of another neighborhood, then S must be a dominating set for G. Thus, in the present case, S is a dominating set if it contains at least one element from each of the *four* neighborhoods $N(a), N(b), N(e)$ and $N(f)$. For the present instance it can be observed in advance that $D = \{c, d\}$ is the unique minimal dominating set for G and, consequently, the required cardinal number is 2. But how can the solution be obtained using only wet-lab operations? And how will proteins play a role in this?

The same plasmid that was used in Section 2 for the solution of the MIS problem was used again as the nanotablet. Writing is done by the CDL technique used in the MIS example. Once again, the six stations of the plasmid tablet are identified here with the six vertices of the graph G. The initial condition of each of the six stations of each plasmid is again considered to represent a 'one'. So. again, *subsegment present* = 1 and *subsegment absent* = 0. However, the interpretation of a '1' at the station associated with a vertex v is now that v *has not* (or not yet) *been chosen* for inclusion in a dominating subset of G. The interpretation of a '0' at the station v is l now that the vertex *has been chosen* for inclusion in a dominating set.

The initial state of our computer is a test tube T containing a vast number of these plasmid tablets. Each of the four neighborhoods that are required for treatment in the present instance is treated in turn following the pattern of treatment of $N(a) = \{a, c\}$: Pour the current tube T into two tubes L and R and treat the neighborhood $N(a)$ by writing a zero at station a in L and a zero at station c in R; unite L and R into (a new) tube T. Note that all tablets in T have a zero at a vertex in $N(a)$. A sequence of three additional steps, one for each of $N(b), N(e)$ and $N(f)$, gives a final tube T that contains only plasmids that have at least one zero in each neighborhood of each vertex in G. An elementary exercise confirms that every minimal dominating subset of V must occur encoded as a set of zeros on plasmids in this final T. Note that plasmids that encode the largest number of *ones* contain solutions to the MIS problem. A reading of the content of T by a gel separation was done just as in the treatment of the MIS problem in Section 2, where again the longest molecular fragments gave the solution: cardinal number = 2. Moreover one can recover the MDS $\{c, d\}$ through sequencing or otherwise, as before. However, the exciting new development in [9, Chap. 4] is *reading from protein output,* as will now be explained:

Recall from Section 2 that the circular plasmid used has an insert INS that has six subsegments to be used as stations at which one can write by the CDL procedure. This INS lies within an open reading frame that provides for transcription into RNA followed by translation into a protein. The lengths of these six subsegments are 36, 36, 36, 36, 45, and 51. Each of these segments consists of DNA code that yields, through transcription into RNA followed by translation, sequences of amino acids of lengths 12, 12, 12, 12, 15 and 17, respectively. (These six polypeptides were wisely chosen to encode epitopes that will allow them, in future work not yet reported, to be read by attachment of antibodies.) The new reading method reported in [9, Chap. 4] and [10] is based on determining the mass of the various remaining INS inserts following the CDL writing operations. For this purpose the candidate plasmids were transformed in an *E. coli* host for protein overexpression. The purified protein molecules produced from the plasmids were analyzed using a technique of mass spectrometry familiar to protein chemists (MALDI-TOF = matrix-assisted laser desorption ionization time-of-flight mass spectrometry). This resulted in a mass spectrum having three sharp peaks, with the maximum at 9292 Da indicating that the minimal dominating set is $\{c, d\}$ for which the predicted mass of the associated protein is 9291.08. The two remaining peaks were at lesser mass values, indicating that either three or four subsegments had been removed.

In other chapters of [9] equally exciting techniques for *reading by using fluorescence* and for *reading using a single-molecule detection system* [15] are described. These new techniques are demonstrated by providing wet lab solutions of a four-variable 3SAT problem, a solution of an instance of the knapsack problem, and a demonstration of the recently developed concept of DNA computing by blocking [14].

7 The Future

Only tiny prototype solutions of algorithmic problems have been successfully carried out by aqueous computing. One can continue to search for aqueous techniques that can be scaled up to larger instances. If large-scale instances are ever to be solved in a practical manner, they will surely be implemented in robotics or in microfluidic circuitry [11, 12]. An alternative path of research is to ask whether the aqueous concept can suggest a technique for carrying out computations within or among living cells. Crucially important medical decisions and actions that require only tiny computations or logical decisions might be taken at the cellular level. Among these is sensing a toxin and signaling its presence; or, better, producing a molecular output that will neutralize the toxin or attack a cell that is producing the toxin. A special significance of the production of a protein as the output of a biomolecular computation (as discussed in Section 6 above, following [9, 10]) is that the protein may act biologically in its environment. The problem may be addressed for transportation through appropriate cellular membranes. Perhaps it will be possible to find a writing scheme by which environmental chemicals write on nanotablets inside cells that, in appropriate circumstances, result in an active protein being read from the nanotablets and in a biologically significant action, either inside or outside the cell.

Note that each of the algorithmic problems discussed here is a standard NP-complete problem. The aqueous approach has features that are very attractive conceptually: the number of computational steps required usually grows only linearly in the length of the descriptions of instances. Virtually all problems are treated in the same, nearly uniform manner. All computations begin with a tube containing a vast number of identical nanotablets. The only variation is in the choice of the scale of vastness, which is based on the scale of the problem instance. The disagreeable features are those shared with virtually all schemes for computing with biomolecules: computational steps are slow and error-prone and, although the number of steps grows slowly as the problem instance grows, the number of DNA molecules having distinct sequences in a tube can grow exponentially. If the final tube from which the result must be read contains a vast number of molecules having distinct structures or distinct sequences, then single-molecule detection [9, 15, Chapter 3] may be required. Perhaps the fruit of all this effort will come from new conceptualizations of computing that can be done in and among living cells.

The emphasis here has been on wet-lab implementations of the aqueous-computing concept. There have also been several additional theoretical explorations and computational designs based on the aqueous concept. See the works of Rani Siromoney, Kamala Krithivasan, and their collaborators [3, 16, 17]. For additional references on aqueous computing, see [7].

References

1. L.M. Adleman (1994) Molecular computation of solutions to combinatorial problems, *Science* **266**:1021–1024.
2. L.M. Adleman (1988) Computing with DNA, *Sci. Amer.* **279**:54–61.
3. M.S. Balan, K. Krithivasan, Y. Sivasubramanyam (2002) Peptide computing – universality and complexity. In: Jonoska, N. and Seeman, N.C. (eds) *DNA Computing – 7th International Workshop on DNA-Based Computers*, June 2001, Lect. Notes Comp. Sci. **2340**, Springer, Heidelberg, Berlin, pp. 290–299.
4. T. Head (2000) Circular suggestions for DNA computing. In: A. Carbone, M. Gromov, P. Prusinkiewicz, (eds.) *Pattern Formation in Biology, Vision and Dynamics*, World Scientific, Singapore & London, pp. 325–335.
5. T. Head, X. Chen, M.J. Nichols, M. Yamamura, S. Gal (2002) Aqueous solutions of algorithmic problems: emphasizing knights on a 3X3. In: N. Jonoska and N.C. Seeman (eds.) *DNA Computing – 7th International Workshop on DNA-Based Computers*, June 2001, LNCS Vol. 2340, Springer, Heidleberg, Berlin, pp. 191–202.
6. T. Head, X. Chen, M. Yamamura, S. Gal (2002) Aqueous computing: a survey with an invitation to participate, *J. Comput. Sci. Technol.*, **17**:672–681.

7. T. Head, S. Gal (2001) Aqueous computing: writing into fluid memory, *Bull. Europ. Assoc. Theor. Comput. Sci.*, **75**:190-198. Updated version reprinted in: Gh. Paun, G. Rozenberg, A. Salomaa, (eds.) *Current Trends in Theoretical Computer Science*, Vol I, World Scientific, New Jersey pp. 493–503.
8. T. Head, G. Rozenberg, G. Bladergroen, C.D.K. Breek, P.H.M. Lomerese, H. Spaink (2000) Computing with DNA by operating on plasmids, *Bio Systems*, **57**:87–93.
9. C. Henkel, (2005) *Experimental DNA computing*, Ph.D. dissertation, Leiden University, The Netherlands.
10. C. Henkel, R. Bladergroen, C. Balog, A. Deelder, T. Head, G. Rozenberg, H. Spaink, (2005) Protein output for DNA computing, *Natural Computing*, **4**:1–10.
11. J.S. McCaskill (2001) Optically programming DNA computing in microflow reactors, *Bio Systems*, **59**:125–138.
12. K. Mogi, S. Kaneda, K. Ono, T. Fukuba, T. Fujii (2004) A microfluidic network for write-in and read-out operations of a molecular memory, *Proceedings of the 8th International Conference on Miniaturized Systems for Chemistry and Life Sciences*, Malmö, Sweden pp. 569–572.
13. Q. Ouyang, P.D. Kaplan, S.M. Liu, A. Libchaber (1997) DNA solution of the maximal clique problem, *Science* **278**:446–449.
14. G. Rozenberg, H. Spaink (2003) DNA computing by blocking, *Theor. Comput. Sci.*, **292**:653-665.
15. K.A. Schmidt, C.V. Henkel, G. Rozenberg, H.P. Spaink (2004) DNA computing using single-molecule hybridization detection, *Nucleic Acids Research* **32**:4962–4968.
16. R. Siromoney, D. Bireswar, (2004) DNA algorithm for breaking propositional logic based cryptosystem, *Bull. Eur. Assoc. Theor. Comput. Sci.*, **79**:170–176.
17. R. Siromoney, B. Das (2004) Plasmids to solve #3SAT. In: N. Jonoska, Gh. Păun, G. Rozenberg, (eds.) *Aspects of Molecular Computing*, Lect. Notes Comp. Sci. **2950**, Springer, Heidelberg, Berlin, pp. 361–366.
18. N. Takahashi, A. Kameda, M. Yamamoto, A. Ohuchi (2005) Aqueous computing with DNA hairpin-based RAM. In: C. Ferretti, G. Mauri, C. Zandron, (eds.) *DNA Computing – 10th International Meeting on DNA Computing*, June 2004, Lect. Notes Comp. Sci. **3384** pp. 355–364.
19. M. Yamamura, T. Head, S. Gal (2000) Aqueous computing – mathematical principles of molecular memory and its biomolecular implementation, Chap. 2. In: H. Kitano, (ed.) *Genetic Algorithms* **4**:49–73 (in Japanese).
20. M. Yamamura, Y. Hiroto, T. Matoba (2002) Another realization of aqueous computing with peptide nucleic acid. In: N. Jonoska, and N.C. Seeman, (eds.) *DNA Computing – 7th International Workshop on DNA-Based Computers*, June 2001, Lect. Notes Comp. Sci. **2340**, Springer, Heidelberg, Berlin, pp. 213–222.

Part VII

Computations Inspired by Cells

Turing Machines with Cells on the Tape

Francesco Bernardini[1], Marian Gheorghe[1],
Natalio Krasnogor[2], and Gheorghe Păun[3]

[1] Department of Computer Science, University of Sheffield
Regent Court, Portobello Street, Sheffield, S1 4DP, UK
F.Bernardini/M.Gheorghe@dcs.shef.ac.uk
[2] School of Computer Science and Information Technology
University of Nottingham
Jubilee Campus, Nottingham, NG8 1BB, UK
Natalio.Krasnogor@nottingham.ac.uk
[3] Institute of Mathematics of the Romanian Academy
PO Box 1-764, 014700 Bucureşti, Romania
and
Research Group on Natural Computing
Department of Computer Science and Artificial Intelligence
University of Sevilla
Avda. Reina Mercedes s/n, 41012 Sevilla, Spain
george.paun@imar.ro, gpaun@us.es

1 Introduction

The present notes start from the remark that a Turing machine works with a tape which is divided into "cells", but the notion of a cell is used in a very restricted and local manner. Thus, a natural question is to consider a sort of Turing machine where these cells are "real cells", that is, membranes containing multisets of symbols, which evolve by means of symport/antiport rules as is well known in membrane computing [5]. In some sense, we get in this way a tissue-like P system, with the cells arranged on a line, but we have several important features which are different from tissue P systems: there is no communication with the environment, but, like in a Turing machine, we consider an "infinite tape", hence an infinite sequence of one-membrane cells; some of these cells (to the left of the "tape") are considered *differentiated*, with specific multisets of objects and specific sets of rules, while all other cells are *non-differentiated*, they have the same contents and the same rules.

An input is provided in the form of symbol-objects introduced in the first cells of the "tape", and the string of these symbols is recognized if the computation halts.

In what follows we prove that such devices – we call them *bio-Turing machines* – with a small number of differentiated cells and symport/antiport rules of small sizes can compute all one-letter recursively enumerable languages. Examples of non-regular and non-context-free languages over two or three letter alphabets which can be recognized by our machines are also given.

Then, we examine Byzantine-like problems: if the input is of the form a^n, for $a \in \{0, 1\}$, but there are cells which cannot correctly read the input, thus falsifying it (but we do not know which cells are in this situation), the problem is to find whether we can find the true input by suitably "programming" the machine. Two variants of the question are considered, with positive answers to the above question.

The chapter has a preliminary character, with many research topics remaining to be considered; several of them are explicitly mentioned in the last section.

2 Prerequisites

We assume the reader to be familiar with the basic elements of computability, and we only specify a few notations, especially related to membrane computing; details can be found, e.g., in [5].

By REG, CF, CS, RE we denote the families of regular, context-free, context-sensitive, and recursively enumerable languages, respectively, without restrictions on the cardinality of the alphabet. When considering only languages over alphabets with at most $n \geq 1$ symbols, then we write $nREG, nCF, nCS, nRE$, respectively.

In what follows we will consider one-membrane cells (like in tissue P systems), communicating through symport/antiport rules. For uniformity, we use only antiport rules, of the form $(i, x; y, j)$, where i, j are labels of cells and x, y are multisets of symbol-objects (as usual in membrane computing, we represent the multisets by strings, with the obvious remark that all permutations of a string represent the same multiset). When using such a rule, the objects indicated by x pass from cell i to cell j and, at the same time, the objects indicated by y pass from cell j to cell i. One of the multisets x, y can be empty (and then we use the empty string λ to denote it), and this corresponds to the case of a symport rule. The maximal length of x, y is called the *weight* of the rule $(i, x; y, j)$.

In the universality proof below we use the notion of a *register machine*. Such a device – in the deterministic version – consists of a given number of registers each of which can hold an arbitrarily large non-negative integer number, and a set of labelled instructions which specify how the numbers stored in registers can change, and which instruction should follow after any used instruction. There are three types of instructions:

- $l_1 : (\mathtt{ADD}(r), l_2)$ (add 1 to register r and then go to the instruction with label l_2),

- $l_1 : (\texttt{SUB}(r), l_2, l_3)$ (if register r is non-empty, then subtract 1 from it and go to the instruction with label l_2, otherwise go to the instruction with label l_3),
- $l_h : \texttt{HALT}$ (the halt instruction).

Thus, formally, a register machine is a construct $M = (m, H, l_0, l_h, I)$, where m is the number of registers, H is the set of instruction labels, l_0 is the start label, l_h is the halt label (assigned to instruction HALT), and I is the set of instructions; each label from H labels only one instruction from I, thus precisely identifying it. A register machine M accepts a number n if, starting with n in the first register and all other registers empty (hence storing the number zero) with the instruction with label l_0, we proceed to apply instructions as indicated by the labels (and made possible by the contents of registers) until reaching the halt instruction, with all registers being empty at that moment. Let $N(M)$ be the set of numbers accepted by M in this way. It is known (see, e.g., [4]) that in this way we can compute all sets of numbers which are Turing computable.

3 Bio-Turing Machines

We pass now to formally define the device we investigate in this paper.

A *bio-Turing machine* is a construct

$$\Pi = (O, \$, k, w_1, \ldots, w_k, w, R),$$

where:

1. O is an alphabet (of objects),
2. \$ is a symbol not in O, used as a marker,
3. $k \geq 1$ is the *degree* of the machine,
4. $w_1, \ldots, w_k$ are strings over O, representing the initial contents of cells $1, 2, \ldots, k$ of the machine,
5. w is the contents of cells $k+1, k+2, \ldots$ (the same for all cells from cell $k+1$ to infinity),
6. R is a finite set of rules of the following two types:
   (a) $(i, x; y, i+1)$, for $1 \leq i \leq k$, $x, y \in O^*$ (usual antiport rules),
   (b) $(*, x; y, *+1)$, for $x, y \in O^*$ (antiport rules without labels of cells).

The intuition is that the tape of the machine is bounded to the left and infinite to the right, and the first k cells from the left are "differentiated", they have different contents and different rules, while all other cells are "non-differentiated", they have the same contents and rules which can be applied irrespective of the cell. That is, any rule $(*, x; y, *+1)$ can be applied for exchanging the multisets x, y among any two neighboring cells $j, j+1$, with $j > k$.

The machine recognizes a string $w = a_1a_2\ldots a_n \in O^*$ of length n as follows: we introduce a_i in cell i, for $1 \leq i \leq n$ (irrespective of the relation between k and n), as well as a marker \$ in cell $n+1$ (that is, these symbols are added to the corresponding multisets w_i, w, respectively). From this *initial configuration*, the computation proceeds as usual in membrane computing, by using the rules from each cell in a non-deterministic maximally parallel way: in each step, a multiset of rules is applied, with the rules and the objects non-deterministically chosen, but in such a way that no further rule can be used at the same time (the objects which possibly remain unused are not enough for applying any further rule). In this way, we pass from a configuration (instantaneous description) of the machine to another configuration. A sequence of transitions forms a computation; if the computation halts (it reaches a configuration where no rule can be used), then the input string $a_1a_2\ldots a_n$ is accepted.

For the readers convenience, in Fig. 1 we give a representation of a bio-Turing machine, in a similar way as the standard Turing machines are represented (here we do not have a read-write head), with the input string also indicated.

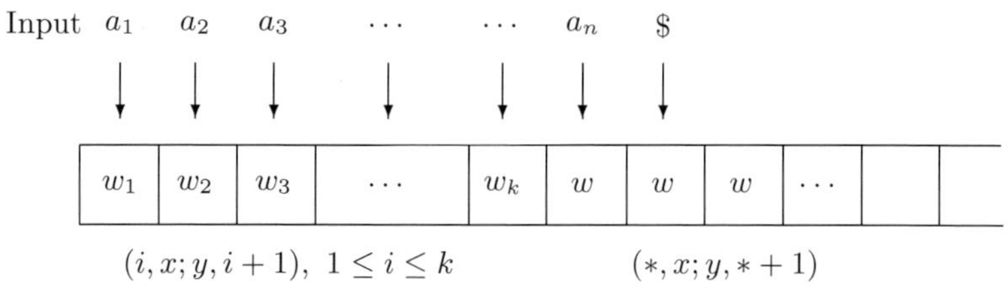

Fig. 1. A representation of a bio-Turing machine

For a machine Π we denote by $L(\Pi)$ the language of all strings recognized as above by Π. The family of all languages $L(\Pi)$ recognized by bio-Turing machines with at most k differentiated cells and using rules of weight at most r is denoted by $LTP_k(anti_r)$. When one of the parameters k and r is not bounded, then we replace it with $*$. When considering only languages over an alphabet with at most $n \geq 1$ symbols we write $nLTP_k(anti_r)$ for the corresponding family.

4 Some Examples

We start with a simple example, illustrating the way a bio-Turing machine works (and also proving that such devices can recognize non-regular languages).

Let us consider the machine

$$\Pi_1 = (O, \$, 2, f, d, \lambda, R),$$

where:

$$\begin{aligned} O &= \{a, b, d, f\}, \\ R &= \{(1, f; \lambda, 2),\ (1, a; d, 2),\ (1, b; d, 2), \\ &\quad (*, \lambda; a, * + 1),\ (*, \lambda; b, * + 1),\ (*, \lambda; \$, * + 1), \\ &\quad (2, \lambda; a, 3),\ (2, \lambda; b, 3), \\ &\quad (1, \lambda; ab, 2), \\ &\quad (2, fa; \$, 3),\ (2, fb; \$, 3),\ (*, f; \lambda, * + 1)\}. \end{aligned}$$

We have two differentiated cells; the first one contains initially the object f, the second one contains the object d. We input a string $w \in \{a, b\}^*$; assume that we have $|w| \geq 3$ (hence initially the dollar symbol is not in cell 3). In the first step, the symbol from cell 1 is moved to cell 2 (and this can happen only once, because the move is done in exchange of d, initially present in cell 2) and, simultaneously, f moves from cell 1 to cell 2. In each step, all symbols a, b from cells $3, 4, \ldots$ of the tape are moved one step to the left. From cell 2 we can move pairs ab to cell 1 (and, because of the maximally parallel mode of using the rules, this should always be done when a pair is present in cell 2). Also the dollar symbol can be moved to the left, but only until reaching cell 3; at that moment it checks whether any copy of a or any copy of b is present in cell 2. If this is the case, then a rule $(2, fa; \$, 3)$, $(2, fb; \$, 3)$ can be used, and the symbol f can move to the right forever (by rule $(*, f; \lambda, * + 1)$), thus preventing the halting of the computation. Therefore, the string w is accepted only if it contains the same number of occurrences of a and of b. (If at the moment when \$ arrives in cell 3 we have a pair ab in cell 2, then we can still use any of the rules $(2, fa; \$, 3)$, $(2, fb; \$, 3)$, and the computation continues forever, but if only the pair ab is present, then it can be moved to cell 1, and the computation halts.)

If the input string is of length two, then the computation stops only if the string is one of ab or ba; the details are left to the reader.

The strings λ, a, b are obviously accepted (none or only one computation step can be done). Consequently,

$$L(\Pi_1) = \{a, b\} \cup \{w \in \{a, b\}^* \mid |w|_a = |w|_b\}.$$

This is a non-regular language, hence $LTP_2(anti_2) - REG \neq \emptyset$.

The previous construction can be easily modified in order to obtain a bio-Turing machine Π_2 accepting the language $L(\Pi_2) = \{a, b, c\} \cup \{w \in \{a, b, c\}^* \mid |w|_a = |w|_b = |w|_c\}$ (we also move c from cell 1 to cell 2 in the first step, and from the right to left, to cell 2, from the rest of the tape; then, the rule $(1, \lambda; abc, 2)$ will check the equality of the number of occurrences of a, b, c; if these occurrences do not match, then f will be released from cell 2 and will move indefinitely to the right). That is, $LTP_2(anti_3) - CF \neq \emptyset$.

It is of interest to observe that for recognizing the language $L(\Pi_1)$ we do not need to use the marker \$. Indeed, let us consider the machine

$$\Pi_1' = (\{a, b, d, f, g\}, \$, 2, f, dg, \lambda, R'),$$

with the rules

$$\begin{aligned} R' = \{&(1, a; d, 2),\ (1, b; d, 2),\\ &(*, \lambda; a, *+1),\ (*, \lambda; b, *+1),\\ &(2, \lambda; a, 3),\ (2, \lambda; b, 3),\\ &(1, \lambda; ab, 2),\\ &(1, f; g, 2),\ (1, g; f, 2),\ (1, f; \lambda, 2),\\ &(2, fa; \lambda, 3),\ (2, fb; \lambda, 3),\ (*, f; \lambda, *+1)\}. \end{aligned}$$

The machine Π_1' works in a similar way as Π works, with the difference of using the symbols f, g for checking the end of the string (and the equality of the number of occurrences of a and b): the symbols f, g oscillate between cells 1 and 2 for an arbitrarily large number of steps, then f comes to cell 2; if there is any a or b without a pair here, then f can go to cell 3, and from here it can move indefinitely to the right of the tape. This ensures that the string contains the same number of occurrences of a and b. (Of course, f can leave cells 1, 2 prematurely, but this just makes the computation take an endless path; if the string is from the language $L(\Pi_1)$, then another computation exists which accepts it.) Note that the marker \$ plays no role in this construction.

In the previous case, the strings are accepted due to a condition about their Parikh image (if a string is accepted, then all permutations of a string are accepted), but this is not always the case with all languages accepted by bio-Turing machines.

To illustrate this assertion, we consider the machine

$$\Pi_3 = (O, \$, 3, \lambda, e, ccddf, \lambda, R),$$

where:

$$\begin{aligned} O = \{&a, b, c, d, e, f\},\\ R = \{&(*, \lambda; a, *+1),\ (*, \lambda; b, *+1),\ (*, \lambda; \$, *+1),\\ &(3, \lambda; a, 4),\ (3, \lambda; b, 4),\ (3, \lambda; \$, 4),\\ &(2, e; f, 3),\ (2, e; ca, 3),\ (2, c; ca, 3),\ (2, c; db, 3),\ (2, d; db, 3),\\ &(1, \lambda; ab, 2),\\ &(2, a; f\$, 3),\ (2, b; f\$, 3),\ (1, \lambda; f, 2),\ (1, f; \lambda, 2)\}. \end{aligned}$$

The functioning of the machine Π_3 is very similar to the functioning of Π_1, with the difference that only strings of the form $a^n b^m$ can be accepted: in order to move symbols a from cell 3 to cell 2 we need copies of c in both cells, and in order to move symbols b we need copies of d in the two cells; initially, c and d are present in cell 3, and we can first bring one c in cell 2 (and this should be done, otherwise the rule $(2, e; f, 3)$ is used and the computation never stops, because f oscillates between cells 1 and 2), and later one copy of d. Checking whether the number of copies of a is equal to the number of copies of b is done like in the first example – but this time we do not examine the first symbol. Therefore, we have

$$L(\Pi_3) = \{a^{n+1}b^n,\ ba^nb^n \mid n \geq 2\} \cup \{aba^nb^{n-1},\ bba^nb^{n-1} \mid n \geq 1\}.$$

This is a non-regular language. Of course, we can also extend this example in such a way as to recognize a non-context-free language, but we leave this exercise to the reader.

5 One-Letter Universality

We briefly investigate now the computational capacity of bio-Turing machines. Directly from the definitions and from the examples given above, we have the following results.

Lemma 1. (i) $LTP_k(anti_r) \subseteq LTP_{k'}(anti_{r'})$, *and* $nLTP_k(anti_r) \subseteq nLTP_{k'}(anti_{r'})$, *for all* $1 \leq k \leq k'$, $1 \leq r \leq r'$, *and* $n \geq 1$.

(ii) $LTP_*(anti_*) \subseteq RE$, *and* $nLTP_*(anti_*) \subseteq nRE$, *for all* $n \geq 1$.

(iii) $2LTP_2(anti_2) - REG \neq \emptyset$, $3LTP_2(anti_3) - CF \neq \emptyset$.

In the previous examples, all the undifferentiated cells were empty, but if we want to recognize more complex languages, then we need a workspace which can be much larger than the number of objects provided by the differentiated cells, hence we need objects also in the undifferentiated cells. This is the case in the following universality result.

Theorem 1. $1RE = 1LTP_k(anti_r)$ *for all* $k \geq 2, r \geq 2$.

Proof. In view of the lemma above, we only need to prove the inclusion $1RE \subseteq 1LTP_2(anti_2)$.

To this aim, let us consider an arbitrary register machine $M = (m, H, l_0, l_h, I)$. We add to it a new register, considered as register 0, as well as the following two new instructions,

$$s_0 : (\mathtt{SUB}(0), s_1, l_0), \quad s_1 : (\mathtt{ADD}(1), s_0),$$

where s_0, s_1 are new labels. Consider the machine obtained in this way, $M' = (m+1, H \cup \{s_0, s_1\}, s_0, l_h, I')$, starting with a number n stored in register 0 and all other registers empty; using the previously added rules, the content of register 0 is moved to register 1, and from now on we follow the instructions of machine M, never again using register 0. Clearly, $N(M') = N(M)$.

We will now construct a bio-Turing machine Π which recognizes strings a_0^n if and only if $n \in N(M)$, which will imply that any language from $1RE$ can be recognized in this way. We take

$$\Pi = (O, \$, 2, de, dfgg, w, R),$$

where:

$$O = E \cup \{a_0, d, e, f, g\},$$
$$E = \{a_i \mid 1 \le i \le m\} \cup \{l, \bar{l}, l', l'', l''', l^{iv} \mid l \in H \cup \{s_0, s_1\}\},$$
w contains all symbols from E exactly once,

and the set R of rules is constructed as follows (together with the groups of rules we give explanations about how they work).

The initial configuration (hence after introducing a_0 in n leftmost cells) starts as follows

a_0de	a_0dfgg	a_0w	a_0w	a_0w	...

The idea of the construction is the following: we bring all copies of a_0 in cell 1, and then we bring arbitrarily many copies of each object from E in cell 2; at some moment, we stop this latter operation and we introduce s_0 in cell 1, triggering in this way a computation in the register machine M'. The contents of each register r of M' will be represented by the number of copies of a_r present in cell 1; initially, we have here the n copies of a_0. The computation in Π will halt only if the computation in M' halts, and, conversely, to each halting computation in M' we can find a halting computation in Π, which will imply that $L(\Pi) = \{a_0^n \mid n \in N(M)\}$.

1. In order to bring all copies of a_0 in cell 1, we use the following rules:

$$(*, \lambda; a_0, *+1),\ (2, \lambda; a_0, 3),\ (1, d; a_0d, 2),$$
$$(*, \lambda; \$, *+1),\ (2, \lambda; \$, 3).$$

After using these rules, the configuration of the machine is as follows:

a_0^nde	$\$dfgg$	w	w	w	...,

hence with \$ in cell 2. Note that a_0 can be moved in cell 1 only if object d is present in both cells 1 and 2.

Because $a_0 \notin E$, the symbol a_0 does not appear in w, hence the process above stops after bringing the input string in cell 1. This was the reason of passing from the initial register machine M to M' – in order to simulate M we need arbitrarily many copies of each symbol a_r, $1 \le r \le m$ (a_1 included), hence, if the input is given as a string a_1^n, then we have to take care not to bring in the first cell more copies of a_1 than in the input.

2. We bring \$ in cell 1, by means of the rule

$$(1, de; \$, 2).$$

The configuration becomes

$a_0^n\$$	$ddefgg$	w	w	w	...

Note that all copies of a_0 are now in cell 1 and that from now on no a_0 can be moved from cell 2 to cell 1, because both copies of d are in cell 2.

3. We now use the object e in order to bring in cell 2 arbitrarily many objects from the set E, and to this aim we use the following rules:

$$(2, ef; \lambda, 3),$$
$$(*, e; \lambda, * + 1),$$
$$(*, \lambda; e\alpha, * + 1), \ (2, \lambda; \alpha, 3), \text{ for all } \alpha \in E.$$

The object f remains in cell 3, while e goes to the right at any distance, comes back with any symbol α, and the process is repeated an arbitrary number of times. All objects α brought to cell 3 pass immediately to cell 2.

4. At some moment, we stop this process, by bringing e back to cell 2, by means of the rule

$$(2, \lambda; e, 3).$$

The configuration looks as follows,

$a_0^n\$$	$ddeggE^\omega$	f	w'_1	w'_2	...,

where $w'_1, w'_2, \ldots$ are any submultisets of w, and E^ω is written to suggest that we have arbitrarily many copies of each object from E (cell 3 contains only f, because all symbols from its initial multiset w were moved to cell 2 by the rule $(2, \lambda; \alpha, 3), \ \alpha \in E$).
Note that e cannot again leave cell 2, because f is in cell 3, hence the rule $(2, ef; \lambda, 3)$ is no longer applicable.

5. We now use object e in order to bring s_0, the starting label of M', in cell 1, and to this aim we use the rules

$$(1, \lambda; es_0, 2), \ (1, \lambda; eg, 2), \ (1, g; g, 2).$$

(If s_0 is not present in cell 2, then a copy of object g is brought in cell 1, and the computation will continue forever by means of the rule $(1, g; g, 2)$.)
The obtained configuration is

$a_0^n es_0\$$	$ddggE^\omega$	f	w'_1	w'_2	...

From now on, no rule from the previous groups can be used, because the necessary pairs of objects are not in the right places.

6. We now start the simulation of the instructions of the register machine M'. An instruction $l_1 : (\mathtt{ADD}(r), l_2) \in I'$ is simulated by using the following rules:

$$(1, l_1; l_2 a_r, 2),$$
$$(1, l_1; g, 2).$$

The label-object l_1 present in cell 1 is exchanged with $l_2 a_r$, which represents the correct simulation of the ADD instruction; if cell 2 does not contain the necessary objects l_2, a_r (that is, the previous phase of the computation has not brought sufficient objects in cell 2), then we have to use the rule $(1, l_1; g, 2)$ and the computation will never halt, because g will oscillate forever between cells 1 and 2 by means of the rule $(1, g; g, 2)$.
7. An instruction $l_1 : (\mathtt{SUB}(r), l_2, l_3) \in I'$ is simulated by using the following rules:

$$(1, l_1; \bar{l}_1 l_1''', 2), \ (1, l_1; g, 2),$$
$$(1, \bar{l}_1 l_1'''; l_1' l_1'', 2), \ (1, \bar{l}_1 l_1'''; g, 2),$$
$$(1, l_1' a_r; l_1''', 2),$$
$$(1, l_1''; l_1^{iv}, 2), \ (1, l_1''; g, 2),$$
$$(1, l_1^{iv} l_1'''; l_2, 2), \ (1, l_1^{iv} l_1'''; g, 2),$$
$$(1, l_1^{iv} l_1'; l_3, 2), \ (1, l_1^{iv} l_1'; g, 2).$$

We start again with l_1 in cell 1. The first two pairs of rules are meant to ensure that l_1''' is present in cell 2; if this is not the case, then the trap object g comes to cell 1 and the computation will not halt. After making sure that l_1''' is present in cell 2, the use of the rule $(1, l_1' a_r; l_1''', 2)$ depends only on the presence of at least one copy of a_r in cell 1 (hence whether or not register r is non-empty). If this is the case, then l_1''' is brought to cell 1. Simultaneously, l_1'' is exchanged with the "checker" l_1^{iv}. If this "checker" finds l_1''' in cell 1, then one brings l_2 in cell 1, otherwise (that is, if the rule $(1, l_1' a_r; l_1''', 2)$ was not used, hence the register was empty) the object l_3 is brought from cell 2 to cell 1. Note that all rules $(1, u; v, 2)$ with the exception of $(1, l_1' a_r; l_1''', 2)$ have companion rules of the form $(1, u; g, 2)$, hence they have to be used, otherwise the computation never stops. This is meant to ensure that if the computation in Π halts, then this is not due to the shortage of necessary objects in cell 2.
The simulation of instructions of M' can be repeated. If the computation in M' stops, hence l_h is introduced in cell 1, then also the computation in Π stops (there is no rule by which we can continue). Conversely, for each halting computation in M' we can find a halting computation in Π, hence the numbers computed by M' are exactly the lengths of strings from $L(\Pi)$.

The observation that the degree of Π is 2 and that the maximal weight of the rules from R is also 2 concludes the proof. □

6 Two Byzantine-Like Problems

Let us consider that some of the cells of a bio-Turing machine are "faulty", in the sense that they cannot correctly "read" the input symbols $a_1, a_2, \ldots, a_n$. More precisely, let us consider the case when the input is composed of only symbols 1 or only symbols 0, and that some of the cells can corrupt the input, so that instead of 1 they will get inside 0, or conversely. The problem is to find bio-Turing machines "input-reliable", in the following sense: we input a string α^n (bounded to the right by \$) as usual, with one symbol α introduced to each cell, and we want to halt with some object a_0 or a_1 in cell 1 telling us whether α was 0 or 1, respectively. That is, *all* computations halts, and the subscript of the object a from cells 1 is precisely the value of α.

Of course, the possibility to construct such a machine depends on the number of faulty cells. If we know that we have at most s such cells, then any input of length at least $2s+1$ can be, in principle, correctly identified: at least $s+1$ values of the input bit remain unchanged, hence it is enough to bring all input objects – as "seen" by the cells, possibly corrupted – in the same cell and compare the number of copies of 0 with the number of copies of 1. The only problem is to "program" this procedure in terms of antiport rules working with the cells of a bio-Turing machine, and this can be done like in the first example from Section 4, with the important difference that this time we always have to halt (hence the use of trap-symbols is no longer allowed).

Because of this difficulty, we give the details of the machine which answers this first Byzantine-like problem. We take

$$\begin{aligned}
\Pi &= (\{1, 0, a_0, a_1, c, d\}, \$, 4, c, d, d, a_0a_1, \lambda, R),\\
R &= \{(1, c\alpha; \lambda, 2),\\
&\qquad (*, \lambda; \alpha, *+1),\ (4, \lambda; \alpha, 5),\ (3, \lambda; \alpha, 4),\ (2, d; d\alpha, 3) \mid \alpha \in \{0, 1\}\}\\
&\cup \{(1, \lambda; 01, 2),\\
&\qquad (*, \lambda; \$, *+1),\ (4, \lambda; \$, 5),\ (3, \lambda; \$, 4),\ (2, d; \$, 3),\\
&\qquad (3, dd; a_0a_1, 4),\ (2, 0; a_0, 3),\ (2, 1; a_1, 3),\ (1, \lambda; a_0, 2),\ (1, \lambda; a_1, 2)\}.
\end{aligned}$$

We introduce a string β^n, with $n \geq 2s+1$ and $\beta \in \{0, 1\}$, in the machine. The bits read by at most s cells can be flipped – but we do not know which cells are doing this. Thus, we bring in cell 2 all bits read by the cells, either correct or corrupted. Pairs 01 are moved to cell 1, hence in the end we will have here either only copies of 0 or only of 1, and the remaining bit is the one equal to β. Both a_0 and a_1 are moved to cell 3 after completing the comparison (after bringing \$ in cell 2, which means that both copies of d are in cell 3), and the object a_i whose i is present in cell 2 is first moved from cell 3 to cell 2 and then to cell 1, thus giving the correct answer. The computation halts at that moment.

Note that the previous system has degree 4 and antiport rules of weight 2.

In the previous setup, the faulty cells were unreliable in what concerns the input bit, which was possibly flipped, but when receiving objects 0 and 1 from the neighboring cells they behaved correctly, always interpreting the received bit correctly and passing it further unchanged. Of course, if such errors were to be considered, then the previous machine will no longer work, and we see no way to cope also with such bilaterally faulty cells – without adding further restrictions.

One such restriction is to consider that the faulty cells are *systematically* flipping the received bit, irrespective from which direction the bit comes, from the "user" as an input, or from a neighboring cell. Somewhat expectedly, in this case we can find the correct input by means of a bio-Turing machine which only checks the two leftmost bits.

Such a machine is the following one:

$$\begin{aligned}
\Pi &= (\{1, 0, a_0, a_1, a_0', a_1', c, d, e, e'\}, \$, 3, ee', da_0'a_1', a_0a_1, \lambda, R),\\
R &= \{(1, \lambda; 1a_1', 2),\ (1, \lambda; 0a_0', 2),\\
&\qquad (1, e00; d, 2),\ (1, e11; d, 2),\ (1, e'01; d, 2),\\
&\qquad (2, ea_0'; a_1, 3),\ (2, ea_1'; a_0, 3),\ (2, e'a_0'; a_0, 3),\ (2, e'a_1'; a_1, 3),\\
&\qquad (1, \lambda; a_0, 2),\ (1, \lambda; a_1, 2)\}.
\end{aligned}$$

Assume that we input $\beta\beta$ in the first two cells; we denote by $\bar{\beta}$ the flipped value of β.

We have four cases: (i) both cells 1 and 2 are faulty; (ii) cell 1 is faulty and cell 2 is not; (iii) cell 1 is not faulty but cell 2 is so; (iv) both cells are not faulty. The bits seen in each case by the cells are $\bar{\beta}\bar{\beta}$, $\bar{\beta}\beta$, $\beta\bar{\beta}$, $\beta\beta$, respectively. The bit from cell 2 is moved to cell 1, together with the associated $a_i', i \in \{0, 1\}$; depending on the type of cell 1, faulty or not, the bit it receives will be flipped or not, hence in cell 1 we will have one of the pairs $\bar{\beta}\beta$, $\bar{\beta}\bar{\beta}$, $\beta\bar{\beta}$, $\beta\beta$, respectively. Therefore, in cases (i) and (iii) the bits we have in cell 1 are different, while in cases (ii) and (iv) the bits from cell 1 are equal. These pairs of cases correspond to the states of cell 2: cell 2 is trustful in cases (ii) and (iv) and faulty in cases (i) and (iii). When the bits from cell 1 are equal we move from here to cell 2 the object e, otherwise we move the object e'.

Note that when moving α from cell 2 to cell 1, we also move the corresponding object a_α', hence what remains is $a_{\bar{\alpha}}'$. Thus, if e finds here $a_{\bar{\alpha}}'$, because e means that cell 2 is trustful, it follows that the input was α; that is why a_α is moved from cell 3 to cell 2, and from here to cell 1. Conversely, if we have e' present in cell 2, because this corresponds to the case when cell 2 is faulty, if a_α' was moved away from cell 2, then this means that $\bar{\alpha}$ was the true input and cell 2 has flipped it; in this case, $a_{\bar{\alpha}}$ is moved from cell 3 to cell 2 and then to cell 1.

In both cases, in four steps we get the correct answer and we halt.

It is worth noting that the result is obtained irrespective of the number of faulty cells, using an input of only two bits, without using the right hand marker \$; the system is of degree 3, and it also uses antiport rules of weight 3.

7 Further Research Topics

The list of questions which remain to be addressed about bio-Turing machines is very large. For instance, the examples from Section 4 and Theorem 1 show that the computing power of these devices is rather large. Can Theorem 1 be extended to languages over arbitrary alphabets? We conjecture that the answer is affirmative. What would be the degree of the system and the weight of antiport rules sufficient for this case? Are there types of bio-Turing machines for which these parameters induce infinite hierarchies? What about the size of "axioms" $w_1, \ldots, w_k$ and w?

Then, a series of variants can be considered. For instance, can the use of the marker symbol \$ be avoided? (This was the case in the first example from Section 4 and in the second Byzantine-like problem.) What about considering deterministic bio-Turing machines? Note that the determinism makes sense, because we work with recognizing machines. Are the deterministic bio-Turing machines strictly less powerful than the non-deterministic ones? (Note that in the cell-like P systems with symport/antiport, the answer to this question depends in general on whether the considered class of systems is universal or not: universality is often obtained for deterministic systems, see, e.g., [1], but there are sub-universal classes of P systems for which the deterministic version is strictly less powerful than non-deterministic systems, see [2].) What corresponds to a linearly bounded automaton? (A possibility is to consider two types of non-differentiated cells, first with a multiset w inside, then, to the right of them to infinity, empty. However, this does not automatically imply that the recognized language is context-sensitive: it remains to prove that – as expected however – we cannot use the empty endless tape for encoding information, for instance, in the distance between two non-empty cells.) What could a push-down automaton mean in this framework? Similar to Turing machines with the tape infinite in both directions, we may consider such bi-infinite tapes also for bio-Turing machines. Another extension can be to consider cells which are not composed of only one membrane – maybe then restricting the weight of rules. It could also be of interest to explicitly consider the size of symport rules, that is, dealing with families of the form $LTP_k(sym_r, anti_q)$, thus playing with all parameters k, r, q (degree, maximal weight of symport rules, maximal weight of antiport rules).

The previous questions can be extended to cellular automata, which also have "cells" which, are not cells. How a cellular automaton with cell-membranes arranged on a grid, containing multisets and exchanging objects (horizontally or vertically) by using antiport rules, can be defined remains to be found. Because we do not create or destroy objects, we either have to confine ourselves to playing a purely combinatorial game, or we need a supply of objects (for instance, by considering as non-differentiated the neighborhood of a finite "central" region, like in a bio-Turing machine).

Also the preliminary results from Section 6 ask for extensions and generalizations. For instance, a natural question is to also identify the faulty cells, while a possible generalization is to consider inputs which can take more than two values.

References

1. R. Freund, Gh. Păun: On deterministic P systems. Submitted, 2004.
2. O.H. Ibarra: On determinism versus nondeterminism in P systems. *Theoretical Computer Science*, to appear (available at `http://psystems.disco.unimib.it`).
3. L. Lamport, R. Shostak, M. Pease: The Byzantine Generals problem. *ACM Transactions on Programming Languages and Systems*, 4, 3 (1982) 382–401.
4. M.L. Minsky: *Computation: Finite and Infinite Machines.* Prentice Hall, Englewood Cliffs, New Jersey, 1967.
5. Gh. Păun: *Membrane Computing. An Introduction.* Springer-Verlag, Berlin, 2002.

Insights into a Biological Computer: Detangling Scrambled Genes in Ciliates

Andre R.O. Cavalcanti and Laura F. Landweber

Department of Ecology and Evolutionary Biology
Princeton University
Princeton, NJ 08544, USA
Andre.Calvacanti@pomona.edu and lfl@princeton.edu

1 Introduction

Ciliates are single-celled eukaryotes characterized by the presence of nuclear dimorphism. Each ciliate cell contains two types of nuclei: a germ-line micronucleus (MIC), used in the exchange of genetic material but mostly transcriptionally silent, and a somatic macronucleus (MAC), where transcription takes place during vegetative life.

Following sexual conjugation, the somatic nucleus of the mother cell is destroyed and a new somatic nucleus forms from a copy of the zygotic micronucleus. The process by which an MIC is converted into an MAC involves massive DNA elimination, rearrangements and amplification (for a review of ciliate biology, see [21]).

The DNA in the germline nucleus is organized in a fashion similar to the DNA of other eukaryotes, with large diploid chromosomes containing many genes and a centromere. Generally, the MAC contains smaller acentric but telomere-carrying chromosomes derived from the MIC chromosomes. The details of the DNA organization in the MAC, however, vary in detail between ciliate lineages: in the oligohymenophoran *Tetrahymena thermophila* approximately 15% of the micronuclear sequences are eliminated during the formation of the new macronucleus, and the chromosomes are fragmented to generate 200 different somatic chromosome types, each with a ploidy of 45 copies;[1] in another oligohymenophoran, Paramecium tetraurelia, approximately 20% of the sequences present in the micronucleus are eliminated, and the macronucleus has approximately 300 chromosomes, each with a ploidy of hundreds of copies; and in spirotrichous ciliates, such as *Sterkiella histriomuscorum* (also known as *Oxytricha trifallax*) and *Stylonychia lemnae*, ~ 95% of the sequences in the micronucleus are eliminated, generating a macronucleus containing ~

[1] In addition to these chromosomes, the macronuclei of ciliates contain a ribosomal DNA (rDNA) chromosome present in thousands of copies.

24,000 different chromosomes, each with a ploidy of thousands of copies — most chromosomes typically contain a single gene, plus flanking short, non-coding subtelomeric regions [21].

The DNA that is eliminated from the micronucleus in the generation of a new macronucleus is referred to as MIC-limited, and the DNA sequences that are retained are called *macronuclear destined sequences* (MDSs). MIC-limited DNA can vary greatly in length, and can be located in intergenic regions or even interrupt genes. MIC-limited DNA sequences that interrupt macronuclear chromosomes are called *internal eliminated sequences* (IESs); because they often interrupt genes, IESs must be removed precisely, and the flanking MDS sequences joined together to produce a functional MAC. In six known genes of spirotrichous ciliates [22, 13], the order of MDSs is different in the macronucleus compared with the micronucleus; these are called *scrambled genes* (Fig. 1 shows some examples of scrambled genes). There are currently three scrambled genes for which several homologues have been studied in multiple spirotrich species: alpha-telomere-binding protein [16], actin I [7] and DNA polymerase alpha [11] (Fig. 1); however, estimates suggest that thousands of scrambled genes probably lurk in the micronuclear genome [21]. Besides having a different order in the MIC, MDS segments in scrambled genes can lie in opposite orientations, reside on different loci [1] and even be intertwined among segments for other genes [13].

We have made available a database containing all scrambled and non-scrambled spirotrich genes for which the micronuclear sequence is known (MDS/IES Db), as well as a program to infer a gene's MDS and IES structures when the macronuclear and the micronuclear version of a gene are known. Both are available online at: `http://oxytricha.princeton.edu/dimorphism/` [3, 2].

Little is known about how ciliates determine the identity of IESs, or the correct order of MDSs in the formation of a new macronucleus. One clue comes from the fact that the end of every MDS contains a nucleotide word that is repeated in the beginning of the next MDS; these repeats are called *pointer sequences.* In the case of simple nonscrambled IESs, this results in IESs flanked by matching pointers; IES removal leaves one pointer sequence behind in the MAC copy, and eliminates one pointer along with the IES. Pointers must somehow facilitate excision of the IES, while at the same time guiding the proper reordering of MDSs in scrambled genes.

In this chapter, we shall discuss how spirotrichous ciliates unscramble their genes in light of recent advances in our understanding of IES excision in *Tetrahymena* and *Paramecium*, but first we shall discuss pointer sequences further and show why they alone cannot guide gene unscrambling.

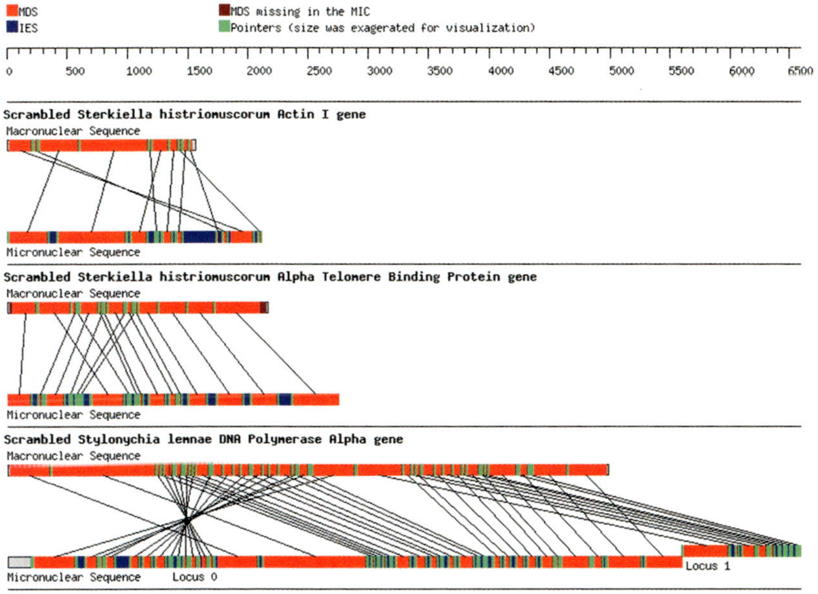

Fig. 1. Examples of the organization of scrambled genes. The top line in each panel represents the macronuclear version and the bottom line the micronuclear version of the genes. Red represents MDSs, blue IESs and green pointer sequences (note that the pointer sequences have been exaggerated). Figure generated by MDS/IES Db (`http://oxytricha.princeton.edu/dimorphism/`) [2].

2 Pointer Sequences

The precise determination of pointer sequences is still under some debate; here we shall use the definition proposed by [14] and implemented in MDS/IES Db. According to this definition, pointers are exact repeats between the ends of MDSs that are consecutive in the MAC; however, in cases where one mismatch leads to the incorporation of three or more matches, the pointers are extended.

In the spirotrich *Euplotes aediculatus* all pointers have the same sequence, TA, and no scrambled genes have been found. In other spirotrichs in general, including another member of the *Euplotes* genus, *E. octocarinatus*, pointer lengths are extremely varied, from a size of 2 (usually the repeat TA, present in many scrambled and non-scrambled genes, albeit usually between non-scrambled MDSs), to a size of 19 in one case, the alpha-telomere-binding protein gene of *Stylonychia mytilus* and *Sterkiella nova* (MDS/IES Db); [2]. Usually, pointers between scrambled MDSs tend to be larger than pointers between nonscrambled MDSs, with average sizes of 9 and 5, respectively [23].

3 IES Excision in Oligohymenophorans

3.1 *Tetrahymena thermophila*

The *T. thermophila* micronucleus contains two types of eliminated sequence: breakage-eliminated sequences (BESs), which are associated with chromosome breakage and telomere addition events that form MAC chromosomes and IESs. The BESs are relatively short (< 50 bp) and flank a conserved, 15 bp chromosome breakage sequence (Cbs) [18]. The second type of eliminated sequence, internal eliminated sequences, has so far been found only in noncoding regions. The genome contains approximately 6000 IESs, ranging in size from ~ 0.5 to > 20 kb. The removal of IESs is coupled to the ligation of the flanking MDSs. The excision of IESs occurs at determinate sites, within a few nucleotides. Some IESs have a specific site of excision, while others can have a few alternative sites. All IESs in *Tetrahymena* are flanked by short repeats (< 10 bp) [20], similar to spirotrich pointer sequences.

It is thought that IES excision in *Tetrahymena* is epigenetically regulated, with a comparison of the newly formed MIC with the old MAC. The sequences present in the old MAC are retained in the new MAC, while sequences absent from the old MAC are eliminated. The first indication that the IES excision mechanism relies on such a comparison was the observation that providing the parent MAC with sequences homologous to a given IES leads to the retention of this IES in the daughter MAC [5].

Mochizuki et al. have proposed that IES excision relies on small RNAs that resemble the genome defense systems of other organisms [17, 26]. This led to the proposal that epigenetic IES excision serves as a system to protect the genome against foreign DNA invasion [26]. In [26] it was shown that a foreign gene that had been integrated into a micronuclear chromosome could be recognized and deleted from the macronucleus of daughter cells. Furthermore, injection of double-stranded RNA (dsRNA) at specific developmental time points led to efficient deletion of homologous regions from the daughter MAC. These deleted regions often extended beyond the boundaries of the injected sequences and were often, but not always, flanked by direct repeats [26].

According to the scan RNA model ([17]; Fig. 3) the entire parental micronucleus is transcribed on both strands, generating a full genome complement of dsRNA. These dsRNA molecules are processed further by an RNAi-like mechanism to produce small dsRNAs of ~ 28 bp [19, 18]. Recently, the enzyme DCL1, a homologue to the Dicer enzymes involved in RNAi mechanisms in other organisms, was implicated in the processing of these dsRNAs [20].

It has been suggested that these small dsRNAs, called scan RNAs (scnRNAs), hybridize with the chromosomes of the old MAC. All the scnRNAs homologous to sequences in the old MAC are destroyed. The remaining scnRNAs, which contain only MIC-limited sequences, would then move to the developing new MAC and tag MIC-limited sequences for excision, by a process similar to heterochromatin formation in other organisms ([17]; Fig. 3).

Experimental evidence supports this model: the MIC, transcriptionally silent during vegetative growth, is transcriptionally active early during conjugation, and bidirectional transcripts containing MIC-limited sequences can be detected [6]. The scnRNAs colocalize with the Twi1p enzyme (a homologue of piwi) in the old MAC, and later in the developing macronucleus, and the proportion of scnRNAs that hybridize to micronuclear versus macronuclear DNA increases as conjugation progresses [18].

In [5], Chalker and Yao showed that the old MAC can influence the genetic complement of the new MAC even in crosses between different cells. If a wild-type cell is crossed with a mutant containing the normally excised M locus, this locus is not excised in the daughter cells of either parent. Recently, [4] showed bidirectional transcription of the M locus in the parental macronucleus, and proposed that these transcripts are the factors that act to winnow out MAC-homologous scnRNAs. These transcripts might never leave the macronucleus and absorb the scnRNAs there, or they might be transported to the cytoplasm where they could play the same role.

3.2 *Paramecium tetraurelia*

Paramecium tetraurelia also has two different kinds of eliminated sequence. Long sequences containing repeated sequences, such as minisatellites or transposons, are imprecisely removed. This imprecise removal can lead to rejoining of the flanking sequences or to the generation of chromosome breaks followed by telomere addition, and it is responsible for the heterogeneity observed among macronuclear chromosomes: telomere positions on macronuclear chromosomes are distributed typically over 0.2 to 2 kb telomere addition regions (TARs). Some MAC chromosomes show alternative TARs, separated by 2 to 10 kb [15]. In addition to these sequences, approximately 60,000 short, non-repetitive IESs, all flanked by TA pointers, are precisely removed from both coding and noncoding regions [10].

Two different epigenetic effects controlling the excision of micronuclear sequences have been described. The first, similar to the one described for *Tetrahymena*, occurs when a sequence that is normally excised from the MAC is inserted into the old MAC of the maternal cell; the new MAC in the daughter cell does not excise this sequence, and it is incorporated into the new MAC [8]. The second effect is the opposite: overloading the macronucleus of the mother cell with a cloned sequence (transgene) in high copy number induces imprecise elimination of this sequence in the daughter cell MAC.

Recently, [9] showed that the second effect is controlled by the expression of the transgene; if the gene is expressed it is not eliminated, whereas if the transgene is not expressed, homologous sequences in the MIC are eliminated. These authors also detected small RNAs of $\sim$ 22 to 23 bp, a slightly smaller size range than that observed for *Tetrahymena* scnRNAs ($\sim$ 28 bp). In the

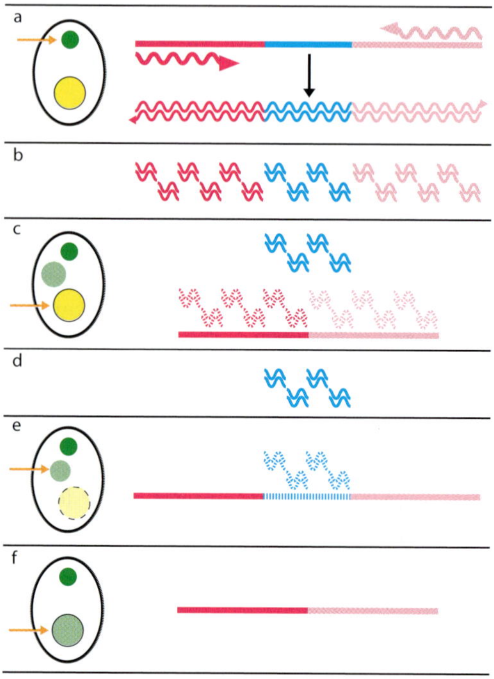

Fig. 3. In [17] a model for IES excision in *T. thermophila* is presented. (**a**) The DNA in the MIC (green circle) is transcribed in both directions. (**b**) Long segments of double-stranded RNA are processed by an RNAi-like mechanism, to form small double-stranded RNAs 28bp long, the scnRNAs (it is not known where this step takes place). (**c**) The scnRNAs are transported to the MAC (yellow circle), where they hybridize to homologous regions in the MAC chromosomes. Following conjugation, the zygotic MIC divides, and one of the copies will become the new MAC. While it is undergoing this transformation, it is called the anlagen (light green circle). (**d**) The scnRNAs that do not find a homologous region in an MAC chromosome are transported back to the developing MAC. (**e**) These sequences now guide the excision of homologous sequences in the anlagen. The old MAC is destroyed apoptotically (light yellow circle). (**f**) The anlagen becomes the new MAC (light green circle).

same paper, they showed that feeding *Paramecium* cells with bacteria producing dsRNA for a fragment of the ND7 gene leads to the elimination of the homologous region from the macronucleus of the daughter cell. This elimination is not precise, but always occurs between two short (2 to 6 bp) direct repeats, further indicating the importance of pointers in the elimination of MIC-limited sequences.

To explain their findings, [9] have proposed a modified version of the scan RNA model described for *Tetrahymena*. They suggest that the small RNAs that they describe have a different origin and function from the scnRNAs:

these small RNAs are produced in the macronucleus and their function is to direct degradation of homologous mRNAs before conjugation. Another class of small RNAs, similar to the scnRNAs in *Tetrahymena*, could be produced by transcription of the micronucleus and might not scan the macronuclear chromosomes, but rather the macronuclear transcripts. This model can explain the seemingly opposite mechanisms of excision: overloading the macronucleus with an excess of a gene would lead to higher production of the ~ 22 to 23 bp RNAs, which would in turn destroy the mRNAs of the gene. This would have the same effect as if the gene were absent from the mother macronucleus, and therefore providing no mRNA transcripts [9].

4 Gene Unscrambling in Spirotrichs

Much less is known about the number, types and distribution of MIC-limited sequences in spirotrichs. The level of fragmentation of the macronucleus in spirotrichs is much higher than in oligohymenophorans. *Oxytricha* has ~ 24,000 to 26,000 different chromosomes in its MAC. Thus the number of eliminated sequences involved in chromosome breakage must be very high. The mean number of IESs in the *Oxytricha* genes sequenced thus far is 14, and an estimate of 26,000 genes would put the number of intragenic IESs at approximately 350,000. The IESs in spirotrichs are also much smaller than those in oligohymenophorans. IESs of size 0 excluding the pointers occur in some scrambled genes, and more than half of all known IESs have sizes < 30 bp. Because most of these IESs occur in coding regions, they must be excised precisely, or else the gene product might contain deleterious deletions and/or frameshifts. In addition, approximately 5% of the micronucleus is composed of ~ 4 kbp transposable elements of the TBE family (T. Doak, personal communication), which are also eliminated as IESs [25].

One should be careful in extrapolating findings in oligohymenophorans to spirotrichs, as their evolutionary distance is very large, as much as 1 Byr. However, the IESs of *Euplotes*, a spirotrich, and *Paramecium* are very similar, suggesting that oligohymenophorans and spirotrichs share an ancestral IES elimination system. Moreover, small RNAs with much the same size and time profile as the *Tetrahymena* scnRNAs were described for *Stylonychia lemnae* [12] and *S. histriomuscorum* (Wong and Landweber, unpublished).

These scnRNAs could be used to eliminate large intergenic DNA by a mechanism similar to that in oligohymenophorans. On the other hand, it is difficult to see how such molecules could drive the elimination of intergenic IESs. First, the size of the IESs in spirotrichs is of the same order as that of the scnRNAs, and many IESs are smaller; if the scnRNAs targeted these regions we would expect the excision to be highly inaccurate. However, recently many cases of imprecise IES excision in developing macronuclei have been observed (Mollenbeck et al. unpublished). This has led [12] to propose a model in which scnRNAs specify sequences to be eliminated by DNA modification, followed by a correction step which could depend on large templates from the old macronucleus.

In particular, the information in scnRNAs seems to be insufficient to guide the excision and reordering needed to detangle scrambled genes. Recently, [24] proposed a model that uses DNA or RNA templates from macronuclear chromosomes to guide the excision of IESs. This model has many advantages over models based exclusively on small RNA, as it provides a precise template for excision; mRNAs could not serve this role, otherwise introns would also be excised from the daughter macronuclei (unless the absence of pointer sequences could block their excision). Full RNA transcripts from the macronuclear chromosome could also serve in such a model. However, no evidence of such templates has been observed in spirotrichs.

Acknowledgments

The authors thank T. Doak, M. Daley and I. McQuillan for suggestions. L.F.L. acknowledges support from NIGMS grant GM59708 and NSF grant 0121422.

References

1. D. H. Ardell, C. A. Lozupone, and L. F. Landweber. Polymorphism, recombination and alternative unscrambling in the DNA polymerase alpha gene of the ciliate *Stylonychia lemnae* (Alveolata; class Spirotrichea). *Genetics*, 165:1761–1777, 2003.
2. A. R. Cavalcanti, T. H. Clarke, and L. F. Landweber. MDS_IES_DB: a database of macronuclear and micronuclear genes in spirotrichous ciliates. *Nucleic Acids Res.*, 33:D396–D398, 2005. (Database issue).
3. A. R. Cavalcanti and L. F. Landweber. Gene unscrambler for detangling scrambled genes in ciliates. *Bioinformatics*, 20:800–802, 2004.
4. D. L. Chalker, Fuller P., and M. C. Yao. Communication between parental and developing genomes during *Tetrahymena* nuclear differentiation is likely mediated by homologous RNAs. *Genetics*, 169:149–160, 2005.
5. D. L. Chalker and M. C. Yao. Non-Mendelian, heritable blocks to DNA rearrangement are induced by loading the somatic nucleus of *Tetrahymena themophila* with germ line-limited DNA. *Mol. Cell Biol.*, 16:3658–3667, 1996.
6. D. L. Chalker and M. C. Yao. Nongenic, bidirectional transcription precedes and may promote developmental DNA deletion in *Tetrahymena thermophila. Genes Dev.*, 15:1287–1298, 2001.
7. M. DuBois and D. M. Prescott. Scrambling of the actin I gene in two *Oxytricha* species. *Proc. Natl. Acad. Sci. USA*, 92:3888–3892, 1995.
8. L. M. Epstein and J. D. Forney. Mendelian and non-mendelian mutations affecting surface antigen expression in *Paramecium tetraurelia. Mol. Cell Biol.*, 4:1583–1590, 1984.
9. O. Garnier, V. Serrano, S. Duharcourt, and E. Meyer. RNA-mediated programming of developmental genome rearrangements in *Paramecium tetraurelia. Mol. Cell Biol.*, 24:7370–7379, 2004.

10. A. Gratias and M. Bétermier. Developmentally programmed excision of internal DNA sequences in *Paramecium aurelia*. *Biochimie*, 83:1009–1022, 2001.
11. D. C. Hoffman and D. M. Prescott. The germline gene encoding DNA polymerase alpha in the hypotrichous ciliate *Oxytricha nova* is extremely scrambled. *Nucleic Acids Res.*, 24:3337–3340, 1996.
12. S. Juranek, S. Rupprecht, J. Postberg, and H. J. Lipps. snRNA specify DNA-sequences but are not sufficient for their correct excision during macronuclear development. Submitted, 2005.
13. S. Kuo, W. J. Chang, and L. F. Landweber. Complex germline architecture: Two genes intertwined on two loci. (Submitted), 2005.
14. L. F. Landweber, T. C. Kuo, and E. A. Curtis. Evolution and assembly of an extremely scrambled gene. *Proc. Natl. Acad. Sci. USA*, 97:3298–3303, 2000.
15. A. Le Mouël, A. Butler, F. Caron, and E. Meyer. Developmentally regulated chromosome fragmentation linked to imprecise elimination of repeated sequences in paramecia. *Eukaryot Cell*, 2:1076–1090, 2003.
16. J. L. Mitcham, A. J. Lynn, and D. M. Prescott. Analysis of a scrambled gene: the gene encoding alpha-telomere-binding protein in *Oxytricha nova*. *Genes Dev.*, 6:788–800, 1992.
17. K. Mochizuki, N. A. Fine, T. Fujisawa, and M. A. Gorovsky. Analysis of a piwi-related gene implicates small RNAs in genome rearrangement in *Tetrahymena*. *Cell*, 110:689–699, 2002.
18. K. Mochizuki and M. A. Gorovsky. RNA polymerase II localizes in *Tetrahymena thermophila* meiotic micronuclei when micronuclear transcription associated with genome rearrangement occurs. *Eukaryot Cell*, 3:1233–1240, 2004.
19. K. Mochizuki and M. A. Gorovsky. Small RNAs in genome rearrangement in *Tetrahymena*. *Curr. Opin. Genet. Dev.*, 14:181–187, 2004.
20. K. Mochizuki and M. A. Gorovsky. A dicer-like protein in *Tetrahymena* has distinct functions in genome rearrangement, chromosome segregation, and meiotic prophase. *Genes Dev.*, 19:77–89, 2005.
21. D. M. Prescott. The DNA of ciliated protozoa. *Microbiol. Rev.*, 58:233–267, 1994.
22. D. M. Prescott. Genome gymnastics: unique modes of DNA evolution and processing in ciliates. *Nat. Rev. Genet.*, 1:191–198, 2000.
23. D.M. Prescott, M. DuBois, Internal eliminated segments (IESs) of Oxytrichidae. *J. Eukariot. Microbiol.* **43** (1996) 432–441.
24. D. M. Prescott, A. Ehrenfeucht, and G. Rozenberg. Template-guided recombination for IES elimination and unscrambling of genes in stichotrichous ciliates. *J. Theor. Biol.*, 222:323–330, 2003.
25. K. Williams, T. G. Doak, and G. Herrick. Developmental precise excision of *Oxytricha trifallax* telomere-bearing elements and formation of circles closed by a copy of the flanking target duplication. *EMBO J.*, 12:4593–4601, 1993.
26. M. C. Yao, P. Fuller, and X. Xi. Programmed DNA deletion as an RNA-guided system of genome defense. *Science*, 300:1581–1584, 2003.

Modelling Simple Operations for Gene Assembly

Tero Harju[1,3], Ion Petre[2,3], and Grzegorz Rozenberg[4]

[1] Department of Mathematics, University of Turku
Turku 20014 Finland
harju@utu.fi
[2] Department of Computer Science, Åbo Akademi University
Turku 20520 Finland
ipetre@abo.fi
[3] Turku Centre for Computer Science
Turku 20520 Finland
[4] Leiden Institute for Advanced Computer Science, Leiden University
Niels Bohrweg 1, 2333 CA Leiden, the Netherlands, and
Department of Computer Science, University of Colorado at Boulder
Boulder, CO 80309-0347, USA
rozenber@liacs.nl

1 Introduction

The *Stichotrichous* ciliates have a very unusual way of organizing their genomic sequences. In the macronucleus, the somatic nucleus of the cell, each gene is a contiguous DNA sequence. Genes are generally placed on their own very short DNA molecules. In the micronucleus, the germline nucleus of the cell, the genes are placed on long chromosomes separated by noncoding material. However, the genes in the micronucleus are organized completely differently than in the macronucleus: a micronuclear gene is broken into pieces called MDSs (macronuclear destined sequences) that are separated by noncoding blocks called IESs (internally eliminated sequences). Moreover, the order of MDSs (compared to their order in the macronuclear version of a given gene) may be shuffled and some MDSs may be inverted. The ciliates may have several copies of the macronucleus (all identical to each other) and several micronuclei (all identical to each other) – the exact number of copies depends on the species. During sexual reproduction, ciliates destroy the old macronuclei and transform a micronucleus into a new macronucleus. In this process, ciliates must assemble all micronuclear genes by placing in the proper (orthodox) order all MDSs to yield a functional macronuclear gene. *Pointers*, short nucleotide sequences that identify each MDS, play an important role in the process. Each MDS M begins with a pointer that is exactly repeated in the end of the MDS preceding M in the orthodox order. The ciliates use the pointers to splice together all MDSs in the correct order.

The intramolecular model for gene assembly, introduced in [10, 28] consists of three operations: ld, hi, and dlad. In each of these operations, the micronuclear chromosome folds on itself so that two or more pointers get aligned and through recombination, two or more MDSs get combined into a bigger composite MDS. The process continues until all MDSs have been assembled. For details related to ciliates and gene assembly we refer to [16, 21, 22, 23, 24, 25, 26, 27]. For details related to the intramolecular model and its mathematical formalizations we refer to [4, 5, 8, 9, 13, 14, 15, 29, 30], as well as to the recent monograph [6]. For a different intermolecular model we refer to [18, 19, 20].

There are no restrictions in general on the number of nucleotides between the two pointers that should be aligned in a certain fold. However, all available experimental data are consistent with restricted versions of our operations, in which between two aligned pointers there is at most one MDS, see [6], [7], and [12]. In this paper we propose a mathematical model that takes this restriction into account by considering "simple" variants of ld, hi, and dlad. The model is formulated in terms of MDS descriptors, signed permutations, and signed double occurrence strings.

2 Mathematical Preliminaries

For an alphabet Σ we denote by Σ^* the set of all finite strings over Σ. For a string u we denote by $\mathsf{dom}(u)$ the set of letters occurring in u. We denote by λ the empty string. For strings u, v over Σ, we say that u is a *substring* of v, denoted $u \leq v$, if $v = xuy$, for some strings x, y (which can be empty).

Let $\Sigma_n = \{1, 2, \ldots, n\}$ and let $\overline{\Sigma_n} = \{\overline{1}, \overline{2}, \ldots, \overline{n}\}$ be a *signed copy* of Σ_n. For any $i \in \Sigma_n$, we say that i is an *unsigned letter*, while $\overline{i}$ is a *signed* letter. For a string $u = a_1 a_2 \ldots a_m$ over $\Sigma_n \cup \overline{\Sigma_n}$, its *inversion* $\overline{u}$ is defined by $\overline{u} = \overline{a}_m \ldots \overline{a}_2 \overline{a}_1$, where $\overline{\overline{a}} = a$, for all $a \in \Sigma_n$.

An *(unsigned) permutation* π over an interval $\Delta = \{i, i+1, \ldots, i+l\}$ is a bijective mapping $\pi : \Delta \to \Delta$. We often identify π with the string $\pi(i)\pi(i+1)\ldots\pi(i+l)$. We say that π is *(cyclically) sorted* if $\pi = k\,(k+1)\ldots i+l\,i\,(i+1)\ldots(k-1)$, for some $i \leq k \leq i+l$. A *signed permutation* over Δ is a string ψ over $\Delta \cup \overline{\Delta}$ such that $\|\psi\|$ is a permutation over Δ, where $\|\cdot\|$ is the mapping defined by $\|k\| = \|\overline{k}\| = k$, for all $i \leq k \leq i+l$. We say that ψ is *(cyclically) sorted* if either ψ, or $\overline{\psi}$ is a sorted unsigned permutation. In the former case we say that ψ is sorted in the *orthodox order*, while in the latter case we say that ψ is sorted in the *inverted order*.

There is a rich literature on sorting (signed and unsigned) permutations, both in connection to their applications to computational biology in topics such as genomic rearrangements or evolutionary distances, and also as a classical topic in discrete mathematics, see, e.g., [1, 2, 11, 17].

3 The Intramolecular Model

We present in this section the intramolecular model: the folds and the recombinations for each of the operations ld, hi, and dlad, as well as their simple variants.

3.1 The Structure of Micronuclear Genes

A micronuclear gene is broken into coding blocks called MDSs (macronuclear destined sequences), separated by non-coding blocks called IESs (internally eliminated sequences). In the macronucleus, however, all MDSs are spliced together into contiguous coding sequences, with no IESs present anymore. It is during gene assembly that ciliates eliminate IES and splice MDSs together. A central role in this process is played by *pointers*, short nucleotide sequences at both ends of each MDS. As it turns out, the pointer at the end of the $(i-1)$st MDS (in the order given by the macronuclear gene sequence), say M_{i-1}, coincides as a nucleotide sequence with the pointer at the beginning of the ith MDS, say M_i, for all i.

Based on these observations, we can represent the micronuclear genes by their sequences of MDSs only. For example, we represent the structure of the micronuclear gene encoding the actin protein in *Sterkiella nova* by the sequence of MDSs $M_3\, M_4\, M_6\, M_5\, M_7\, M_9\, \overline{M}_2\, M_1\, M_8$, where we indicate that the second MDS, M_2, is inverted in the micronucleus. Moreover, in some cases, we represent each MDS by its pair of pointers: we denote by i the pointer at the beginning of the ith MDS M_i. Thus, MDS M_i can be represented by its pair of pointers as $(i, i+1)$. The first and the last MDSs are special, and so M_1 is represented by $(b, 2)$ and M_k by (k, e), where b and e are special beginning/ending markers. In this case, the gene in Fig. 1 is represented as $(3,4)(4,5)(6,7)(5,6)(7,8)(9,e)(\overline{3},\overline{2})(b,2)(8,9)$. One more simplification can also be made. The gene may be represented by the sequence of its pointers only, thus ignoring the markers and the parenthesis above – this representation still gives enough information to trace the gene assembly process. Details on model forming can be found in [6].

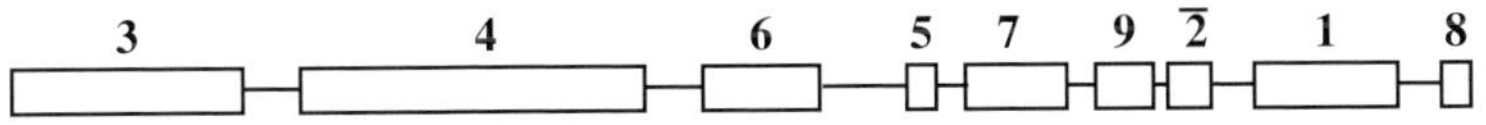

Fig. 1. Structure of the micronuclear gene encoding actin protein in *Sterkiella nova*.

3.2 Three Molecular Operations

Three molecular operations, ld, hi, dlad, were conjectured in [10] and [28] for gene assembly. In each of them, the micronuclear genome folds on itself in such a way that certain types of folds may be formed and recombination may take place, see Fig. 2. It is important to note that all foldings are aligned by pointers. We refer for more details to [6].

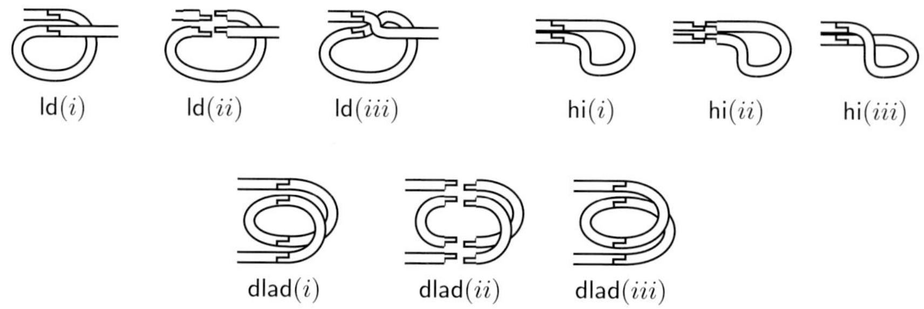

Fig. 2. Illustration of the ld, hi, dlad molecular operation showing in each case: (i) the folding, (ii) the recombination, and (iii) the result.

It is known that ld, hi, and dlad can assemble any gene pattern or, in other words, any sequence of MDSs can be transformed into an assembled MDS (b, e) (in which case we say that it has been assembled in the *orthodox* order) or $(\overline{e}, \overline{b})$ (we say it has been assembled in the *inverted* order), see [6] and [7] for formal proofs.

3.3 Simple Operations for Gene Assembly

Note that all three operations ld, hi, dlad are *intramolecular*, that is, a molecule folds on itself to rearrange its coding blocks. For a different, intermolecular model for gene assembly, see, [18], [19], and [20].

Since ld excises one circular molecule, that molecule can only contain noncoding blocks (or, in a special case, contain the entire gene, see [6] for details on the boundary ld): we say that ld must always be *simple* in a successful assembly. As such, the effect of ld is that it will combine two consecutive MDSs into a bigger composite MDS. For example, consider that M_iM_{i+1} is a part of the molecule, i.e., MDS M_{i+1} succeeds M_i being separated by one IES I. Thus, the pointer $i + 1$ has two occurrences that flank I: one at the end of MDS M_i and the other one at the beginning of MDS M_{i+1}. Then ld makes a fold as in Fig. 2:ld(i) aligned by the pointer $i + 1$, excises IES I as a circular molecule and combines M_i and M_{i+1} into a longer coding block as shown in Fig. 2:ld(ii)–ld(iii).

In the case of hi and dlad, the rearranged sequences may be arbitrarily large. For example, in the actin I gene in *S. nova*, see Fig. 1, pointer 3 has two occurrences: one at the beginning of M_3 and one, inverted, at the end of M_2. Thus, hi is applicable to this sequence with the hairpin aligned on pointer 3, even though five MDSs separate the two occurrences of pointer 3. Similarly, dlad is applicable to the MDS sequence $M_2M_8M_6M_5M_1M_7M_3M_{10}M_9M_4$, with the double loops aligned on pointers 3 and 5. Here the first two occurrences of pointers 3, 5 are separated by two MDSs (M_8 and M_6) and their second occurrences are separated by four MDSs (M_3, M_{10}, M_9, M_4).

It turns out, however, that all available experimental data, see [3], are consistent with applications of the so-called "simple" hi and dlad: particular instances of hi and dlad where the folds, and thus the rearranged sequences contain only one MDS. We define the simple operations in the following.

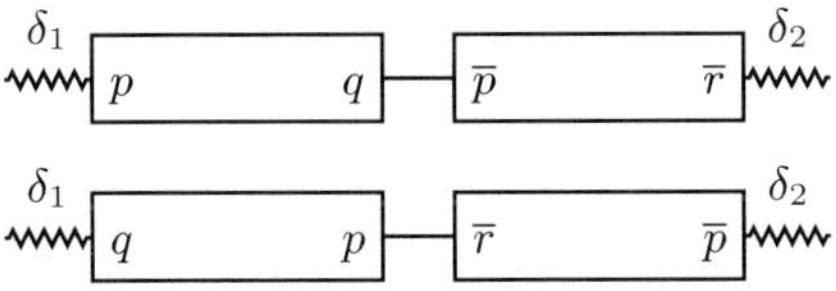

Fig. 3. The MDS/IES structures where the *simple hi*-rule is applicable. Between the two MDSs there is only one IES.

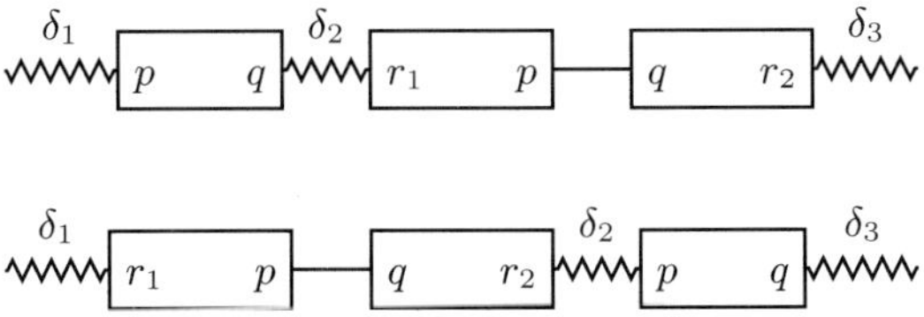

Fig. 4. The MDS/IES structures where *simple dlad*-rules are applicable. The straight line denotes one IES.

An application of the hi-operation on pointer p is *simple* if the part of the molecule that separates the two copies of p in an inverted repeat contains only one MDS and one IES. We have here two cases, depending on whether the first occurrence of p is incoming or outgoing. The two possibilities are illustrated in Fig. 3, where the MDSs are indicated by rectangles and their flanking pointers are shown.

An application of dlad on pointers p, q is *simple* if the sequence between the first occurrences of p, q and the sequence between the second occurrences of p, q consist of either one MDS or one IES. We have again two cases, depending on whether the first occurrence of p is incoming or outgoing. The two possibilities are illustrated in Fig. 4.

Recall that an operation ld is always simple (by definition) in the intramolecular model so that no coding sequence is lost.

One immediate property of simple operations is that they are not universal, i.e., there are sequences of MDSs that cannot be assembled by simple operations. One such example is the sequence $\overline{M_1}\, M_4\, M_3\, M_2$. Indeed, neither ld, nor simple hi, nor simple dlad is applicable to this sequence.

4 Formal Models for Simple Operations

We introduce in this section a formal model for simple operations. The model is formulated on three levels of abstraction: MDS descriptors, signed permutations, and signed double occurrence strings.

4.1 Modelling by MDS Descriptors

As noted above, micronuclear gene patterns may be represented by the sequence of their MDSs, while MDSs may be represented only by the pair of their flanking pointers, ignoring the rest of the sequences altogether. Indeed, since all the folds required by gene assembly are aligned on pointers, and the splicing of MDSs takes place through pointers, the whole process can be traced even with this (remarkable) simplification. Thus, an MDS M_i is represented as $(i, i+1)$, while its inversion is denoted as $(\overline{i+1}, \overline{i})$. A sequence of such pairs will be called an *MDS descriptor* and will be used to represent the structure of micronuclear genes. We define the notion formally in the following.

Let $\mathcal{M} = \{b, e, \overline{b}, \overline{e}\}$ be the set of markers and their inversions, and $\Pi_\kappa = \{2, 3, \ldots, \kappa\} \cup \{\overline{2}, \overline{3}, \ldots, \overline{\kappa}\}$ the set of pointers and their inversions, where κ is the number of MDS in the gene of interest. In the following, κ is an arbitrary but fixed nonnegative integer.

Let then

$$\begin{aligned}\Gamma_\kappa =& \{\, (b,e), (\overline{e},\overline{b}), (b,i), (\overline{i},\overline{b}), (i,e), (\overline{e},\overline{i}) \mid 2 \le i \le \kappa \,\} \\ & \cup \; \{\, (i,j) \mid 2 \le i < j \le \kappa \,\}.\end{aligned}$$

For each $x \in \Pi_\kappa \cup \mathcal{M}$, let

$$\widehat{x} = \begin{cases} 1, & \text{if } x \in \{b, \overline{b}\}, \\ \kappa+1, & \text{if } x \in \{e, \overline{e}\}, \\ \|x\|, & \text{if } x \in \Pi_\kappa. \end{cases}$$

For each $\delta = (x, y) \in \Gamma_\kappa$, let $\widehat{\delta} = [\min\{\widehat{x}, \widehat{y}\}, \max\{\widehat{x}, \widehat{y}\} - 1]$.

Example 1. Let $\delta = (4,5)(\overline{8},\overline{6})(b,4)(8,e)(5,6)$. Then the pairs occurring in δ have the following values: $\widehat{(4,5)} = [4,4]$, $\widehat{(\overline{8},\overline{6})} = [6,7]$, $\widehat{(b,4)} = [1,3]$, $\widehat{(8,e)} = [8,8]$ and $\widehat{(5,6)} = [5,5]$. □

Consider $\delta \in \Gamma_\kappa^*$, $\delta = \delta_1\delta_2\ldots\delta_n$, with $\delta_i \in \Gamma_\kappa$ for each i. We say that δ is an *MDS descriptor* if the intervals $\widehat{\delta_i}$, for $i = 1, 2, \ldots, n$, form a partition of the interval $[1, \kappa+1]$.

For each micronuclear gene pattern, its associated MDS descriptor is obtained by denoting each MDS by its pair of pointers or markers.

Example 2. The MDS descriptor associated to gene actin in *S. nova*, see Fig. 1, is $(3,4)\,(4,5)\,(6,7)\,(5,6)\,(7,8)\,(9,e)\,(\overline{3},\overline{2})\,(b,2)\,(8,9)$.

We can now define the simple operations as rewriting rules on MDS descriptors in accordance with the molecular model shown in Fig. 3 and 4.

(1) For each pointer $p \in \Pi_\kappa$, the ld-*rule* for p is defined as follows:

$$\mathsf{ld}_p(\delta_1(q,p)(p,r)\delta_2) = \delta_1(q,r)\delta_2, \qquad (\ell 1)$$
$$\mathsf{ld}_p((p,m_1)(m_2,p)) = (m_2,m_1), \qquad (\ell 2)$$

where $q,r \in \Pi_\kappa \cup \mathcal{M}$, $m_1,m_2 \in \mathcal{M}$ and $\delta_1,\delta_2 \in \Gamma_\kappa^*$.

(2) For each pointer $p \in \Pi_\kappa$, the sh-*rule* for p is defined as follows:

$$\mathsf{sh}_p(\delta_1(p,q)(\overline{p},\overline{r})\delta_2) = \delta_1(\overline{q},\overline{r})\delta_2, \qquad \text{(h1)}$$
$$\mathsf{sh}_p(\delta_1(q,p)(\overline{r},\overline{p})\delta_2) = \delta_1(q,r)\delta_2, \qquad \text{(h2)}$$

where $q,r \in \Pi_\kappa \cup \mathcal{M}$, and $\delta_i \in \Gamma_\kappa^*$, for each $i = 1,2,3$.

(3) For each pointers $p,q \in \Pi_\kappa$, the sd-*rule* for p,q is defined as follows:

$$\mathsf{sd}_{p,q}(\delta_1(p,q)\delta_2(r_1,p)(q,r_2)\delta_2) = \delta_1\delta_2(r_1,r_2)\delta_3, \qquad \text{(d1)}$$
$$\mathsf{sd}_{p,q}(\delta_1(r_1,p)(q,r_2)\delta_2(p,q)\delta_3) = \delta_1(r_1,p)(q,r_2)\delta_2(p,q)\delta_3, \qquad \text{(d2)}$$

where $r_1,r_2 \in \Pi_\kappa \cup \mathcal{M}$, and $\delta_i \in \Gamma_\kappa^*$, for each $i = 1,2,3$.

For an MDS descriptor δ and operations $\varphi_1,\ldots,\varphi_n$, $n \geq 1$, a composition $\varphi = \varphi_\kappa \ldots \varphi_1$ is an *assembly strategy* for δ, if φ is applicable to δ. Also, φ is *successful* for δ if either $\varphi(\delta) = (b,e)$ (in which case we say that δ has been assembled in the *orthodox order*) or $\varphi(\delta) = (\overline{e},\overline{b})$ (and we say that δ has been assembled in the *inverted order*).

Example 3. The actin gene in *S. nova* may be assembled by simple operations as follows. If $\delta = (3,4)\,(4,5)\,(6,7)\,(5,6)\,(7,8)\,(9,e)\,(\overline{3},\overline{2})\,(b,2)\,(8,9)$, then

$$\mathsf{ld}_4(\delta) = (3,5)\,(6,7)\,(5,6)\,(7,8)\,(9,e)\,(\overline{3},\overline{2})\,(b,2)\,(8,9)$$
$$\mathsf{sd}_{5,6}(\mathsf{ld}_4(\delta)) = (3,7)\,(7,8)\,(9,e)\,(\overline{3},\overline{2})\,(b,2)\,(8,9)$$
$$\mathsf{ld}_7(\mathsf{sd}_{5,6}(\mathsf{ld}_4(\delta))) = (3,8)\,(9,e)\,(\overline{3},\overline{2})\,(b,2)\,(8,9)$$
$$\mathsf{sd}_{8,9}(\mathsf{ld}_7(\mathsf{sd}_{5,6}(\mathsf{ld}_4(\delta)))) = (3,e)\,(\overline{3},\overline{2})\,(b,2)$$
$$\mathsf{sh}_3(\mathsf{sd}_{8,9}(\mathsf{ld}_7(\mathsf{sd}_{5,6}(\mathsf{ld}_4(\delta))))) = (\overline{e},\overline{2})\,(b,2)$$
$$\mathsf{sh}_2(\mathsf{sh}_3(\mathsf{sd}_{8,9}(\mathsf{ld}_7(\mathsf{sd}_{5,6}(\mathsf{ld}_4(\delta)))))) = (\overline{e},\overline{b}).$$

4.2 Modelling by Signed Permutations

The gene structure of a ciliate can also be represented as a signed permutation, denoting the sequence and orientation of each MDS, while omitting all IESs. For example, the signed permutation associated to gene actin I in *S. nova* is $3\,4\,6\,5\,7\,9\,\overline{2}\,1\,8$. The rearrangements made by ld, hi, dlad at the molecular level leading to bigger composite MDSs correspond to permutations that combine two already sorted blocks into a longer sorted block. Thus, in the framework of permutations, assembling a gene is equivalent to sorting the permutation associated to the micronuclear gene as explained below. Indeed, the gene is assembled once all MDSs are placed in the correct order.

When formalizing the gene assembly as a sorting of permutations we will effectively ignore the operation ld observing that once such an operation becomes applicable to a gene pattern, it can be applied at any later step of the assembly, see [4] and [8] for a formal proof. In particular, we can assume that all ld operations are applied in the last stage of the assembly, once all MDSs are sorted in the correct order. In this way, the process of gene assembly can indeed be described as a process of sorting the associated signed permutation, i.e., arranging the MDSs in the proper order, be that orthodox or inverted.

It is worth noting that the signed permutations are equivalent to the MDS descriptors as far as their expressibility is concerned. Indeed, the mapping ψ defined so that $\psi(i) = (i, i+1)$, for all $1 < i < \kappa$, $\psi(1) = (b, 2)$, and $\psi(\kappa) = (\kappa, e)$ is a bijective morphism between the set of signed permutations and the set of MDS descriptors. Some differences do exist when modelling gene assembly with descriptors or permutations. For example, modelling the assembly with MDS descriptors is a rewriting process of eliminating pointers, leading ultimately to assembled descriptors with no pointers. On this level, we can keep track of every pointer in the gene assembly – this is often useful. The downside is that the descriptors introduce a tedious mathematical notation and reasoning about them is typically involved. The signed permutations on the other hand represent an elegant, classical topic in mathematics and a large literature about them exists. Gene assembly on permutations becomes a process of sorting signed permutations, a topic that is well-studied in the literature. An additional technical advantage here is that the base alphabet of the permutation does not change through the process as is the case with the descriptors. The downside of the signed permutations is that they do not denote the pointers explicitly.

The molecular model of simple operations in Fig. 3 and 4 can be formalized as a sorting of signed permutations as follows.

(2′) For each $p \geq 1$, sh_p is defined as follows:

$$\mathsf{sh}_p(x\,(p+1)\ldots(p+k+1)\,\overline{p}\,y) = x\,\overline{(p+k+1)}\ldots\overline{(p+1)}\,\overline{p}\,y,$$
$$\mathsf{sh}_p(x\,\overline{p}\ldots\overline{(p-k)}(p+1)\,y) = x\,(p-k)\ldots p\,(p+1)\,y,$$
$$\mathsf{sh}_p(x\,p\,\overline{(p+k+1)}\ldots\overline{p+1}\,y) = x\,p\,(p+1)\ldots(p+k+1)\,y,$$
$$\mathsf{sh}_p(x\,\overline{(p+1)}(p-k)\ldots p\,y) = x\,\overline{(p+1)}\,\overline{p}\ldots\overline{(p-k)}\,y,$$

where $k \geq 0$ and x, y, z are signed strings over Σ_n. We denote $\mathsf{Sh} = \{\mathsf{sh}_i \mid 1 \leq i \leq n\}$.

(3′) For each p, $2 \leq p \leq n-1$, sd_p is defined as follows:

$$\mathsf{sd}_p(x\,(p-i)\ldots p\,y\,(p-i-1)\,(p+1)\,z) = x\,y\,(p-i-1)\,(p-i)\ldots p\,(p+1)\,z,$$
$$\mathsf{sd}_p(x\,(p-i-1)\,(p+1)\,y\,(p-i)\ldots p\,z) = x\,(p-i-1)\,(p-i)\ldots p\,(p+1)\,y\,z,$$

where $i \geq 0$ and x, y, z are signed strings over Σ_n. We also define $\mathsf{sd}_{\overline{p}}$ as follows:

$$\mathsf{sd}_{\overline{p}}(x\,(\overline{p+1})(\overline{p-i-1})\,y\,\overline{p}\ldots(\overline{p-i})\,z) = x\,(\overline{p+1})\,\overline{p}\ldots(\overline{p-i})\,(\overline{p-i-1})\,y\,z,$$
$$\mathsf{sd}_{\overline{p}}(x\,\overline{p}\ldots(\overline{p-i})\,y\,(\overline{p+1})\,(\overline{p-i-1})\,z) = x\,y\,(\overline{p+1})\,\overline{p}\ldots(\overline{p-i})\,(\overline{p-i-1})\,z,$$

where $i \geq 0$ and x, y, z are signed strings over Σ_n. We denote $\mathsf{Sd} = \{\mathsf{sd}_i, \mathsf{sd}_{\overline{i}} \mid 1 \leq i \leq n\}$.

We say that a signed permutation π over the set of integers $\{i, i+1, \ldots, i+l\}$ is *sortable* if there are operations $\phi_1, \ldots, \phi_k \in \mathsf{Sh} \cup \mathsf{Sd}$ such that $(\phi_1 \circ \ldots \circ \phi_k)(\pi)$ is a (cyclically) sorted permutation. We also say in this case that $\phi_1 \circ \ldots \circ \phi_k$ is a *sorting strategy* for π. We say that π is Sh-*sortable* if $\phi_1, \ldots, \phi_k \in \mathsf{Sh}$ and we say that π is Sd-*sortable* if $\phi_1, \ldots, \phi_k \in \mathsf{Sd}$. A composition ϕ is called an *unsuccessful strategy* for π if $\phi(\pi)$ is an unsortable permutation.

Example 4. (i) The permutation $\pi_1 = 3\,\overline{4}\,\overline{5}\,6\,\overline{1}\,2$ is sortable and a sorting strategy is $\mathsf{sh}_1(\mathsf{sh}_4(\mathsf{sh}_3(\pi_1))) = 3\,4\,5\,6\,1\,2$. The permutation $\pi_1' = 3\,4\,5\,6\,\overline{1}\,\overline{2}$ is unsortable. Indeed, no sh operations and no sd operation is applicable to π_1'.

(ii) The permutation $\pi_2 = 1\,3\,4\,2\,\overline{5}$ is sortable and it has only one sorting strategy: $\mathsf{sh}_4(\mathsf{sd}_2(\pi_2)) = 1\,2\,3\,4\,5$.

(iii) There exist permutations with several successful strategies, even leading to different sorted permutations. One such permutation is $\pi_3 = 3\,5\,1\,2\,4$. Indeed, $\mathsf{sd}_3(\pi_3) = 5\,1\,2\,3\,4$, while $\mathsf{sd}_4(\pi_3) = 3\,4\,5\,1\,2$.

(iv) The simple operations yield a nondeterministic process: there are permutations having both successful and unsuccessful sorting strategies. One such permutation is $\pi_4 = 1\,3\,5\,7\,9\,2\,4\,6\,8$. Note that $\mathsf{sd}_3(\mathsf{sd}_5(\mathsf{sd}_7(\pi_4))) = 1\,9\,2\,3\,4\,5\,6\,7\,8$ is an unsortable permutation. However, π_4 can be sorted, e.g., by the following strategy: $\mathsf{sd}_2(\mathsf{sd}_4(\mathsf{sd}_6(\mathsf{sd}_8(\pi_4)))) = 1\,2\,3\,4\,5\,6\,7\,8\,9$.

(v) The permutation $\pi_5 = 1\,3\,5\,2\,4$ has both successful and unsuccessful sorting strategies. Indeed, $\mathsf{sd}_3(\pi_5) = 1\,5\,2\,3\,4$, an unsortable permutation. However, $\mathsf{sd}_2(\mathsf{sd}_4(\pi_5)) = 1\,2\,3\,4\,5$ is sorted.

(vi) Applying a cyclic shift to a permutation may render it unsortable. Indeed, permutation $2\,1\,4\,3\,5$ is sortable, while $5\,2\,1\,4\,3$ is not.

(vii) Consider the signed permutation $\pi_7 = 1\,11\,3\,9\,5\,7\,2\,4\,13\,6\,15\,8\,10\,12\,14\,16$. Operation sd may be applied to π_7 on integers 3, 6, 9, 11, 13, and 15. Doing that, however, leads to an unsortable permutation:

$$\mathsf{sd}_3(\mathsf{sd}_6(\mathsf{sd}_9(\mathsf{sd}_{11}(\mathsf{sd}_{13}(\mathsf{sd}_{15}(\pi_7)))))) = 1\,5\,6\,7\,2\,3\,4\,8\,9\,10\,11\,12\,13\,14\,15\,16.$$

However, omitting sd_3 from the above composition leads to a sorting strategy for π_7: let

$$\pi_7' = \mathsf{sd}_6(\mathsf{sd}_9(\mathsf{sd}_{11}(\mathsf{sd}_{13}(\mathsf{sd}_{15}(\pi_7))))) = 1\,3\,5\,6\,7\,2\,4\,8\,9\,10\,11\,12\,13\,14\,15\,16.$$

Then $\mathsf{sd}_2(\mathsf{sd}_4(\pi_7'))$ is a sorted permutation.

(viii) Consider the signed permutation π_8 associated to the actin gene in *S. nova*, $\pi_8 = 3\,4\,6\,5\,7\,9\,\overline{2}\,1\,8$. A sorting strategy for π_8 is shown below (compare it with Example 3):

$$\begin{aligned}
\mathsf{sd}_5(\pi_8) &= 3\,4\,5\,6\,7\,9\,\overline{2}\,1\,8\\
\mathsf{sd}_8(\mathsf{sd}_5(\pi_8)) &= 3\,4\,5\,6\,7\,8\,9\,\overline{2}\,1\\
\mathsf{sh}_2(\mathsf{sd}_8(\mathsf{sd}_5(\pi_8))) &= \overline{9}\,\overline{8}\,\overline{7}\,\overline{6}\,\overline{5}\,\overline{4}\,\overline{3}\,\overline{2}\,1\\
\mathsf{sh}_1(\mathsf{sh}_2(\mathsf{sd}_8(\mathsf{sd}_5(\pi_8)))) &= \overline{9}\,\overline{8}\,\overline{7}\,\overline{6}\,\overline{5}\,\overline{4}\,\overline{3}\,\overline{2}\,\overline{1}.
\end{aligned}$$

4.3 Modelling by Signed Double Occurrence Strings

The structure of a gene may be simplified by representing only the sequence of its pointers, see [4], [6], and [8]. Indeed, since the assembled gene has no pointers anymore and all the operations are based on the sequence and orientation of pointers, such a simplification is possible. The strings we obtain are called signed double occurrence strings and are defined in the following.

Let Σ be an alphabet and $\overline{\Sigma}$ its signed copy. A string $v \in (\Sigma \cup \overline{\Sigma})^*$ is a *signed double occurrence string* if for every letter $a \in \mathsf{dom}(v)$, v has exactly two occurrences from the set $\{a, \overline{a}\}$. We also say then that v is a *legal* string. If v contains both substrings a and $\overline{a}$, then a is *positive* in u; otherwise, a is *negative* in u.

Example 5. Consider the signed string $u = 2\,4\,3\,\overline{2}\,\overline{5}\,3\,4\,5$ over Δ_5. Clearly, u is legal. Pointers 2 and 5 are positive in u, while 3 and 4 are negative in u. On the other hand, the string $w = 2\,4\,3\,\overline{2}\,\overline{5}\,3\,5$ is not legal, since 4 has only one occurrence in w.

We can associate a unique legal string to any gene pattern by writing its sequence of pointers only. Formally, we can define the following mapping: $\mu((i,j)) = i\,j$, for all $2 \leq i < j \leq \kappa$, $\mu((b,i)) = i$, $\mu((i,e)) = i$, for all $2 \leq k \leq \kappa$, and $\mu((b,e)) = \lambda$. Then μ defines a morphism from the set of MDS descriptors to the set of legal strings. We say that $\mu(\delta)$ is the legal string associated to δ.

Example 6. The MDS descriptor associated to the actin gene in *S. nova* is $\delta = (3,4)\,(4,5)\,(6,7)\,(5,6)\,(7,8)\,(9,e)\,(\overline{3},\overline{2})\,(b,2)\,(8,9)$. Its legal string, obtained from δ by writing only the sequence of pointers and their orientations is $3\,4\,4\,5\,6\,7\,5\,6\,7\,8\,9\,\overline{3}\,\overline{2}\,2\,8\,9$.

We refer to [6] for the formalization of the intramolecular model on the level of general legal strings. We only define in the following the simple operations as rewriting rules on legal strings. Without risk of confusion, we will use the notation ld, sh and sd also for legal strings.

The simple operations can be defined as rewriting rules on legal strings as follows.

(1″) For each pointer $p \in \Pi_\kappa$, the ld-*rule* for p is defined by

$$\mathsf{ld}_p(u_1\, p\, p\, u_2) = u_1\, u_2.$$

where $u_1, u_2 \in (\Sigma \cup \overline{\Sigma})^*$. Let $\mathsf{Ld} = \{\mathsf{ld}_p \mid p \in \Pi_\kappa,\ \kappa \geq 2\}$.

(2″) For each pointer $p \in \Pi_\kappa$, the sh-*rule* for p is defined by

$$\mathsf{sh}_p(u_1\, p\, u_2\, \overline{p}\, u_3) = u_1\, \overline{u}_2\, u_3,$$

where $u_1, u_2, u_3 \in (\Sigma \cup \overline{\Sigma})^*$ and $|u_2| \leq 1$. Let $\mathsf{Sh} = \{\mathsf{sh}_p \mid p \in \Pi_\kappa,\ \kappa \geq 2\}$.

(3″) For each pointers $p, q \in \Pi_\kappa$, the sd-*rule* for p, q is defined by

$$\mathsf{sd}_{p,q}(u_1\, p\, q\, u_2\, p\, q u_3) = u_1\, u_2\, u_3,$$

where $u_1, u_2, u_3 \in (\Sigma \cup \overline{\Sigma})^*$. Let $\mathsf{Sd} = \{\mathsf{sd}_{p,q} \mid p, q \in \Pi_\kappa,\ \kappa \geq 2\}$.

A composition $\varphi = \varphi_n \ldots \varphi_1$ of operations from $\mathsf{Ld} \cup \mathsf{Sh} \cup \mathsf{Sd}$ is a *string reduction* of u, if φ is applicable to u. Also, φ is *successful* for u if $\varphi(u) = \lambda$, the empty string.

Example 7. The signed double occurrence string associated to the actin gene in *S. nova* is $u = 3\,4\,4\,5\,6\,7\,5\,6\,7\,8\,9\,\overline{3}\,\overline{2}\,2\,8\,9$. Here is a successful reduction of u using only simple operations (compare it with Examples 3 and 4(viii)):

$$\mathsf{ld}_4(u) = 3\,5\,6\,7\,5\,6\,7\,8\,9\,\overline{3}\,\overline{2}\,2\,8\,9$$
$$\mathsf{sd}_{5,6}(\mathsf{ld}_4(u)) = 3\,7\,7\,8\,9\,\overline{3}\,\overline{2}\,2\,8\,9$$
$$\mathsf{ld}_7(\mathsf{sd}_{5,6}(\mathsf{ld}_4(u))) = 3\,8\,9\,\overline{3}\,\overline{2}\,2\,8\,9$$
$$\mathsf{sd}_{8,9}(\mathsf{ld}_7(\mathsf{sd}_{5,6}(\mathsf{ld}_4(u)))) = 3\,\overline{3}\,\overline{2}\,2$$
$$\mathsf{sh}_3(\mathsf{sd}_{8,9}(\mathsf{ld}_7(\mathsf{sd}_{5,6}(\mathsf{ld}_4(u))))) = \overline{2}\,2$$
$$\mathsf{sh}_2(\mathsf{sh}_3(\mathsf{sd}_{8,9}(\mathsf{ld}_7(\mathsf{sd}_{5,6}(\mathsf{ld}_4(u)))))) = \lambda.$$

5 Discussion

In this paper we introduced a molecular model of the so-called simple operations, a restricted variant of the intramolecular model for gene assembly. In simple operations the type of fold that a micronuclear chromosome has to make during an assembly is very restricted: only one MDS is relocated during the subsequent recombination. While this variant is not universal anymore, it is still powerful enough to assemble all known micronuclear gene patterns. A number of questions (research topics) considering simple operations are natural and worth investigating. One of the most important ones is: what are the gene patterns that can be assembled using the simple operations? Also, we noticed that, while the simple model is not universal, it remains non-deterministic: there are gene patterns that have both successful and unsuccessful assembly strategies. Deciding if a given pattern may be assembled by simple operations and finding/characterizing its successful strategies is another important problem. From a computational point of view, a study of the complexity of the simple assemblies seems very interesting. A detailed study of simple operations was already initiated in [12].

Acknowledgement. The authors gratefully acknowledge support from the European Union project MolCoNet, IST-2001-32008, the European Research Training Network SegraVis, the Academy of Finland (project 39802 to TH and project 203667 to IP), and the NSF (grant 0121422 to GR).

References

1. P. Berman, S. Hannenhalli, Fast sorting by reversals. *Combinatorial Pattern Matching, Lecture Notes in Comput. Sci.* **1072** (1996) 168–185.
2. A. Caprara, Sorting by reversals is difficult. In S. Istrail, P. Pevzner and M. Waterman (eds.) *Proceedings of the 1st Annual International Conference on Computational Molecular Biology* (1997) 75–83.
3. A. Cavalcanti, T.H. Clarke, L. Landweber, MDS_IES_DB: a database of macronuclear and micronuclear genes in spirotrichous ciliates. *Nucleic Acids Research* **33** (2005) 396–398.
4. A. Ehrenfeucht, T. Harju, I. Petre, D.M. Prescott, G. Rozenberg, Formal systems for gene assembly in ciliates. *Theoret. Comput. Sci.* **292** (2003) 199–219.
5. A. Ehrenfeucht, T. Harju, I. Petre, G. Rozenberg, Characterizing the micronuclear gene patterns in ciliates. *Theory of Comput. Syst.* **35** (2002) 501–519.
6. A. Ehrenfeucht, T. Harju, I. Petre, D.M. Prescott, G. Rozenberg, *Computation in Living Cells: Gene Assembly in Ciliates*, Springer (2003).
7. A. Ehrenfeucht, I. Petre, D.M. Prescott, G. Rozenberg, Universal and simple operations for gene assembly in ciliates. In: V. Mitrana and C. Martin-Vide (eds.) *Words, Sequences, Languages: Where Computer Science, Biology and Linguistics Meet*, Kluwer Academic, Dortrecht (2001) 329–342.
8. A. Ehrenfeucht, I. Petre, D.M. Prescott, G. Rozenberg, String and graph reduction systems for gene assembly in ciliates. *Math. Structures Comput. Sci.* **12** (2001) 113–134.
9. A. Ehrenfeucht, I. Petre, D.M. Prescott, G. Rozenberg, Circularity and other invariants of gene assembly in cliates. In: M. Ito, Gh. Păun and S. Yu (eds.) *Words, Semigroups, and Transductions*, World Scientific, Singapore (2001) 81–97.
10. A. Ehrenfeucht, D.M. Prescott, G. Rozenberg, Computational aspects of gene (un)scrambling in ciliates. In: L. F. Landweber, E. Winfree (eds.) *Evolution as Computation*, Springer, Berlin, Heidelberg, New York (2001) pp. 216–256.
11. S. Hannenhalli, P.A. Pevzner, Transforming cabbage into turnip (Polynomial algorithm for sorting signed permutations by reversals). In: *Proceedings of the 27th Annual ACM Symposium on Theory of Computing* (1995) 178–189.
12. T. Harju, I. Petre, V. Rogojin G. Rozenberg, Simple operations for gene assembly. In: *Preproceedings of the 11th International Meeting on DNA-Based Computers*, 2005, to appear.

13. T. Harju, I. Petre, C. Li G. Rozenberg, Parallelism in gene assembly In: *Proceedings of DNA-Based Computers 10*, LNCS 3384 Springer, Berlin, Heidelberg (2005) 138-147.
14. T. Harju, I. Petre, G. Rozenberg, Gene assembly in ciliates: molecular operations I, In: Gh. Paun, G. Rozenberg, A. Salomaa (Eds.) *Current Trends in Theoretical Computer Science* World Scientific Pub. Co. Inc. (2004).
15. T. Harju, I. Petre, G. Rozenberg, Gene assembly in ciliates: formal frameworks II. In: Gh. Paun, G. Rozenberg, A. Salomaa (Eds.) *Current Trends in Theoretical Computer Science* World Scientific Pub. Co. Inc. (2004).
16. C.L. Jahn, L.A. Klobutcher, Genome remodeling in ciliated protozoa. *Ann. Rev. Microbiol.* **56** (2000) 489–520.
17. H. Kaplan, R. Shamir, R.E. Tarjan, A faster and simpler algorithm for sorting signed permutations by reversals. *SIAM J. Comput.* **29** (1999) 880–892.
18. L. Kari, L.F. Landweber, Computational power of gene rearrangement. In: E. Winfree and D. K. Gifford (eds.) *Proceedings of DNA Based Computers V*, American Mathematical Society (1999) 207–216.
19. L.F. Landweber, L. Kari, The evolution of cellular computing: Nature's solution to a computational problem. In: *Proceedings of the 4th DIMACS Meeting on DNA-Based Computers*, Philadelphia, PA (1998) 3–15.
20. L.F. Landweber, L. Kari, Universal molecular computation in ciliates. In: L. F. Landweber and E. Winfree (eds.) *Evolution as Computation*, Springer, Berlin Heidelberg New York (2002).
21. D.M. Prescott, *Cells: Principles of Molecular Structure and Function*, Jones and Barlett, Boston (1988).
22. D.M. Prescott, Cutting, splicing, reordering, and elimination of DNA sequences in hypotrichous ciliates. *BioEssays* **14** (1992) 317–324.
23. D.M. Prescott, The unusual organization and processing of genomic DNA in hypotrichous ciliates. *Trends in Genet.* **8** (1992) 439–445.
24. D.M. Prescott, The DNA of ciliated protozoa. *Microbiol. Rev.* **58**(2) (1994) 233–267.
25. D.M. Prescott, The evolutionary scrambling and developmental unscrambling of germlike genes in hypotrichous ciliates. *Nucl. Acids Res.* **27** (1999)1243–1250.
26. D.M. Prescott, Genome gymnastics: unique modes of DNA evolution and processing in ciliates. *Nat. Rev. Genet.* 1(3) (2000) 191–198.
27. D.M. Prescott, M. DuBois, Internal eliminated segments (IESs) of Oxytrichidae. *J. Eukariot. Microbiol.* **43** (1996) 432–441.
28. D.M. Prescott, A. Ehrenfeucht, G. Rozenberg, Molecular operations for DNA processing in hypotrichous ciliates. *Europ. J. Protistology* **37** (2001) 241–260.
29. D.M. Prescott, G. Rozenberg, How ciliates manipulate their own DNA – A splendid example of natural computing. *Natural Computing* **1** (2002) 165–183.
30. D.M. Prescott, G. Rozenberg, Encrypted genes and their reassembly in ciliates. In: M. Amos (ed.) *Cellular Computing*, Oxford University Press, Oxford (2003).

Part VIII

Appendix

Publications by Nadrian C. Seeman

1. N.C. Seeman, MIND3: A FORTRAN Patterson Superposition Program for the IBM 1130. In: *Crystallographic Computing*, F.R. Ahmed, ed., Munksgaard, Copenhagen 87–89 (1970).
2. B. Berking and N.C. Seeman, The Crystal and Molecular Structure of 1,6;2,3-Dianhydro-?-D-gulopyranose. *Acta Crystallographica* **B27** 1752–1760 (1971).
3. N.C. Seeman, J.L. Sussman, H.M. Berman and S.-H. Kim, Nucleic Acid Conformation: The Crystal Structure of a Naturally Occurring Dinucleoside Phosphate (UpA). *Nature New Biology* **233** 90–92 (1971).
4. N.C. Seeman, E.L. McGandy and R.D. Rosenstein, The Crystal and Molecular Structure of Pyrazole-3-L-Alanine. *Journal of the American Chemical Society* **94** 1717–1720 (1972).
5. J.J. Madden, E.L. McGandy and N.C. Seeman, The Crystal Structure of the Orthorhombic Form of L-(+)-Histidine. *Acta Crystallographica* **B28** 2377–2382 (1972).
6. J.J. Madden, E.L. McGandy, N.C. Seeman, M.M. Harding and A. Hoy, The Crystal Structure of the Monoclinic Form of L-(+)-Histidine. *Acta Crystallographica* **B28** 2382–2389 (1972).
7. J.L. Sussman, N.C. Seeman, S.-H. Kim and H.M. Berman, The Crystal Structure of a Naturally Occurring Dinucleoside Phosphate: Uridylyl-3',5'-adenosine. Model for RNA Chain Folding. *Journal of Molecular Biology* **66** 403–421 (1972).
8. N.C. Seeman, Determinata la Struttura Cristallina di un Dinucleosido-fosfato Naturale (UpA). In: *Enciclopedia Della Scienze e Della Tecnica, Annuario Della EST*, Mondadori, Milano pp. 68–69 (1972) (in Italian).
9. S.-H. Kim, H.M. Berman, N.C. Seeman and M.D. Newton, Seven Basic Conformations of Nucleic Acid Structural Units. *Acta Crystallographica* **B29** 703–710 (1973).
10. J.M. Rosenberg, N.C. Seeman, J.-J.P. Kim, F.L. Suddath, H.B. Nicholas and A. Rich, Double Helix at Atomic Resolution. *Nature* **243** 150–154 (1973).

11. R.O. Day, N.C. Seeman, J.M. Rosenberg and A. Rich, A Crystalline Fragment of the Double Helix: The Structure of the Dinucleoside Phosphate Guanylyl-3',5'-cytidine. *Proceedings of the National Academy of Sciences* (USA) **70** 849–853 (1973).
12. S.-H. Kim, F.L. Suddath, G.J. Quigley, A. McPherson, J.L. Sussman, A.H.-J. Wang, N.C. Seeman and A. Rich, The Three Dimensional Tertiary Structure of Transfer RNA. *Science* **185** 435–440 (1974).
13. S.-H. Kim, J.L. Sussman, G.J. Quigley, F.L. Suddath, A. McPherson, A.H.-J. Wang, N.C. Seeman and A. Rich, A Generalized Structure for Transfer RNA. *Proceedings of the National Academy of Sciences* (USA) **71** 4970–4974 (1974).
14. N.C. Seeman, R.O. Day and A. Rich, Nucleic Acid-Mutagen Interactions: The Crystal Structure of Adenylyl-3',5'-uridine Plus 9-Aminoacridine. *Nature* **253** 324–326 (1975).
15. G.J. Quigley, A.H.-J. Wang, N.C. Seeman, F.L. Suddath, A. Rich, J.L. Sussman and S.-H. Kim, Hydrogen Bonding in Yeast Phenylalanine Transfer RNA. *Proceedings of the National Academy of Sciences* (USA) **72** 4866–4870 (1975).
16. G.J. Quigley, N.C. Seeman, A.H.-J. Wang, F.L. Suddath and A. Rich, Yeast Phenylalanine Transfer RNA: Atomic Coordinates and Torsion Angles. *Nucleic Acids Research* **2** 2329–2341 (1975).
17. S.-H. Kim, F.L. Suddath, G.J. Quigley, A. McPherson, J.L. Sussman, A.H.-J. Wang, N.C. Seeman and A. Rich, The Tertiary Structure of Yeast Phenylalanine Transfer RNA. In: *Structure and Conformations of Nucleic Acids and Protein-Nucleic Acid Interactions*, ed. by M. Sundaralingam and S.T. Rao, University Park Press, Baltimore (1975) pp. 7–14.
18. N.C. Seeman, J.M. Rosenberg and A. Rich, Sequence Specific Recognition of Double Helical Nucleic Acids by Proteins. *Proceedings of the National Academy of Sciences* (USA) **73** 804–808 (1976).
19. J.M Rosenberg, N.C. Seeman, R.O. Day and A. Rich, RNA Double Helices Derived from Studies of Small Fragments. *Biochemical and Biophysical Research Communications* **69** 979–987 (1976).
20. A. Rich and N.C. Seeman, RNA Double Helices at Atomic Resolution. In: *Handbook of Biochemistry and Molecular Biology*, vol. II, 3rd edition, ed. by G. Fasman, CRC Press, Cleveland (1976) pp. 463-469.
21. N.C. Seeman, J.M Rosenberg, F.L. Suddath, J.-J.P. Kim and A. Rich, RNA Double Helical Fragments at Atomic Resolution I: The Crystal and Molecular Structure of Adenylyl-3',5'-uridine Hexahydrate. *Journal of Molecular Biology* **104** 109–144 (1976).
22. J.M. Rosenberg, N.C. Seeman, R.O. Day and A. Rich, RNA Double Helical Fragments at Atomic Resolution II: The Crystal and Molecular Structure of Guanylyl-3',5'-cytidine Nonahydrate. *Journal of Molecular Biology* **104** 145–167 (1976).

23. A. Rich, N.C. Seeman and J.M. Rosenberg, Protein Recognition of Base Pairs in a Double Helix. In: *Nucleic Acid-Protein Recognition*, ed. by H.G. Vogel, Academic Press, New York (1977) pp. 361–374.
24. N.H. Woo, N.C. Seeman and A. Rich, The Crystal Structure of Putrescine Diphosphate: A Model for Amine-Nucleic Acid Interactions. *Biopolymers* **18** 539–552 (1979).
25. N.C. Seeman, Single Crystal Crystallography. In: *Stereodynamics of Molecular Systems*, ed. by R.H. Sarma, Pergamon Press, New York (1979) pp. 75–109.
26. N.C. Seeman, Single Crystal Crystallography for Nucleic Acid Structural Studies. In: *Geometry and Dynamics of Nucleic Acids*, ed. by R.H. Sarma, Pergamon Press, New York (1980) pp. 47–82.
27. N.C. Seeman, Crystallographic Investigation of Oligonucleotide Structure. In: *Geometry and Dynamics of Nucleic Acids*, ed. by R.H. Sarma, Pergamon Press, New York (1980) pp. 109–142.
28. J.F. Hyde, J.A. Zubieta and N.C. Seeman, The Crystal and Molecular Structure of ((Mo(S-CH2-CH2-S-CH2-CH2-S)2)2Ag)PF6.1/2 DMF, A Thiolate-Bridged Mixed Metal Cluster. *Inorganica Chemica Acta* **54** L137–L139 (1981).
29. J.A. DiVerdi, S.J. Opella, R.-I. Ma, N.R. Kallenbach and N.C. Seeman, 31P NMR of DNA in Eukaryotic Chromosomal Complexes. *Biochemical and Biophysical Research Communications* **102** 885–890 (1981).
30. N.C. Seeman, Nucleic Acid Junctions: Building Blocks for Genetic Engineering in Three Dimensions. In: *Biomolecular Stereodynamics*, ed. by R.H. Sarma, Adenine Press, New York (1981) pp. 269–277.
31. N.C. Seeman and B.H. Robinson, Simulation of Double Stranded Branch Point Migration. In: *Biomolecular Stereodynamics*, ed. by R.H. Sarma, Adenine Press, New York (1981) pp. 279–300.
32. N.C. Seeman, Nucleic Acid Junctions and Lattices. *Journal of Theoretical Biology* **99** 237–247 (1982).
33. R.D. Rosenstein, M. Oberding, J. Hyde, J.A. Zubieta, K.D. Karlin and N.C. Seeman, The Crystal Structure of Hypoxanthinium Nitrate Monohydrate: Intercalation of Water Between Purines. *Crystal Structure Communications* **11** 1507–1513 (1982).
34. N.C. Seeman and N.R. Kallenbach, Nucleic Acid Junctions, The Tensors of Life? In: *Nucleic Acids: The Vectors of Life*, ed. by B. Pullman and J. Jortner, D. Reidel, Dordrecht (1983) pp. 183–200.
35. N.C. Seeman and N.R. Kallenbach, Design of Immobile Nucleic Acid Junctions. *Biophysical Journal* **44** 201–209 (1983).
36. N.R. Kallenbach, R.-I. Ma and N.C. Seeman, An Immobile Nucleic Acid Junction Constructed from Oligonucleotides. *Nature* **305** 829–831 (1983).
37. N.R. Kallenbach, R.-I. Ma, A.J. Wand, G.H. Veeneman, J.H. van Boom and N.C. Seeman, Fourth Rank Immobile Nucleic Acid Junctions. *Journal of Biomolecular Structure and Dynamics* **1** 158–168 (1983).

38. N.C. Seeman, M.F. Maestre, R.-I. Ma and N.R. Kallenbach, Physical Characterization of a Nucleic Acid Junction. In: *Progress in Clinical and Biological Research, Vol. 172A: The Molecular Basis of Cancer*, ed. by R. Rein, Alan R. Liss Inc., New York (1985) pp. 99–108.
39. N.C. Seeman, The Interactive Manipulation and Design of Macromolecular Architecture Utilizing Nucleic Acid Junctions. *Journal of Molecular Graphics* **3** 34–39 (1985).
40. N.C. Seeman, Macromolecular Design, Nucleic Acid Junctions and Crystal Formation. *Journal of Biomolecular Structure and Dynamics* **3** 11–34 (1985).
41. D.E. Wemmer, A.J. Wand, N.C. Seeman and N.R. Kallenbach, NMR Analysis of DNA Junctions: Imino Proton NMR Studies of Individual Arms and Intact Junction. *Biochemistry* **24** 5745–5749 (1985).
42. N.R. Kallenbach and N.C. Seeman, Stable Branched DNA Structures: DNA Junctions. *Comments on Cellular and Molecular Biophysics* **4** 1–16 (1986).
43. R.D. Sheardy and N.C. Seeman, The Construction of an Oligonucleotide Containing 5-Iodo-Deoxyuridine. *Journal of Organic Chemistry* **51** 4301–4303 (1986).
44. R.-I. Ma, N.R. Kallenbach, R.D. Sheardy, M.L. Petrillo and N.C. Seeman, 3-Arm Nucleic Acid Junctions Are Flexible. *Nucleic Acids Research* **14** 9745–9753 (1986).
45. B.H. Robinson and N.C. Seeman, Simulation of Double Stranded Branch Point Migration. *Biophysical Journal* **51** 611–626 (1987).
46. L.A. Marky, N.R. Kallenbach, K.A. McDonough, N.C. Seeman and K.J. Breslauer, The Melting Behavior of a Nucleic Acid Junction: A Calorimetric and Spectroscopic Study. *Biopolymers* **26** 1621–1634 (1987).
47. B.H. Robinson and N.C. Seeman, The Design of a Biochip: A Self-assembling Molecular-Scale Memory Device. *Protein Engineering* **1** 295–300 (1987).
48. N.C. Seeman, Physical Models for Exploring DNA Topology. *Journal of Biomolecular Structure and Dynamics* **5** 997–1004 (1988).
49. N.C. Seeman and N.R. Kallenbach, Nucleic Acid Junctions: A Successful Experiment in Macromolecular Design. In: *Molecular Structure: Chemical Reactivity and Biological Activity*, ed. by J.J. Stezowski, J.-L. Huang and M.-C. Shao, Oxford University Press, Oxford (1988) pp. 189–194.
50. M.E.A. Churchill, T.D. Tullius, N.R. Kallenbach and N.C. Seeman, A Holliday Recombination Intermediate is Twofold Symmetric. *Proceedings of the National Academy of Sciences* (USA) **85** 4653–4656 (1988).
51. M.L. Petrillo, C.J. Newton, R.P. Cunningham, R.-I. Ma, N.R. Kallenbach and N.C. Seeman. The Ligation and Flexibility of 4-Arm DNA Junctions. *Biopolymers* **27** 1337–1352 (1988).
52. J.-H. Chen, M.E.A. Churchill, T.D. Tullius, N.R. Kallenbach and N.C. Seeman, Construction and Analysis of Mono-Mobile DNA Junctions. *Biochemistry* **27** 6032–6038 (1988).

53. J.-H. Chen, N.C. Seeman and N.R. Kallenbach, Tracts of A-T Nucleotide Pairs Retard the Electrophoretic Mobility of Short DNA Duplexes. *Nucleic Acids Research* **16** 6803–6812 (1988).
54. J.E. Mueller, B. Kemper, R.P. Cunningham, N.R. Kallenbach and N.C. Seeman, T4 Endonuclease VII Cleaves the Crossover Strands of Holliday Junction Analogs. *Proceedings of the National Academy of Sciences* (U.S.A.) **85** 9441–9445 (1988).
55. N.C. Seeman, J.-H. Chen and N.R. Kallenbach, Gel Electrophoretic Analysis of DNA Branched Junctions. *Electrophoresis* **10** 345–354 (1989).
56. Q. Guo, N.C. Seeman and N.R. Kallenbach, Site-Specific Interaction of Intercalating Drugs with a Branched DNA Molecule. *Biochemistry* **28** 2355–2359 (1989).
57. N.C. Seeman, Nanoscale Assembly and Manipulation of Branched DNA: A Biological Starting Point for Nanotechnology, *NANOCON Proceedings*, ed. by J. Lewis and J.L. Quel, NANOCON, P.O. Box 40176, Bellevue, WA 98004, pp. 101–123; transcript of oral presentation, pp. 30–36 (1989). http://www.halcyon.com/nanojbl/NanoConProc/nanocon3.html
58. J.-H. Chen, N.R. Kallenbach and N.C. Seeman, A Specific Quadrilateral Synthesized from DNA Branched Junctions. *Journal of the American Chemical Society* **111** 6402–6407 (1989).
59. M. Lu, Q. Guo, N.C. Seeman and N.R. Kallenbach, DNase I Cleavage of Branched DNA Molecules, *Journal of Biological Chemistry* **264** 20851–20854 (1989).
60. F. Jensch, H. Kosak, N.C. Seeman and B. Kemper, Cruciform Cutting Endonucleases from S. cerevisiae and Phage T4 Show Conserved Reactions With Branched DNAs, *EMBO Journal* **8** 4325–4334 (1989).
61. N.C. Seeman, J.E. Mueller, J.-H. Chen, M.E.A. Churchill, A. Kimball, T.D. Tullius, B. Kemper, R.P. Cunningham and N.R. Kallenbach, Immobile Junctions Suggest New Features of the Structural Chemistry of Recombination, In: *Structure & Methods: Human Genome Initiative & DNA Recombination*, Vol. 1, R.H. Sarma and M.H. Sarma, eds., Adenine Press, New York, pp. 137–156 (1990).
62. Q. Guo, M. Lu, N.C. Seeman and N.R. Kallenbach, Drug Binding by Branched DNA Molecules: Analysis by Chemical Footprinting of Intercalation into an Immobile Junction, *Biochemistry* **29** 570–578 (1990).
63. M. Lu, Q. Guo, R.F. Pasternack, D.J. Wink. N.C. Seeman and N.R. Kallenbach, Drug Binding by Branched DNA: Selective Interaction of Tetrapyridyl Porphyrins with an Immobile Junction, *Biochemistry* **29** 1614–1624 (1990).
64. A. Kimball, Q. Guo, M. Lu, N.R. Kallenbach, R.P. Cunningham, N.C. Seeman and T.D. Tullius, Conformational Isomers of Holliday Junctions, *Journal of Biological Chemistry* **265** 6544–6547 (1990).
65. M. Lu, Q. Guo, N.C. Seeman and N.R. Kallenbach, Drug Binding by Branched DNA: Selective Interaction of the Dye Stains-All with an Immobile Junction, *Biochemistry* **29** 3407–3412 (1990).

66. J.E. Mueller, C.J. Newton, F. Jensch, B. Kemper, R.P. Cunningham, N.R. Kallenbach and N.C. Seeman, Resolution of Holliday Junction Analogs by T4 Endonuclease VII Can Be Directed by Substrate Structure, *Journal of Biological Chemistry* **265** 13918–13924 (1990).
67. N.C. Seeman, De Novo Design of Sequences for Nucleic Acid Structure Engineering, *Journal of Biomolecular Structure and Dynamics* **8** 573–581 (1990).
68. M. Lu, Q. Guo, J.E. Mueller, B. Kemper, F.W. Studier, N.C. Seeman and N.R. Kallenbach, Characterization of a Bimobile DNA Junction, *Journal of Biological Chemistry* **265** 16778–16785 (1990).
69. J. Chen and N.C. Seeman, The Synthesis from DNA of a Molecule with the Connectivity of a Cube, *Nature* **350** 631–633 (1991).
70. Y. Wang, J.E. Mueller, B. Kemper, and N.C. Seeman, The Assembly and Characterization of 5-Arm and 6-Arm DNA Junctions, *Biochemistry* **30** 5667–5674 (1991).
71. M. Lu, Q. Guo, N.C. Seeman and N.R. Kallenbach, Parallel and Antiparallel Holliday Junctions Differ in Structure and Stability, *Journal of Molecular Biology* **221** 1419–1432 (1991).
72. J.E. Mueller, S.M. Du and N.C. Seeman, The Design and Synthesis of a Knot from Single-Stranded DNA, *Journal of the American Chemical Society* **113** 6306–6308 (1991).
73. J. Chen and N.C. Seeman, The Electrophoretic Properties of a DNA Cube and its Sub-Structure Catenanes, *Electrophoresis* **12** 607–611 (1991).
74. N.C. Seeman, DNA Structural Engineering Utilizing Immobile Junctions, *Current Opinion in Structural Biology* **1** 653–661 (1991).
75. N.C. Seeman, The Construction of 3-D Stick Figures from Branched DNA, *DNA and Cell Biology* **10** 475–486 (1991).
76. N.C. Seeman, The Use of Branched DNA for Nanoscale Fabrication, *Nanotechnology* **2** 149–159 (1991).
77. M. Lu, Q. Guo, L.A. Marky, N.C. Seeman and N.R. Kallenbach, Thermodynamics of DNA Chain Branching, *Journal of Molecular Biology* **223** 781–789 (1992).
78. Y. Zhang and N.C. Seeman, A Solid-Support Methodology for the Construction of Geometrical Objects from DNA, *Journal of the American Chemical Society* **114** 2656–2663 (1992).
79. S.M. Du and N.C. Seeman, The Synthesis of a DNA Knot Containing both Positive and Negative Nodes, *Journal of the American Chemical Society* **114** 9652–9655 (1992).
80. S.M. Du, S. Zhang and N.C. Seeman, DNA Junctions, Antijunctions and Mesojunctions, *Biochemistry* **31** 10955–10963 (1992).
81. N.C. Seeman, The Design of Single-Stranded Nucleic Acid Knots, *Molecular Engineering* **2** 297–307 (1992).
82. T.-J. Fu and N.C. Seeman, DNA Double Crossover Structures, *Biochemistry* **32** 3211–3220 (1993).

83. N.C. Seeman, Nanoengineering with DNA, Biomolecular Materials: *Materials Research Society Symposium Proceedings* **292** 123–134 (1993).
84. H. Wang, S.M. Du and N.C. Seeman, Tight Single-Stranded DNA Knots, *Journal of Biomolecular Structure and Dynamics* **10** 853–863 (1993).
85. N.C. Seeman, Branched DNA: A 3-D Structural Design System, *Clinical Chemistry* **39** 722–724 (1993).
86. N.C. Seeman, J. Chen, S.M. Du, J.E. Mueller, Y. Zhang, T.-J. Fu, H. Wang, Y. Wang, and S. Zhang, Synthetic DNA Knots and Catenanes, *New Journal of Chemistry* **17** 739–755 (1993).
87. V.N. Morozov, J. Sherman, N.R. Kallenbach, S.M. Du and N.C. Seeman, A Scanning Tunneling Microscopy Study of the Formation and Chemical Activation of Step Defects on the Basal Plane of Pyrolytic Graphite, *Journal of Microscopy* **170** pt. 3, 237–245 (1993).
88. S. Zhang, T.-J. Fu and N.C. Seeman, Construction of Symmetric, Immobile DNA Branched Junctions, *Biochemistry* **32** 8062–8067 (1993).
89. N.C. Seeman and N.R. Kallenbach, DNA Branched Junctions, *Annual Review of Biophysics and Biomolecular Structure* **23** 53–86 (1994).
90. S.M. Du and N.C. Seeman, The Construction of a Trefoil Knot from a DNA Branched Junction Motif, *Biopolymers* **34** 31–37 (1994).
91. V.N. Morozov, N.C. Seeman and N.R. Kallenbach, New Methods for Depositing and Imaging Molecules in Scanning-Tunneling Microscopy, *Scanning Microscopy International* **7** 757–779 (1994).
92. N.C. Seeman, Structural Control and Engineering of Nucleic Acids, In: *Concepts in Protein Engineering and Design*, P. Wrede and G. Schneider, eds., Walter-de-Gruyter, Berlin, 319–343 (1994).
93. T-J. Fu, Y.-C. Tse-Dinh, and N.C. Seeman, Holliday Junction Crossover Topology, *Journal of Molecular Biology* **236** 91–105 (1994).
94. N.C. Seeman, Y. Zhang, and J. Chen, DNA Nanoconstructions, *Journal of Vacuum Science and Technology* **A12** 1895–1903 (1994).
95. T.-J. Fu, B. Kemper and N.C. Seeman, Endonuclease VII Cleavage of DNA Double Crossover Molecules, *Biochemistry* **33** 3896–3905 (1994).
96. N.C. Seeman, Y. Zhang, T.-J. Fu, S. Zhang, Y. Wang and J. Chen, Chemical Synthesis of Nanostructures, Biomolecular Materials by Design: *Materials Research Society Symposium Proceedings* **330** 45–56 (1994).
97. S. Zhang, and N.C. Seeman, Symmetric Holliday Junction Crossover Isomers, *Journal of Molecular Biology* **238** 658–668 (1994).
98. B. Liu, N.B. Leontis and N.C. Seeman, Bulged 3-arm DNA Branched Junctions as Components for Nanoconstruction, *Nanobiology* **3** 177–188 (1994).
99. N.C. Seeman, Y. Zhang, S.M. Du, H. Wang, J.E. Mueller, Y. Wang, B. Liu, J. Qi and J. Chen, The Control of DNA Structure and Topology: An Overview, *Molecularly Designed Ultrafine/Nanostructured Materials: Materials Research Society Symposium Proceedings* **351** 57–66 (1994).
100. Y. Zhang and N.C. Seeman, The Construction of a DNA Truncated Octahedron, *Journal of the American Chemical Society* **116** 1661–1669 (1994).

101. S.M. Du, H. Wang, Y.C. Tse-Dinh and N.C. Seeman, Topological Transformations of Synthetic DNA Knots, *Biochemistry* **34** 673–682 (1995).
102. H. Wang and N.C. Seeman, Structural Domains of DNA Mesojunctions, *Biochemistry* **34** 920–929 (1995).
103. S.M. Du, B.D. Stollar, and N.C. Seeman, A Synthetic DNA Molecule in Three Knotted Topologies, *Journal of the American Chemical Society* **117** 1194–1200 (1995).
104. D.M.J. Lilley, R.M. Clegg, S. Diekmann, N.C. Seeman, E. von Kitzing and P. Hagerman, A Nomenclature of Junctions and Branchpoints in Nucleic Acids, *European Journal of Biochemistry* **230** 1-2 (1995); *Nucleic Acids Research* **23** 3363-3364 (1995); *Journal of Molecular Biology* **255** 554–555 (1996).
105. N.C. Seeman, Molecular Craftwork with DNA, *The Chemical Intelligencer* **1** (3) 38–47 (1995).
106. N.C. Seeman, Y. Zhang, S.M. Du and J. Chen, Construction of DNA Polyhedra and Knots Through Symmetry Minimization, In: *Supramolecular Stereochemistry,* ed. by J. Siegel, NATO Advanced Research Workshops Series, Kluwer, Dordrecht, 27–32 (1995).
107. E. Flapan and N.C. Seeman, A Topological Rubber Glove Obtained from a Synthetic Single Stranded DNA Molecule, *Journal of the Chemical Society, Chemical Communications*, 2249–2250 (1995).
108. J.A. Sekiguchi, N.C. Seeman and S. Shuman, Asymmetric Resolution of Holliday Junctions by Eukaryotic DNA Topoisomerase I. *Proceedings of the National Academy of Sciences* (USA) **93** 785–789 (1996).
109. N.C. Seeman, H. Wang, J. Qi., X.J. Li, X.P. Yang, Y. Wang, H. Qiu, B. Liu, Z. Shen, W. Sun, F. Liu, J.J. Molenda, S.M. Du, J. Chen, J.E. Mueller., Y. Zhang, T.-J. Fu, and S. Zhang, DNA Nanotechnology and Topology. In: *Biological Structure and Dynamics*, ed. by R.H. Sarma and M.H. Sarma, Adenine Press, New York, vol. 2, pp. 319–341 (1996).
110. J. Qi, X. Li, X. Yang and N.C. Seeman, The Ligation of Triangles Built from Bulged Three-Arm DNA Branched Junctions, *Journal of the American Chemical Society* **118** 6121–6130 (1996).
111. X. Li, X. Yang, J. Qi, and N.C. Seeman, Antiparallel DNA Double Crossover Molecules as Components for Nanoconstruction, *Journal of the American Chemical Society* **118** 6131–6140 (1996).
112. H. Wang, R.J. Di Gate, and N.C. Seeman, An RNA Topoisomerase, *Proceedings of the National Academy of Sciences* (USA) **93** 9477-9482 (1996).
113. N.C. Seeman, Design and Engineering of Nucleic Acid Nanoscale Assemblies, *Current Opinion in Structural Biology* **6** 519–526 (1996). http://biomednet.com/cgi-bin/fulltext/fulltext.pl?uid=sb6402#sec5
114. C. Mao, W. Sun and N.C. Seeman, Assembly of Borromean Rings from DNA, *Nature* **386** 137–138 (1997).
115. H. Qiu, J.C. Dewan and N.C. Seeman, A DNA Decamer with a Sticky End: The Crystal Structure of d-CGACGATCGT. *Journal of Molecular Biology* **267** 881–898 (1997).

116. X. Li, H. Wang and N.C. Seeman, Direct Evidence for Holliday Junction Crossover Isomerization. *Biochemistry* **36** 4240–4247 (1997).
117. N.C. Seeman, J. Chen, Y. Zhang, T.-J. Fu, X. Li, X. Yang, Y. Wang, J. Qi, B. Liu and F. Liu, Control of Structure and Topology in DNA Nanotechnology. In: *Molecular Nanotechnology: Biological Approaches and Novel Applications*, ed. by S.A. Minden, IBC Libraries, Southborough, MA, Chapter 2.2 (31 pages) (1997).
118. N.C. Seeman, DNA Components for Molecular Architecture. *Accounts of Chemical Research* **30** 357–363 (1997).
119. N.C. Seeman, J. Qi, X. Li, X. Yang, N.B. Leontis, B. Liu, Y. Zhang, S.M. Du and J. Chen, The Control of DNA Structure: From Topological Modules to Geometrical Modules. In: *Modular Chemistry*, ed. by J. Michl, Kluwer, 95–104 (1997).
120. X. Yang, A. Vologodskii, B. Liu, B. Kemper and N.C. Seeman, Torsional Control of Double Stranded DNA Branch Migration. *Biopolymers* **45** 69–83 (1998).
121. N.C. Seeman, DNA Nanotechnology: Novel DNA Constructions. Annual Review of *Biophysics and Biomolecular Structure* **27** 225–248 (1998).
122. N.C. Seeman, DNA Nanotechnology, In: *WTEC Workshop Report on R&D Status and Trends in Nanoparticles, Nanostructured Materials, and Nanodevices in the United States*, ed. by R.W. Siegel, E. Hu and M.C. Roco, International Technology Research Institute, Baltimore (1998).
123. A. V. Vologodskii, X. Yang and N.C. Seeman, Non-complementary DNA Helical Structure Induced by Positive Torsional Stress, *Nucleic Acids Research* **26** 1503–1508 (1998).
124. H. Wang, R.J. Di Gate and N.C. Seeman, The Construction of an RNA Knot and Its Role in Demonstrating that E. coli DNA Topoisomerase III Is an RNA Topoisomerase. *Structure, Motion, Interaction and Expression of Biological Macromolecules*, ed. by R.H. Sarma and M.H. Sarma, Adenine Press, New York, pp. 103–116 (1998).
125. E. Winfree, F. Liu, L. A. Wenzler, and N.C. Seeman, Design and Self-assembly of Two-Dimensional DNA Crystals, *Nature* **394** 539–544 (1998).
126. N.C. Seeman, Directing the Structure of Matter Through Nanotechnology, Proceedings of IEEE *International Joint Symposia on Intelligence and Systems IEEE Computer Society*, Los Alamitos, CA, 146–150 (1998).
127. W. Sun, C. Mao, F. Liu and N.C. Seeman, Sequence Dependence of Branch Migratory Minima. *Journal of Molecular Biology* **282** 59–70 (1998).
128. N.C. Seeman, H. Wang, X. Yang, F. Liu, C. Mao, W. Sun, L. Wenzler, Z. Shen, R. Sha, H. Yan, M.H. Wong, P. Sa-Ardyen, B. Liu, H. Qiu, X. Li, J. Qi, S.M. Du, Y. Zhang, J.E. Mueller, T.-J. Fu, Y. Wang, and J. Chen, New Motifs in DNA Nanotechnology, *Nanotechnology* **9** 257–273 (1998). http://www.foresight.org/Conferences/MNT05/Papers/Seeman/ index.html.
129. N.C. Seeman, J. Chen, Y. Zhang, B. Liu, H. Qiu, T.-J. Fu, Y. Wang, X. Li, X. Yang, J. Qi, F. Liu, L.A. Wenzler, S, Du, J.E. Mueller, H. Wang,

C. Mao, W. Sun, Z. Shen, M.H. Wong, H, Yan and R. Sha, A Bottom-up Approach to Nanotechnology Using DNA, In: *Biological Molecules in Nanotechnology*, ed. by S.C. Lee, IBC Libraries, Southborough, MA, pp. 45–58 (1998).

130. X. Yang, L.A. Wenzler, J. Qi, X. Li and N.C. Seeman, Ligation of DNA Triangles Containing Double Crossover Molecules, *Journal of the American Chemical Society* **120** 9779-9786 (1998).
131. E. Winfree, X. Yang and N.C. Seeman, Universal Computation via Self-assembly of DNA: Some Theory and Experiments, In: *DNA Based Computers II,* ed. by L.F. Landweber and E.B. Baum, Am. Math. Soc., Providence, pp. 191–213 (1998).
132. N.C. Seeman, H. Wang, B. Liu, J. Qi, X. Li, X. Yang, F. Liu, W. Sun, Z. Shen, Y. Wang, R. Sha, C. Mao, S. Zhang, T.-J. Fu, S.M. Du, J.E. Mueller, Y. Zhang and J. Chen, The Perils of Polynucleotides: The Experimental Gap Between the Design and Assembly of Unusual DNA Structures, In: *DNA Based Computers II,* ed. by L.F. Landweber and E.B. Baum, Am. Math. Soc., Providence, pp. 215–233 (1998).
133. N.C. Seeman, Nucleic Acid Nanostructures and Topology. Angewandte Chemie. 110, 3408-3428 (1998); *Angewandte Chemie International Edition* **37** 3220–3238 (1998).
134. C. Mao, W. Sun, Z. Shen and N.C. Seeman, A DNA Nanomechanical Device Based on the B-Z Transition, *Nature* **397** 144–146 (1999).
135. F. Liu, R. Sha and N.C. Seeman, Modifying the Surface Features of Two-Dimensional DNA Crystals, *Journal of the American Chemical Society* **121** 917–922 (1999).
136. R. Sha, F. Liu, M.F. Bruist and N.C. Seeman, Parallel Helical Domains in DNA Branched Junctions Containing 5', 5' and 3', 3' Linkages, *Biochemistry* **38** 2832–2841 (1999).
137. C. Mao, W. Sun and N.C. Seeman, Designed Two-Dimensional DNA Holliday Junction Arrays Visualized by Atomic Force Microscopy, *Journal of the American Chemical Society* **121** 5437–5443 (1999).
138. N.C. Seeman, Synthetic DNA Topology, Molecular Catenanes, *Rotaxanes and Knots*, ed. by J.-P. Sauvage and C. Dietrich-Buchecker, Wiley-VCH, Weinheim pp. 323–356 (1999).
139. F. Liu, H. Wang and N.C. Seeman, Short Extensions to Sticky Ends for DNA Nanotechnology and DNA-Based Computation, *Nanobiology* **4** 257–262 (1999).
140. N.C. Seeman, DNA Engineering and Its Application to Nanotechnology, Trends in *Biotechnology* **17**, 437–443 (1999).
141. W. Sun, C. Mao, H. Iwasaki, B. Kemper and N.C. Seeman, No Braiding of Holliday Junctions in Positively Supercoiled DNA Molecules, *Journal of Molecular Biology* **294** 683–699 (1999).
142. T. LaBean, H. Yan, J. Kopatsch, F. Liu, E. Winfree, J.H. Reif and N.C. Seeman, The Construction, Analysis, Ligation and Self-assembly of DNA

Triple Crossover Complexes, *Journal of the American Chemical Society* **122** 1848–1860 (2000).

143. N.C. Seeman, F. Liu, C. Mao, X. Yang, L.A. Wenzler, R. Sha, W. Sun, Z. Shen, X. Li, J. Qi, Y. Zhang, T.-J. Fu, J. Chen and E. Winfree, Two Dimensions and Two States in DNA Nanotechnology, *Proceedings of the 11th Conversation in Biomolecular Stereodynamics*, ed. by R.H. Sarma and M.H. Sarma, Adenine Press, New York, pp. 253–262 (2000).
144. N.C. Seeman, DNA Nanotechnology: From Topological Control to Structural Control, in *Pattern Formation in Biology, Vision and Dynamics*, ed. by A. Carbone, M. Gromov, P.Prusinkiewicz, World Scientific Publishing Company, Singapore, pp. 271–309 (2000).
145. N.C. Seeman, C. Mao, F. Liu, R. Sha, X. Yang, L. Wenzler, X. Li, Z. Shen, H. Yan, P. Sa-Ardyen, X. Zhang, W. Shen, J. Birac, P. Lukeman, Y. Pinto, J. Qi, B. Liu, H. Qiu, S.M. Du, H. Wang, W. Sun, Y. Wang, T.-J. Fu, Y. Zhang, J.E. Mueller and J. Chen, Nicks, Nodes, and New Motifs for DNA Nanotechnology, *Frontiers of Nano-optoelectronic Systems*, ed. by L. Pavesi and E. Buzanova, Kluwer, Dordrecht, pp. 177–198 (2000).
146. R. Sha, F. Liu and N.C. Seeman, Direct Evidence for Spontaneous Branch Migration in Antiparallel DNA Holliday Junctions, *Biochemistry* **39** 11514–11522 (2000).
147. R. Sha, H. Iwasaki, F. Liu, H. Shinagawa and N.C. Seeman, Cleavage of Symmetric Immobile DNA Junctions by Escherichia coli Ruv C, *Biochemistry* **39** 11982–11988 (2000).
148. R. Sha, F. Liu, D.P. Millar and N.C. Seeman, Atomic Force Microscopy of Parallel DNA Branched Junction Arrays, *Chemistry & Biology* **7** 743–751 (2000).
149. A. Podtelezhnikov, C. Mao, N.C. Seeman and A. Vologodskii, Multimerization-Cyclization of DNA Fragments as a Method of Conformational Analysis, *Biophys. J.* **79** 2692–2704 (2000).
150. C. Mao, T. LaBean, J.H. Reif and N.C. Seeman, Logical Computation Using Algorithmic Self-assembly of DNA Triple Crossover Molecules, *Nature* **407** 493–496 (2000); Erratum: *Nature* **408** 750-750 (2000).
151. N.C. Seeman, In the Nick of Space: Generalized Nucleic Acid Complementarity and the Development of DNA Nanotechnology, *Synlett* 2000 1536–1548 (2000).
152. N.C. Seeman, DNA Nicks and Nodes and Nanotechnology, *NanoLetters* **1** 22–26 (2001).
153. J.H. Reif, T.H. LaBean and N.C. Seeman, Challenges and Applications for Self-assembled DNA Nanostructures, *Sixth International Workshop on DNA-Based Computers, DNA 2000*, Leiden, The Netherlands, (June, 2000) ed. A. Condon, G. Rozenberg. Springer-Verlag, Berlin, Heidelberg, Lecture Notes in Computer Science **2054** 173–198 (2001).
154. H. Yan, X. Zhang, Z. Shen and N.C. Seeman, A Robust DNA Mechanical Device Controlled by Hybridization Topology, *Nature* **415** 62–65 (2002).

155. N.C. Seeman, DNA Nanotechnology: Life's Central Performer in a New Role, *Biological Physics Newsletter* **2** (1) 2–6 (2002) .
156. N.C. Seeman, It Started with Watson and Crick, But it Sure Didn't End There: Pitfalls and Possibilities Beyond the Classic Double Helix, *Natural Computing* **1** 53-8–4 (2002).
157. N.C. Seeman and A.M. Belcher, Emulating Biology: Nanotechnology from the Bottom Up, *Proceedings of the National Academy of Sciences* (USA) **99** (supp. 2), 6451–6455 (2002).
158. R. Sha, F. Liu and N.C. Seeman, Atomic Force Measurement of the Interdomain Angle in Symmetric Holliday Junctions, *Biochemistry* **41** 5950–5955 (2002).
159. N.C. Seeman, Key Experimental Approaches in DNA Nanotechnology, Current Protocols in Nucleic Acid Chemistry, Unit 12.1, John Wiley & Sons, New York (2002).
160. R. Sha, F. Liu, H. Iwasaki and N.C. Seeman, Parallel Symmetric Immobile DNA Junctions as Substrates for E. coli RuvC Holliday Junction Resolvase. *Biochemistry* **41** 10985–10993 (2002).
161. A. Carbone and N.C. Seeman, Circuits and Programmable Self-assembling DNA Structures, *Proc. Nat. Acad. Sci.* (USA) **99** 12577–12582 (2002).
162. X. Zhang, H. Yan, Z. Shen and N.C. Seeman, Paranemic Cohesion of Topologically-Closed DNA Molecules, *J Am. Chem. Soc.* **124** 12940–12941 (2002).
163. S. Xiao, F. Liu, A. Rosen, J.F. Hainfeld, N.C. Seeman, K.M. Musier-Forsyth and R.A. Kiehl, Self-assembly of Nanoparticle Arrays by DNA Scaffolding, *J. Nanoparticle Research* **4** 313–317 (2002).
164. A. Carbone and N.C. Seeman, Fractal Designs Based on DNA Parallelogram Structures, *Natural Computing* **1** 469–480 (2002).
165. L. Zhu, O. dos Santos, N.C. Seeman and J.W. Canary, Reaction of N3-Benzoyl-3, 5-O-(di-tert-butylsilanediyl)uridine with Hindered Electrophiles: Intermolecular N3 to 2'-O Protecting Group Transfer, *Nucleosides, Nucleotides & Nucleic Acids* **21** 723-7-35 (2002).
166. N.C. Seeman, DNA in a Material World, *Nature* **421** 427–431 (2003).
167. N.C. Seeman, Structural DNA Nanotechnology: A New Organizing Principle for Advanced Nanomaterials, *Materials Today* **6** (7) 24–29 (2003).
168. P. Sa-Ardyen, N. Jonoska and N.C. Seeman, Self-assembling DNA Graphs, *DNA-Based Computers VIII*, LNCS **2568** Springer-Verlag, Berlin, 1–9 (2003).
169. H. Yan and N.C. Seeman, Edge-Sharing Motifs in DNA Nanotechnology. Journal of *Supramolecular Chemistry* **1** 229–237 (2003).
170. P. Sa-Ardyen, A.V. Vologodskii and N.C. Seeman, The Flexibility of DNA Double Crossover Molecules. *Biophysical Journal* **84** 3829–3837 (2003).
171. N. Jonoska, P. Sa-Ardyen and N.C. Seeman, Computation by Self-assembly of DNA Graphs, *J. Genetic Programming and Evolvable Machines* **4** 123–137 (2003).

172. N.C. Seeman, Biochemistry and Structural DNA Nanotechnology: An Evolving Symbiotic Relationship, *Biochemistry* **42** 7259–7269 (2003).
173. L. Zhu, P.S. Lukeman, J. Canary and N.C. Seeman, Nylon/DNA: Single-Stranded DNA with Covalently Stitched Nylon Lining, *J. Am. Chem. Soc.* **125** 10178–10179 (2003).
174. A. Carbone and N.C. Seeman, Coding and Geometrical Shapes in Nanostructures: a Fractal DNA-Assembly, *Natural Computing* **2** 133–151 (2003).
175. N.C. Seeman, DNA: Beyond the Double Helix, *Macromolecular Symposia* **201** 237–244 (2003).
176. N.C. Seeman, At the Crossroads of Chemistry, Biology and Materials: Structural DNA Nanotechnology, *Chemistry & Biology* **10** 1151–1159 (2003).
177. P. Sa-Ardyen, N. Jonoska and N.C. Seeman, Self-assembling DNA Graphs, *Natural Computing* **2** 427–438 (2003).
178. N.C. Seeman, DNA Nanostructures for Mechanics and Computing: Non-linear Thinking with Life's Central Molecule. *NanoBiotechnology.* Editors, Chad Mirkin and Christof Niemeyer, Wiley-VCH Verlag GmbH & Co., Weinheim, Chapter 20, pp. 308–318 (2004).
179. Z. Shen, H. Yan, T. Wang and N.C. Seeman, Paranemic Crossover DNA: A Generalized Holliday Structure with Applications in Nanotechnology *J. Am. Chem. Soc.* **126** 1666–1674 (2004).
180. N. Jonoska, S. Liao, and N.C. Seeman, Transducers with Programmable input by DNA Self-assembly, in: *Aspects of Molecular Computing*, N. Jonoska, G. Paun, G. Rozenberg eds. Lecture Notes in Computer Science **2340**, Springer-Verlag, Berlin, 219–240 (2004).
181. A. Carbone and N.C. Seeman, Molecular Tiling and DNA Self-assembly, in: *Aspects of Molecular Computing*, N. Jonoska, G. Paun, G. Rozenberg eds. Lecture Notes in Computer Science **2340**, Springer-Verlag, Berlin, 61–83 (2004).
182. S. Liao, C. Mao, J.J. Birktoft, S. Shuman and N.C. Seeman, Resolution of Undistorted Symmetric Immobile DNA Junctions by Vaccinia Topoisomerase I, *Biochemistry* **43** 1520–1531 (2004).
183. N.C. Seeman. Nanotechnology and the Double Helix, *Scientific American* **290** (6) 64–75 (2004).
184. P. Sa-Ardyen, N. Jonoska and N.C. Seeman, The Construction of Graphs Whose Edges Are DNA Helix Axes, *J. Am. Chem. Soc.*, **126** 6648–6657 (2004).
185. W.B. Sherman and N.C. Seeman, A Precisely Controlled DNA Bipedal Walking Device, *NanoLetters* **4** 1203-1207 (2004); Erratum, **4** 1801–1801.
186. P.S. Lukeman, A. Mittal, and N.C. Seeman, Two Dimensional PNA/DNA Arrays: Estimating the Helicity of Unusual Nucleic Acid Polymers, *Chemical Communications 2004*, pp. 1694–1695 (2004).
187. N.C. Seeman, DNA Nanotechnology, *Encyclopedia of Supramolecular Chemistry*, Marcel Dekker, New York, pp. 475–483 (2004).

188. A. Carbone, C. Mao, P. Constantinou, B. Ding, J. Kopatsch, W.B. Sherman and N.C. Seeman, 3D Fractal DNA Assembly from Coding, Geometry and Protection, *Natural Computing* **3** 235–252 (2004).
189. B. Ding, R. Sha and N.C. Seeman, Pseudohexagonal 2D DNA Crystals from Double Crossover Cohesion, *J. Am. Chem. Soc.* **126** 10230–10231 (2004).
190. P. Paukstelis, J. Nowakowski, J.J. Birktoft and N.C. Seeman, The Crystal Structure of a Continuous Three-Dimensional DNA Lattice, *Chemistry and Biology* **11** 1119–1126 (2004).
191. W. Shen, M. Bruist, S. Goodman and N.C. Seeman, A Nanomechanical Device for Measuring the Excess Binding Energy of Proteins That Distort DNA, *Angew. Chem. Int. Ed.* **43** 4750–4752 (2004); *Angew. Chem.* **116** 4854–4856 (2004).
192. S. Liao and N.C. Seeman, Translation of DNA Signals into Polymer Assembly Instructions, *Science* **306** 2072–2074 (2004).
193. J.D. Le, Y. Pinto, N.C. Seeman, K. Musier-Forsyth, T.A. Taton and R.A. Kiehl, Self-assembly of Nanoelectronic Component Arrays by In Situ Hybridization to 2D DNA Scaffolding, *NanoLetters* **4** 2343–2347 (2004).
194. G. Serrano and N.C. Seeman, Nanotecnologia Basada en ADN, *Revista de Quimica* **19** 11-20 (2004), in Spanish.
195. N.C. Seeman and P.S. Lukeman, Nucleic Acid Nanostructures, *Reports on Progress in Physics* **68** 237–270 (2005).
196. N.C. Seeman, B. Ding, S. Liao, T. Wang, W.B. Sherman, P.E. Constantinou, J. Kopatsch, C. Mao, R. Sha, F. Liu, H. Yan and P.S. Lukeman, Experiments in Structural DNA Nanotechnology: Arrays and Devices, *Proc. SPIE; Nanofabrication: Technologies, Devices and Applications* **5592**, 71–81 (2005).
197. N.C. Seeman, From Genes to Machines: DNA Nanomechanical Devices, *Trends in Biochemical Sciences* **30** 119–235 (2005).
198. N.C. Seeman. Structural DNA Nanotechnology: An Overview. *Methods in Molecular Biology 303: Bionanotechnology Protocols*, Editors, Sandra J. Rosenthal and David W. Wright, Humana Press, Totowa, NJ, pp. 143–166 (2005).
199. F. Mathieu, S. Liao, C. Mao, J. Kopatsch, T. Wang, and N.C. Seeman, Six-Helix Bundles Designed from DNA, *NanoLetters* **5** 661–665 (2005).
200. R.Sha, X.Zhang, S.Liao, P.E.Constantinou, B.Ding, T.Wang, A.V.Garibotti, H.Zhong, L.B.Israel, X.Wang, G.Wu, B.Chakraborty, J.Chen, Y. Zhang, C. Mao, H.Yan, J.Kopatsch, J.Zheng, P.S.Lukeman, W.B.Sherman, N.C. Seeman, Motifs and Methods in Structural DNA Nanotechnology, Proc. Intl. Conf. Nanomaterials, NANO 2005, July 13-15, 2005, Mepco Schlenk Engineering College, Srivakasi, India, V. Rajendran, ed., pp. 3–10 (2005).
201. M. Cavaliere, N. Jonoska, S. Yogev, R. Piran, E. Keinan, N.C. Seeman *Biomolecular Implementation of Computing Devices with Unbounded*

Memory (G. Ferretti et al., eds.) Springer Lecture Notes in Computer Science **3384** (2005) 35–49.

202. R. Sha, X. Zhang, S. Liao, P.E. Constantinou, B. Ding, T. Wang, A.V. Garibotti, H.Zhong, L.B.Israel, X.Wang, G.Wu, B.Chakraborty, J.Chen, Y.Zhang, H.Yan, Z.Shen, W.Shen, P.Sa-Ardyen, J.Kopatsch, J.Zheng, P.S.Lukeman, W.B.Sherman, C. Mao, N. Jonoska, N.C. Seeman, Structural DNA Nanotechnology: Molecular Construction and Computation, in Unconventional Computing 2005 (C. Calude et al., eds.) Lecture Notes in Computer Science **3699** 20–31 (2005).
203. C. Mao, P.E. Constantinou, F. Liu, Y. Pinto, J. Kopatsch, P.S. Lukeman, T. Wang, B. Ding, H. Yan, J.J. Birktoft, R. Sha, H.Zhong, L. Foley, L.A. Wenzler, R. Sweet, M. Becker and N.C. Seeman, The Design of Self-assembled 3D DNA Networks, Proc. Intl. Symp. on Nanoscale Devices, Materials, and Biological Systems, 206th Meeting of the Electrochemical Society, Honolulu, PV 2004-XX, Editors: M. Cahay, M. Urquidi-Macdonald, S. Bandyopadhyay, P. Guo, H. Hasegawa, N. Koshida, J.P. Leburton, D.J. Lockwood, S. Seal, and A. Stella. in press (2005).
204. M. Endo, N.C. Seeman and T. Majima, DNA Tube Structures Controlled by a Four-Way-Branched DNA Connector, *Angew. Chemie, Int. Ed.* **44** 6074–6077 (2005).
205. N.C. Seeman, DNA Enables Nanoscale Control of the Structure of Matter, Quart. Rev. Biophys., in press (2005).
206. Y. Baryshnikov, E. Coffman, N.C. Seeman, T. Yimwadsana, Self-correcting Self-assembly: Growth Models and the Hammersley Process, Proceedings of DNA-11, in press (2005).

Natural Computing Series

W.M. Spears: **Evolutionary Algorithms. The Role of Mutation and Recombination.** XIV, 222 pages, 55 figs., 23 tables. 2000

H.-G. Beyer: **The Theory of Evolution Strategies.** XIX, 380 pages, 52 figs., 9 tables. 2001

L. Kallel, B. Naudts, A. Rogers (Eds.): **Theoretical Aspects of Evolutionary Computing.** X, 497 pages. 2001

G. Păun: **Membrane Computing. An Introduction.** XI, 429 pages, 37 figs., 5 tables. 2002

A.A. Freitas: **Data Mining and Knowledge Discovery with Evolutionary Algorithms.** XIV, 264 pages, 74 figs., 10 tables. 2002

H.-P. Schwefel, I. Wegener, K. Weinert (Eds.): **Advances in Computational Intelligence. Theory and Practice.** VIII, 325 pages. 2003

A. Ghosh, S. Tsutsui (Eds.): **Advances in Evolutionary Computing. Theory and Applications.** XVI, 1006 pages. 2003

L.F. Landweber, E. Winfree (Eds.): **Evolution as Computation.** DIMACS Workshop, Princeton, January 1999. XV, 332 pages. 2002

M. Hirvensalo: **Quantum Computing.** 2nd ed., XI, 214 pages. 2004 (first edition published in the series)

A.E. Eiben, J.E. Smith: **Introduction to Evolutionary Computing.** XV, 299 pages. 2003

A. Ehrenfeucht, T. Harju, I. Petre, D.M. Prescott, G. Rozenberg: **Computation in Living Cells. Gene Assembly in Ciliates.** XIV, 202 pages. 2004

L. Sekanina: **Evolvable Components. From Theory to Hardware Implementations.** XVI, 194 pages. 2004

G. Ciobanu, G. Rozenberg (Eds.): **Modelling in Molecular Biology.** X, 310 pages. 2004

R.W. Morrison: **Designing Evolutionary Algorithms for Dynamic Environments.** XII, 148 pages, 78 figs. 2004

R. Paton†, H. Bolouri, M. Holcombe, J.H. Parish, R. Tateson (Eds.): **Computation in Cells and Tissues. Perspectives and Tools of Thought.** XIV, 358 pages, 134 figs. 2004

M. Amos: **Theoretical and Experimental DNA Computation.** XIV, 170 pages, 78 figs. 2005

M. Tomassini: **Spatially Structured Evolutionary Algorithms.** XIV, 192 pages, 91 figs., 21 tables. 2005

G. Ciobanu, G. Paun, M.J. Pérez-Jiménez (Eds.): **Applications of Membrane Computing.** X, 441 pages, 99 figs., 24 tables. 2006

K. V. Price, R. M. Storn, J. A. Lampinen: **Differential Evolution.** XX, 538 pages, 292 figs., 48 tables and CD-ROM. 2006

J. Chen, N. Jonoska, G. Rozenberg: **Nanotechnology: Science and Computation.** XII, 385 pages, 126 figs., 10 tables. 2006

A. Brabazon, M. O'Neill: **Biologically Inspired Algorithms for Financial Modelling.** XVI, 275 pages, 92 figs., 39 tables. 2006